Reconfiguring the Global Governance of Climate Change

This book charts the course and causes of UN, G7 and G20 governance of climate change through the crucial period of 2015–2021. It provides a careful, comprehensive and reliable description of the individual and interactive contributions of the G7, G20 and UN summits and analyses their results.

The authors explain these contributions and results by considering the impacts of causal candidates, such as a changing physical ecosystem and international political system and the actions of individual leaders of the world's most systemically significant countries. They apply and improve an established, compact causal model, grounded in international relations theory, to guide these tasks.

By developing, prescribing and implementing immediate, realistic actionable policy solutions to cope with the urgent, existential challenge of controlling climate change, this volume will appeal to scholars of international relations, global governance and global environmental governance.

John J. Kirton, PhD, is a professor of political science and the founder and director of the G7 and G20 Research Groups, and co-director of the BRICS Research Group and the Global Health Diplomacy Program, based at the University of Toronto, Canada.

Ella Kokotsis, PhD, is the director of accountability of the G7 and G20 Research Groups, based at the University of Toronto, Canada.

Brittaney Warren, MES, is the lead researcher on climate change for the G7 and G20 Research Groups as well as the BRICS Research Group, based at the University of Toronto, Canada.

Global Governance

Series Editor: John J. Kirton, University of Toronto, Canada

Global governance is growing rapidly to meet the compounding challenges of a globalized 21st-century world. Many issues once dealt with largely at the local, national or regional level are now going global, in the economic, social and political-security domains. In response, new and renewed intergovernmental institutions are arising and adapting, multilevel governance is expanding and sub-national actors are playing a greater role, creating complex combinations and private partnerships to this end.

This series focuses on the new dynamics of global governance in the 21st century by

- Addressing the changes in the structure, operation and impact of individual intergovernmental institutions, above all their innovative responses to the growing global challenges they confront.
- Exploring how they affect, are affected by and relate to non-state actors of global relevance and reach.
- Examining the processes of cooperation, competition and convergence among international institutions and the many global governance gaps where global challenges such as terrorism, transnational crime and energy do not confront powerful international institutions devoted to their control.
- Dealing with how global institutions govern the links among key issues such as climate change and health.

In all cases, the series focuses on the central questions of how global governance institutions and processes generate the effective, legitimate, accountable results required to govern today's interconnected, complex, uncertain and crisis-ridden world.

For more information about this series, please visit: https://www.routledge .com/Global-Governance/book-series/ASHSER1420

Sovereignty and Illicit Social Order
Christopher Marc Lilyblad

The Crises of Legitimacy in Global Governance
Edited by Gonca Oguz Gok and Hakan Mehmetcik

Institutionalised Summits in International Governance
Promoting and Limiting Change
Daniel Odinius

Reconfiguring the Global Governance of Climate Change
John J. Kirton, Ella Kokotsis and Brittaney Warren

Reconfiguring the Global Governance of Climate Change

John J. Kirton, Ella Kokotsis and
Brittaney Warren

Routledge
Taylor & Francis Group

LONDON AND NEW YORK

First published 2022
by Routledge
4 Park Square, Milton Park, Abingdon, Oxon OX14 4RN

and by Routledge
605 Third Avenue, New York, NY 10158

Routledge is an imprint of the Taylor & Francis Group, an informa business

© 2022 John J. Kirton, Ella Kokotsis and Brittaney Warren

British Library Cataloguing-in-Publication Data
A catalogue record for this book is available from the British Library

Library of Congress Cataloging-in-Publication Data
A catalog record has been requested for this book

ISBN: 9780367151768 (hbk)
ISBN: 9781032227368 (pbk)
ISBN: 9780429055485 (ebk)

DOI: 10.4324/9780429055485

Typeset in Times New Roman
by Deanta Global Publishing Services, Chennai, India

Contents

About the Authors

John J. Kirton, PhD, is a professor of political science and the founder and director of the G7 and G20 Research Groups, and co-director of the BRICS Research Group and the Global Health Diplomacy Program, based at the University of Toronto, Canada. He is also a visiting professor of international relations and public affairs at the Shanghai International Studies University and a non-resident senior fellow at the Chongyang Institute for Financial Studies in Renmin University of China. He is the author of *G20 Governance for a Globalized World* (2013) and *China's G20 Leadership* (2016) and co-author with Ella Kokotsis of *The Global Governance of Climate Change: G7, G20 and UN Leadership* (2015), as well as many articles and chapters, and editor and co-editor of numerous books in the Global Governance series and Global Environmental Governance series published by Routledge. He holds a PhD in international studies from the School of Advanced International Studies at Johns Hopkins University, Washington DC, United States.

Ella Kokotsis, PhD, is the director of accountability of the G7 and G20 Research Groups, based at the University of Toronto, Canada. An expert on summit compliance, she has consulted with various governmental and non-governmental organizations, including the Canadian government's National Round Table on the Environment and the Economy, the Council on Foreign Relations on its African development agenda, the Russian government on global health issues, the Canadian government during its 2010 G8 and G20 summits, and Chinese scholars on climate change in the Arctic. Her scholarly research methodology for assessing summit compliance continues to be the basis for the annual accountability reports produced by the G7, G20 and BRICS Research Groups. Over the past 25 years, Ella has consulted broadly on various aspects of summitry with numerous governmental organizations, academic institutions, think tanks and non-governmental organizations. She is the author of *Keeping International Commitments: Compliance, Credibility and the G7 1988–1995* (1999) and co-author, with John J. Kirton, of *The Global Governance of Climate Change: G7, G20 and UN Leadership* (2015). She holds a PhD in international politics from the University of Toronto.

Brittaney Warren, MES, is the lead researcher on climate change for the G7 and G20 Research Groups as well as the BRICS Research Group, based at the University of Toronto, Canada. She has published on the G20 and G7's governance of climate change and the environment, including its links to health, and on accountability measures to improve summit performance. She holds a master's degree in environmental studies from York University.

Preface and Acknowledgements

This book was conceived as a follow-up to *The Global Governance of Climate Change: G7, G20 and UN Leadership*, published just before the historic Paris Summit on climate change in 2015. This new book was inspired by the dramatic developments in the years following the publication of the 2015 volume. These developments began with the hope many placed in the Paris Agreement, the struggle to implement and improve it, the withdrawal of Donald Trump's United States announced in June 2017, the steady approach toward critical thresholds of greenhouse gas emissions and atmospheric concentrations and the relentless rise in deadly, destructive, climate change–related extreme weather events. This cadence seemed to defy the hope, highlighted in the earlier book, that G7 and G20 leadership in global climate governance could succeed where that of the United Nations had failed. After several years in which all three global governance bodies failed, some were now left to rely on actions by other actors to solve the crisis that anthropogenic climate change had now become. But the crisis was so large and so urgent that a solution still required all hands on deck, working in the same direction, led by the formidable material, coercive, regulatory and normative power of the most powerful leaders and countries in the world. Understanding how and why they struggled from 2015 to 2020, and started to improve together in 2021, provides a foundation for identifying how they can do even better in a world in which there is no time left to lose.

These seven years allow for a stronger scholarly test of the concert equality model of G7 governance and the systemic hub model of G20 governance that guided the earlier book. Such a test highlights the importance of previously underestimated causes, most notably the arrival of a climate change–denying leader in the United States, and the global proliferation of the deadly COVID-19 pandemic in 2020 and 2021.

In producing this work, we owe a huge debt of gratitude to the many individuals who made this book possible. Cindy Xinying Ou assisted with the initial literature review and much research on subsequent chapters. Ana Zotovic conducted archival research, Sarah Saqib and Zhenglin Liu were instrumental in numerous sections, and Matthew McIntosh also provided support on Chapter 3. Members of the G7 and G20 Research Groups and volunteers for the Global Governance Program produced many important individual compliance assessments.

A special word of thanks goes to Madeline Koch, who wears two hats. As the executive director of the Global Governance Program, she organized the field teams at the annual G7 and G20 summits, which allowed us to experience first-hand the dynamics and nuance of these capstone events and secure the otherwise fugitive information available only at the summits themselves. As a professional copy editor, she helped us turn multiple drafts of this manuscript into something publishable.

John Kirton's colleagues in the Department of Political Science, led by professors Steven Bernstein and Matthew Hoffman with their pioneering work on global environmental governance and later Jessica Green and Kate Neville, were a constant source of inspiration. Maria Banda, in the International Relations Program, provided critical insights into the legal dimensions of controlling climate change. Professors Louis Pauly and Aurel Braun provided essential support throughout.

Beyond the University of Toronto, Marina Larionova at the Russian Academy of National Economy and Public Administration in Moscow, Miranda Schreurs of the Bavarian School of Public Policy at the Technical University of Munich, Chiara Oldani at the Università degli Studi della Tuscia in Rome, Jan Wouters of the Leuven Centre for Global Governance Studies at KU Leuven, and Naoki Tanaka at the Center for International Public Policy Studies in Tokyo have been valued partners. Dennis Snower of the Council for Global Problem Solving and our colleagues in the Think 20 constantly enriched our work.

Financial support for our summit work came from Steven Toope as director of the Munk School of Global Affairs and Public Policy, Louis Pauly as chair of the Department of Political Science, Randall Hansen as director of the Centre for European, Russian and Eurasian Studies, as well as the Asian Institute at the Munk School, Jackman Foundation, Centre for International Governance Innovation, Insurance Bureau of Canada, the Deutscher Akademischer Austauschdienst, Environment and Climate Change Canada, and Global Affairs Canada.

A quick word about how to get the most out of this book: This text refers to dimensions of performance, causes of performance and judgements of a summit's performance. These are all explained in detail in the introduction.

Acknowledgements

In producing this book, we are particularly grateful for the broader professional and personal support of several special individuals.

John Kirton dedicates this book to Barbara Eastman, PhD, who has supported his work in studying and shaping G7 summits since 1987, including generous contributions to the endowment of the Global Governance Program at Trinity College in the University of Toronto for on-site and ongoing summit research. A pioneer of environmentally sound and renewable energy financing and development across North America and internationally, Barbara has long understood and acted on the urgency of combatting climate change on a global scale.

Ella Kokotsis dedicates this book to her late father, Panos Kokotsis, who was a source of unwavering love and support throughout her life; a loving, compassionate and generous husband to Barbara, devoted father to Margarete and father-in-law to Rob Maccarone; and a caring grandfather to Ella and Rob's children, Max and Zara. Although Ella lost her father in April 2019 during the writing of this book, his beloved memory and courageous battle with Alzheimer's disease inspired her to bring this important project to completion.

Brittaney Warren dedicates this book to her parents, Donnie Warren and Bernice Wallace. Donnie, with his love of nature and the outdoors, has been a constant source of inspiration and of the financial support that enabled her to co-author this book. Bernice also inspired her appreciation of nature, introducing her as a child to the beauty of "down home" in Canada's East Coast and now developing with her a family home in the majestic Muskoka woods.

A final word of thanks and remembrance to Professor Joseph Daniels, who tragically lost his life in February 2020. Joe's seminal work on G7 summit compliance serves as an ongoing source of inspiration to all of us.

John J. Kirton, Ella Kokotsis and Brittaney Warren

Toronto, August 2021

Abbreviations and Acronyms

3Rs	reduce, reuse, recycle
AfDB	African Development Bank
AOSIS	Alliance of Small Island States
APEC	Asia-Pacific Economic Cooperation
ASEAN	Association of Southeast Asian Nations
AU	African Union
B20	Business 20
B3W initiative	Build Back Better World initiative
BASIC	Brazil, South Africa, India and China
BRICS	Brazil, Russia, India, China and South Africa
CARICOM	Caribbean Community
CCE	circular carbon economy
CDM	Clean Development Mechanism
CDR	common but differentiated responsibilities
CDU	Christian Democratic Union
CFSG	Climate Finance Study Group
COP	Conference of the Parties
CSU	Christian Social Union
D10	Democratic 10 (G7 plus Australia, India and Korea)
D11	Democratic 11 (G7 plus Australia, India, Korea and South Africa)
ECCC	Environment and Climate Change Canada
ECIU	Energy and Climate Intelligence Unit
ENB	*Earth Negotiations Bulletin*
FSB	Financial Stability Board
G7	Group of Seven (Canada, France, Germany, Italy, Japan, the United Kingdom, the United States and the European Union)
G8	Group of Eight (G7 plus Russia)
G20	Group of 20 (Argentina, Australia, Brazil, Canada, China, France, Germany, India, Indonesia, Italy, Japan, Korea, Mexico, Russia, Saudi Arabia, South Africa, Turkey, the United Kingdom, the United States and the European Union)
G77	Group of 77 (developing countries)

GDP	gross domestic product
HFCs	hydrofluorocarbons
ICT	information and communications technologies
IEA	International Energy Agency
ILO	International Labour Organization
IMF	International Monetary Fund
IMO	International Maritime Organization
INDC	intended nationally determined contribution
IPBES	Intergovernmental Science-Policy Platform on Biodiversity and Ecosystem Services
IPCC	Intergovernmental Panel on Climate Change
IRENA	International Renewable Energy Agency
ISIS	Islamic State in Iraq and Syria
LDC	least developed country
LMDCs	like-minded developing countries
MDB	multilateral development bank
MEF	Major Economies Forum on Energy and Climate
MIKTA	Mexico, Indonesia, Korea, Turkey and Australia
NATO	North Atlantic Treaty Organization
NDC	nationally determined contribution
NEPAD	New Partnership for Africa's Development
OECD	Organisation for Economic Co-operation and Development
PMO	Prime Minister's Office
ppm	parts per million
PSI	plurilateral summit institution
SDGs	Sustainable Development Goals
SDM	Sustainable Development Mechanism
SIDS	small island developing states
SPD	Social Democratic Party
TCFD	Task Force on Climate-related Financial Disclosures
UNEP	United Nations Environment Programme
UNFCCC	United Nations Framework Convention on Climate Change
UNGA	United Nations General Assembly
UNSC	United Nations Security Council
W20	Women 20
WHO	World Health Organization
WMO	World Meteorological Organization
WTO	World Trade Organization

1   Introduction

The Challenge

On August 9, 2021, the world's 234 leading climate scientists and 195 governments, including all those of the G7 and G20 countries, agreed on several shocking fundamental facts (Intergovernmental Panel on Climate Change [IPCC] 2021). Global warming could not be stopped from intensifying over the next 30 years. In 20 years, the world would be 1.5°C warmer than pre-industrial levels, even with deep emissions cuts starting today. Every global region was already suffering severe damage from climate change, which would steadily intensify for decades. Confirmation came weeks before the IPCC reported these stark certainties, as unprecedented heatwaves and wildfires devastated the United States, Canada, Russia, Greece and Turkey; historic floods hit Germany, Europe and China; and droughts killed many more beyond. United Nations secretary general António Guterres called the IPCC report a "code red warning" (UN 2021). The world looked in desperation to the G20's Rome Summit in October and, above all, the UN's long-awaited climate summit in Glasgow in November to avert this "hell on earth" (*Financial Times* 2021f).

As 2021 unfolded, new hope arose that the global summit governance of climate change would finally take the necessary steps to avert the climate catastrophe. The diversionary shock of COVID-19 was diminishing in some parts of the world, replaced by the soaring shocks of climate change everywhere. New leaders arrived to chair and shape the G7 and G20 summits, Boris Johnson of the United Kingdom and Mario Draghi of Italy, respectively. The UN prepared to hold the 26th Conference of the Parties (COP) to the United Nations Framework Convention on Climate Change (UNFCCC) in November 2021. Moreover, the special ad hoc, climate-focused summits arising since 2017 expanded, with Joe Biden, the new, climate-committed US president hosting the first Leaders Summit on Climate on April 22–23, with leaders from over 40 major emitting countries. This unprecedented sequence of global summits, all working closely together for the first time, brought the reconfiguration of global climate governance to new heights.

DOI: 10.4324/9780429055485-1

Debate

The scholarly debate about the course and causes of global climate governance after 2015 revolved around the respective contributions of the formal multilateral organizations of the UN galaxy, the old informal plurilateral summits of the G7 and G20, the 10 newer special summit supplements since 2017, and the effectiveness of the centralized or fragmented, top-down or bottom-up regime complexes, including non-state actors that the UNFCCC fostered (Gupta 2016; Kahler 2017). Among the many possible combinations, several major schools of thought stood out.

The first looked to the promising multilateral UNFCCC as central and prospectively effective. Hall (2016) showed how a broad array of UN functional organizations became involved in climate change in support of the UNFCCC and how the UN's September 2015 summit that adopted the Sustainable Development Goals (SDGs) introduced a new, much more inclusive, integrative focus for the climate cause. Betsill et al. (2015) pointed to the UNFCCC possibly providing the needed overall coordination. Floyd (2015) argued for the integrative role of the United Nations Security Council (UNSC) and its human security concept in overcoming the dangerous fragmentation of the climate security regime created by the intrusion of other organizations. Engelbrekt (2016) and Scott (2015) similarly believed in the UNSC's integrative power.

The second school saw little independent role for informal, global plurilateral summit institutions (PSIs). Rinaldi and Martuscelli (2016) noted the disposition of the BRICS of Brazil, Russia, India, China and South Africa to act in multilateral forums but highlighted the constraints on a consensus among them.

The third school saw UN failure, including prospectively at COP26 in 2021, due to China's domestic coal addiction and the international linkage of climate cooperation with China's key geopolitical demands (Erickson and Collins 2021). This could, however, be overcome if the G7 democracies with Australia and Korea formed a "competition club" to impose a carbon price at home and carbon border adjustments on China abroad. As these democracies accounted for about half of global gross domestic product (GDP) in 2019, and members of the Organisation for Economic Co-operation and Development had almost 75% of global GDP and 35% of global carbon emissions, they could force an export-dependent China to adjust.

The fourth school saw plurilateral promise in a more UN-supportive way, with such bodies operating from the bottom up to reinforce the central UNFCCC. Slaughter (2020, 74, 93–94) saw the G20 as a potential UN supporter, as G20 summits addressed climate change from the start, and did so quite well at Hamburg in 2017, despite the economic interests of the key fossil fuel–producing countries of Australia, Russia and Saudi Arabia. For Slaughter, this meant the G20 should shift from negotiating outstanding issues of the Paris Agreement to technical implementation, as this was a G20 strength.

A variant of the fourth school saw fading climate leadership from the G20, G7 and BRICS since 2017, due to the arrival of climate change–denying Donald

Trump in the United States and Jair Bolsonaro in Brazil, and the rise of nationalist, populist parties in Europe. Bauer et al. (2019, 2020, 115) argued that the G20 summits in 2015 and 2016 provided strong support for the Paris Agreement and the SDGs. However, the G20 then struggled to provide leadership. Still, with its globally predominant share of economic capabilities and global emissions, its small club, greater procedural flexibility than the UNFCCC's COPs and strong engagement groups, the G20 should be able to lead, as it did increasingly before 2016.

The fifth school saw a broader role for informal PSIs. Falkner (2016) identified how climate clubs in the context of multilateral negotiations usefully enhanced political dialogue, created club benefits that improved mitigation strategies, reduced free riding on coalitions of the willing and helped legitimate global climate governance, despite the slowing multilateral progress caused by major power shifts.

The sixth school identified the G20's potential and performance in global energy governance and in climate governance itself. Heubaum and Biermann (2015) saw the historically limited policy interaction between the climate and energy regimes improving due to the International Energy Agency's expanding agenda embracing renewables and its bottom-up new partnerships with the International Renewable Energy Agency and the UNFCCC. Sainsbury and Wurf (2016) agreed. Downie (2015a, b, 2020) showed how the G20 had worked with multilateral organizations on energy to play a significant role. Van de Graaf and van Asselt (2017) discussed its role on energy subsidies. He (2016) argued that the G20 and China as its summit host in 2016 should lead in global energy governance.

The seventh school saw greater G20 climate performance, due to growing Chinese and Indian leadership. Kirton (2016a) identified China's G20 leadership on energy and climate change. Rashmi (2020) argued that a more climate-committed India could lead emerging powers through the G20 to control climate change, especially with the postponement of COP26 in 2020.

The eighth school emphasized the G20's poor performance on climate change, relative to its core, early focus on economic issues. Kirton and Warren argued that this was due to the G20's governance of climate change in a separate siloed, rather than a synergistic way that linked to closely connected subjects such as health, rather than simply the economy (Kirton and Warren 2020a, b; Warren 2020; Kirton and Wang 2021). Berger et al. (2019, 502) saw declining performance, arguing that the G20, not being a "club of like-minded," found it increasingly challenging to reach consensus on "fundamental issues such as ... the amelioration of dangerous climate change." Ambumozhi (2018, 88) observed that the G20 "tends to focus on less controversial issues affecting the financial architecture ... such as climate change" due to members' diverse mix of energy resources and nationally determined contributions commitments under the UNFCCC.

The ninth school highlighted the G7's potential, displayed at the 2015 Elmau Summit and the UNSC. Hall (2016, 69–70) suggested that the G7, UN and others could productively work together, due to "increased scientific and political consensus," and called for continued action on "financing, multilateral organizations, heads of state and scientific research" plus "from civil society." Kirton et al. (2018) found that between 2013 and 2016, the G7 performed better than the G20 on climate change, with both having generally rising commitments and compliance.

The 10th school emphasized G7 failure. Kirton et al. (2019) concluded that the G7's 2017 Taormina Summit largely failed on climate change, due to a recalcitrant Trump. Andrione-Moylan et al. (2019, 172) suggested that Trump spurred China's Xi Jinping to assume global climate leadership. He (2016) saw such leadership arising even before Trump arrived.

The 11th school advised the G20 to seize control of climate change governance from the UN, due to the cumulating crises coupled with the failure of the UN's multilateral organizations in response. Bishop and Payne (2020, 134) said "the broad remit of the G20" should be expanded even if it means "seizing directional control of the politics of an issue like climate change, from the UN system."

The 12th school saw false rhetoric and real failure everywhere, as the G20, UN and other intergovernmental institutions all consistently failed to match their deliberative and direction-setting words with appropriately ambitious decisions and above all implementing delivery. This was due to inadequate accountability mechanisms and insufficient civil society involvement, despite the compounding climate crisis. Stevenson (2021) emphasized the G20 summit's unfulfilled promises to phase out fossil fuel subsidies since 2009, its decline in ambitious climate promises with Trump's arrival, and the leading global rank of G20 members China, the United States, Russia, the European Union and India in subsidizing fossil fuels. Others argued that reversing climate change required all hands on deck (Hale 2016; Chan et al. 2019).

The 13th school saw supplementary special climate summits failing, due to fading US power. Matthews (2021, 12) recalled that Barack Obama had to design the 2015 Paris Agreement without the need for Senate treaty ratification, and although climate change was Biden's top priority, his team might overestimate "the leverage that the United States retains for initiatives that depend on its example, such as the global summits the president wants to convene on climate change."

Puzzles

Despite their important contributions, these schools present numerous puzzles. First, several offer prescriptions or describe possible contributions, rather than explain what the central global institutions for governing climate change have done and why. Many do not specify the unique contribution of summitry to spurring the authoritative and comprehensive action that climate change demands (cf. Karlsson-Vinkhuyzen et al. 2016). Few treat the full role of the G20 or touch on the G7, despite their increasing if intermittent emphasis on climate change. None deal directly with the important interaction among the G7, G20 and UN. Nor do any comprehensively cover developments in the critical years from 2015 to 2021 (Kirton and Kokotsis 2015; Engelbrekt 2016). The major work with a comprehensive, systematic, theoretically informed and analytically guided model for such a task was published before the Paris Agreement was signed (Kirton and Kokotsis 2015). No schools traced and assessed a central feature of the reconfiguration of global climate governance — the increasing number of new, ad hoc, special climate plurilateral summits, starting with the One Planet Summit in 2017. None put together the three central global summit systems of the G7, G20 and UN, and the special climate summits at the centre of the ever-growing and diversifying regime complex for climate change, or assessed how and why each tried to integrate all the key actors into a whole-of-global-governance approach.

Purpose

This book thus has four central purposes. The first is to provide a careful, comprehensive, detailed, reliable description of the individual and interactive contributions of the G7, G20 and UN summits and the special climate summits in the global governance of climate change from 2015 to mid 2021. This critical period stretches from the 2015 UN Paris Summit to the lead-up to the successor UN Glasgow Summit in 2021. The second purpose is to explain these contributions and results, by considering the impact of causal candidates at all levels of analysis, from that of a changing physical ecosystem and international political system to the individual leaders of the world's most systemically significant countries. The third purpose is to apply and improve an established, compact causal model, grounded in international relations theory, to guide these tasks (see Appendix A). This will assist others in developing, prescribing and implementing immediate, realistic actionable policy solutions to cope with the urgent, existential challenge of climate change, in a world rapidly running out of time. Its fourth purpose is to continue the work begun by Kirton and Kokotsis (2015) in *The Global Governance of Climate Change: G7, G20 and UN Leadership*, which told the story up to just before the 2015 Paris Agreement.

Argument

This book argues that G7, G20 and UN performance has varied widely, individually and interactively, from 2015 to 2021, and that the addition of 10 ad hoc, plurilateral special climate summits since 2017 reconfigured global summit climate governance to spur much stronger performance in 2021 (see the section titled "Overall Performance" on page 11 for an explanation of the categories of performance). This assessment is based on a comparison of summits to their predecessors. Given the climate crisis, regular and special summitry has fallen far short of contributing to restoring and sustaining climate and ecological balance.

Four phases in this reconfiguration stand out. The first phase, producing Paris in 2015, explored in Chapter 2, saw the G7's strong performance at Elmau, but not the G20's small performance at Antalya, contribute to the solid performance of the UN's Paris Summit and Agreement in December (see Appendix B).

The second phase, relying on Paris from 2016 to 2019, examined in Chapters 3–7, saw G7 and G20 leadership rotate among the G7's substantial performance at Ise-Shima in 2016, the G20's strong performance at Hamburg in 2017, the G7's significant performance at Charlevoix in 2018 and the G20's significant performance at Osaka in 2019, while the UN ministerial COPs produced a solid performance in 2017 and small ones otherwise. The G7 and G20 summits were increasingly supplemented by the One Planet Summit's significant performance in 2017, substantial performance in 2018 and small performance in 2019, and the significant performance at the UN Climate Action Summit in 2019.

The third phase, the COVID-19 crowd-out in 2020, analysed in Chapter 7, saw G7, G20 and UN summit climate governance virtually disappear, as the scheduled G7 and UN summits failed to take place and the G20 Riyadh Summit produced only a small performance in November. Yet, the reconfiguration strengthened with three special climate summits that year: the UN's High-Level Roundtable on Climate Action and Summit on Biodiversity with their small performances in September and the Climate Ambition Summit with a significant performance in December.

The fourth phase, combined leadership in 2021, detailed in Chapter 8, saw the established G7, G20 and COP summits and the new expanding plurilateral climate ones work together to produce a strong performance at two of its summits during the first half of the year, hoping to spur success at Glasgow near the end of 2021. The solid performance of the One Planet Summit on Biodiversity, the small performance of the Climate Adaptation Summit in January and the strong performance of the Leaders Summit on Climate in April joined the strongly performing G7 Cornwall Summit in June to propel a prospective substantial performance at both the G20 Rome Summit in October and the UN's Glasgow Summit in November.

These four phases were largely created by the changing conditions of the key causes highlighted in the systemic hub model of G20 governance (Kirton 2013; Kirton and Kokotsis 2015). The most potent propellers were the steadily increasing climate shocks and the recent relative decline of diversionary ones, the growing failure of multilateral organizations to meet the increasing ecological need, and the equalization of capability and convergence of democracy and climate policy among leading G7 and G20 members. Above all, summit hosting by domestically strong, globally connected, climate-committed G7 and G20 leaders mattered, including Angela Merkel, Shinzo Abe, Justin Trudeau and Emmanuel Macron from 2015 to 2020, and Johnson, Draghi and Biden in 2021, compared to other leaders such as Trump, Mauricio Macri and King Salman bin Abdulaziz Al Saud.

In the producing Paris phase, the strong performance at Elmau and the significant performance at Paris were propelled by the increasing climate shocks that exposed the vulnerability of G7, G20 and UN members. Multilateral organizational failure grew, as the ministerial-level UNFCCC and its 2009 Copenhagen Accord became increasingly inadequate to meet the growing global need. This increased the demand for plurilateral summit governance. Globally predominant and internally equalizing capability slowly passed from the G7 to the G20, which, with rising coal-committed China and India and fossil fuel–dependent Russia and Saudi Arabia, could not close the gap. Converging characteristics on open democracy and climate policies were much stronger in the G7 than in the G20 containing China and Russia, which had recently annexed the Crimean region of Ukraine. Leaders' domestic political control was also stronger in the G7, as Merkel brought exceptional experience as a G7 host — a former environment minister with scientific expertise. The G7's position as its leaders' valued club at the hub of a growing network of global summitry was strongly enhanced by the contraction back to seven from eight countries with Russia's suspension, and the intimacy and trust among the many veteran leaders at Elmau.

In the second phase, from 2016 to 2019, climate shocks, vulnerabilities and scientific warnings steadily grew, but remained largely unrecognized by G7 and G20 leaders. The ministers managing the COP process still failed to meet the growing need. Global predominance, equalizing economic capabilities and emissions shifted from the G7 toward the G20, where increasingly less democratic, more climate-skeptical leaders from Russia, China and Saudi Arabia were reinforced by the arrival of a climate change–denying Trump and Bolsonaro to constrain ambitious climate action. Significant or strong climate performance only arose at summits hosted by democratic G7 leaders with high experience, climate conviction and domestic support, as under Merkel at the 2017 G20 in Hamburg, Trudeau at the 2018 G7 in Charlevoix, Abe at the 2019 G20 in Osaka, and Macron at his inaugural One Planet Summit in 2017.

During the third phase, in 2020, the exceptional shock of COVID-19 diverted almost all G7 and G20 summit action away from climate change and delayed the UN's Glasgow Summit for a full year. China's capabilities and emissions began to return to pre-pandemic levels, while its democratic characteristics continued to decline. The G7 and G20 were crippled by Trump's presidential re-election preoccupations, Merkel's fading political power, the United States and Saudi Arabia as summit hosts, and the interruption of the in-person summitry that helped make the G7 and G20 summits their leaders' valued clubs. Global summit climate governance was kept alive, if barely, by the UN's High-Level Roundtable on Climate Action and Summit on Biodiversity and more so by the Paris Climate Ambition Summit, hosted by Macron.

Combined leadership came for the first time in the fourth phase in 2021. The reconfiguration of global climate summit governance reached a new stage, with the deadly diversionary darkness of the COVID-19 shock giving way to the dawn of green global governance. After the two special climate summits in January, the UK-hosted virtual G7 ad hoc summit in February was followed by Biden's Leaders Summit on Climate in April and the G7's Cornwall Summit in June, and then by the G20's Rome Summit in October and the UN's Glasgow Summit in November. Every step of this unprecedentedly intense sequence of summits addressing climate change was hosted by the G7's democratic members, led by the renewed climate-committed United States and the United Kingdom. They were reacting to the severe climate shocks and science showing that atmospheric concentrations of emissions were reaching critical thresholds, by increasing and sustained pressure from their citizens, and by the inability of the much more inclusive COP26 preparations to escape the degraded digital diplomacy still imposed by COVID-19. Further pushes came from the rapidly rising capabilities, democratic character and environmental performance of the United States and the United Kingdom, led by climate-committed leaders with solid summit experience and strong domestic political support. At the core of the G7 and G20 clubs now returning to in-person summitry were the G7's Johnson and Draghi as co-hosts of COP26. The global summit network radiating out from this hub was significantly expanded by Biden's new summits of the Quadrilateral and the Leaders Summit on Climate. Despite this, the transformational systemic change climate science urgently called for remained elusive.

The Enhanced Systemic Hub Model

The analysis in this book is guided by an enhanced version of the systemic hub model of G7 and G20 governance (see Appendix A) (Kirton 2013, 2016a; Kirton and Kokotsis 2015).

The model still begins with the six dimensions of summit performance applicable to all PSIs: domestic political management; private and public deliberation; principled and normative direction setting; collective decision making; members' delivery of these decisions; and the institutional development of global governance inside and outside the PSI. The model continues with six causes of this performance, again applicable to all PSIs: shock-activated vulnerability; multilateral organizational failure; predominant equalizing capability; converging political characteristics; domestic political cohesion; and the club at the hub of a global summit network.

Dimensions of Performance

Domestic Political Management

This performance dimension recognizes that leaders go abroad to promote their policies and political standing back home. This starts by securing the status and prestige that come from attending summits and being seen in the exclusive top-tier group of powerful leaders in their collective photo ops. It extends to the summit leaders' formal, collective endorsement of a specific country's contribution through explicit compliments in the communiqués — the summit's collectively endorsed outcome documents — and the attention and approval the leader's summit performance gets back home.

Domestic political management is first measured by whether the leaders deem the club important enough to attend the summit and its sessions on climate change. The second measure is how much the attending members' efforts to advance climate action, either within or outside the summit format, is publicly recognized in favourable terms in communiqué compliments. Additional measures are public approval through the attention and endorsement of the summit and the leader's performance in the elite media, legislature, civil society, public opinion and eventual re-election results.

Deliberation

The second dimension is leaders' deliberation, through both private conversations at the summit and in public conclusions recorded in the communiqués.

Private deliberations through leaders' confidential conversations are measured by the length, intensity and quality of their personal, face-to-face exchanges, especially when they are alone together at their leaders-only dinners and sessions, ceremonial events and bilateral encounters whether scheduled or spontaneous.

Public consensus is measured, both overall and on specific subjects such as climate change, by the amount and portion of attention leaders give in the communiqués.

Direction Setting

The third dimension is direction setting, through affirmations of consensus principles and norms in the communiqués. Principles are statements of fact, causation and rectitude; norms are statements of prescription, proscription and permission (Krasner 1983).

The first measure of this ideational dimension of performance is the number of times the institution's distinctive foundational mission is mentioned in the communiqués. For the G7, that is democracy and human rights and, for the G20, financial stability and making globalization work for all. The second measure is the links made between climate change and other subjects, as a cause, consequence or co-existing condition of climate change. The third measure is the prominence given to climate change through its placement in the preamble of the communiqué.

Decision Making

Decision making is expressed through public, collective, precise, future-oriented, politically binding commitments (Abbott and Snidal 2000). This is measured first by the number and portion of climate change commitments, and second by the balance between climate commitments that are highly binding (where leaders said they "will" or "shall" do new things, or do more than they have done before) and low binding ones (where leaders reaffirm, reiterate or promise to continue what they have previously promised to do). A third measure is the balance between subject-specific siloed commitments (which address only climate change) and synergistic ones (which are linked in the commitment to other subjects, especially in ways that put climate change first and identify the co-benefits of climate action).

A commitment can contain other catalysts or compressors of performance, especially the delivery of these decisions through members' compliance with their commitments (Warren 2021b).

Delivery

The fifth performance dimension is the delivery of these decisions through members' compliance with the commitments before the next summit, as their governments act individually to implement them in the collectively specified or consistent way (Kokotsis 1999; Kirton and Larionova 2018b). They do this using the standard array of instruments available to their governments. These include the increasingly ambitious mechanisms and actions of issuing verbal reaffirmations, providing and sharing data and best practices, assigning budgets and personnel, creating offices and programs, changing regulations and policies, passing laws or even changing their country's constitution.

Such compliance-compatible actions come with a weak causal claim. They do not assume that the commitment is the exclusive cause of the subsequent compliance-consistent behaviour. Rather, they assume that the commitment shapes and constrains a government's subsequent action. Even if a member has already agreed to a commitment and intends to implement it, the commitment helps compel the member to do so, making it more difficult to downgrade or defect. Similar commitments accepted by members at other summits increase this constraining effect.

Compliance is measured by the total all-summit, all-subject, all-member average for the summit based on all commitments assessed for compliance, and by the average by year, by subject and by member. Performance on the subset of climate change commitments assessed for compliance is similarly measured.

Development of Global Governance

The sixth performance dimension is the institutional development of global governance both inside and outside the G7 or G20. It is measured by the number, breadth and balance of communiqué references to specific international institutions, by name, in the passages on climate change. It is also relevant whether these references present the G7 or G20 as leading, following supportively or simply noting these other institutions neutrally. References to bodies of the G7, G20 and UN are particularly important.

The balance of references to inside and outside institutions indicates how inward or outward looking the summit is. The number and level of bodies on either the same or different subjects, from leaders through ministerial bodies to official-level bodies are also relevant.

Overall Performance

Together, these six dimensions allow for an overall judgement of a summit's performance, with the greatest weight generally given to the number and quality of commitments and subsequent compliance. Following a schema introduced by Putnam and Bayne (1987) for measuring the performance of the annual G7 summits, this book uses the following scale:

- very strong for A
- strong for A–
- significant for B+
- substantial for B
- solid for B–
- small for C
- slender for D and
- failure for F, defined as a summit that did nothing or made things worse.

Causes of Performance

Performance along these six dimensions is caused by six forces, which interact sequentially. The core hypothesis is that a strong summit performance is caused first by a high level of shock-activated vulnerability, to which, second, the core multilateral organizations fail to respond adequately, leaving, third, a small group of countries with globally predominant and internally equalizing capability to fill the global governance gap in a collective way. Fourth, performance is strengthened by their upwardly converging characteristics of democracy and environmental performance at home. Fifth, performance increases if leaders have summit experience, a personal desire to control climate change, political support from their executive and legislative branches, other politically consequential actors and favourable public opinion. Sixth, they produce high performance if their summit is their personally valued club, hosted by a committed and skilled leader, at the hub of an extensive and expanding network of global summit governance that radiates influence out and brings in valuable inputs and legitimacy from others.

Shock-Activated Vulnerability

Shock-activated vulnerability assumes that the most powerful leaders of the world's most powerful or systemically significant countries are largely slow, cybernetic and thus reactive learners, rather than proactive, prescient, preventive, rationally optimizing ones. They are spurred to act when a severe, second or subsequent, tightly spaced shock, similar in its subject or source to previous shocks, activates their awareness of their country's vulnerability and the inadequacy of their unilateral action in response.

Shock-activated vulnerability is measured first by the shocks and vulnerabilities that leaders collectively recognize in their communiqués, especially when they themselves make the causal connection by recording their performance due to a particular shock and vulnerability.

A second measure, with less physical and temporal proximity to their summit performance, are the newsworthy stories highlighted on the front pages or in major newscasts of the elite media that leaders and their politically consequential publics consume. This media-highlighted, shock-activated vulnerability is especially important in making democratic leaders aware and arousing them to action. It is measured by the subject of articles appearing on the front pages of elite daily newspapers that leaders still rely on, although social media attention can be relevant too (Anderson and Huntington 2017; Fownes et al. 2018).

A third measure is the physical shock-activated vulnerability and the degree of death and damage produced. A fourth, less proximate measure is scientifically reported shock-activated vulnerability — the results of scientific reports from authoritative sources. A fifth is forecasts of future top risks, ranked by probability and payoff, by the influential global elite business community.

In all cases, it is relevant if the shock and vulnerabilities arise or affect inside or outside G7 and G20 members. Leaders are expected to perform more quickly and strongly to shocks and vulnerabilities arising within or affecting their own country or the countries of their colleagues in the group, especially if such shocks are similar in subject or source to those they have recently experienced.

Diversionary shock-activated vulnerabilities are also important, and may involve subjects seen as very different and distant in many ways from the subject of climate change. The impact of shock-activated vulnerability on a specific subject thus relates to the shock-activated vulnerability on the many other subjects that demand global governance at a specific summit and time.

Multilateral Organizational Failure

The second cause, highlighted by both realist and liberal-institutionalist scholars, is multilateral organizational failure, above all by the bodies from the Bretton Wood–UN galaxy constructed since 1944 and then from the other PSIs filling the remaining global governance gaps. This failure flows first from the absence of a multilateral organization with a mandate and mission to address the particular new problems that globalization brings, with climate change the most fully globalized and existential problem of all.

A second component is the fragmentation and resulting lack of coordination or the competition and buck passing among the many bodies dedicated to the same subject (Freytag and Kirton 2017; Downie 2020). A third component is their weakness in leadership, budgets, staff, programs and instruments to meet the task at hand. Relevant measures include elite and mass perceptions of the democracy, effectiveness and fairness of relevant multilateral organizations (Verhaegen et al. 2021). The failures in multilateral organizations' existence, centrality and strength, especially in response to shock-activated vulnerability, prompt the G7 and G20 to fill the gap with high performance of their own.

Predominant Equalizing Capabilities

The third cause, featured by realist scholars, is the globally predominant and internally equalizing capabilities of PSI members, both overall and in the specialized capabilities most relevant to the subject of the shock. Their collective predominance makes them more likely to shape global governance as a whole, as outside countries will follow the PSIs' lead in beneficent, coercive or voluntary ways. Internal equalization reinforces the PSIs' ability to have any member lead, including the most shocked, vulnerable and committed ones.

Overall capability is generally measured by members' GDP at current exchange rates. Specialized capabilities are defined by the instruments that work best against the shocks at hand.

Capabilities should be assessed in relation to relative vulnerabilities. In the case of climate change, G7 and even G20 members are the least vulnerable to its effects, relative to poorer, less powerful countries. However, rich countries' collective vulnerability has risen over the years. Inside these groups, relatively small island states and countries with their political and economic capitals on coasts are most vulnerable to sea level rises, hurricanes and typhoons. Here, the United Kingdom, Japan and then the United States stand out, even though all G7 and G20 countries have oceanic coasts.

Converging Political Characteristics

The fourth cause is converging political characteristics at a high level on open democracy and domestic environmental performance. Highly convergent democracy promotes action to control climate change (Fiorino 2018). The shifting level and convergence of G7, G20 and UN members' democratic openness is measured by several standard data sets. Political and policy convergence on climate change is assessed by PSI members' policy performance at home. The Germanwatch Climate Change Performance Index, which ranks 60 countries, offers the most useful data set.

Domestic Political Cohesion

The fifth cause is domestic political cohesion. It is composed, first, at the state level, of PSI leaders' control of their executive, legislature, regulatory agencies, judiciaries and sub-federal governments; second, their popularity and proximity to the next election; and third, at the individual level, their personal expertise, experience and convictions about the G7 and G20 and climate change. It is important to identify particular combinations of these components, as leaders with high political control but climate change–denying convictions constrain summit success.

Club at the Network Hub

The sixth cause is the PSI's position as the club at the network hub, specifically, as the leaders' valued club at the compact hub of an extensive and expanding network of global PSIs. First, the host leader at the core of the hub can influentially shape a summit's agenda, and does so in conjunction with the other PSIs in which they participate. Second, the compact, controlled participation in the PSI as a club at the hub fosters low transaction costs, transparency, trust and status, giving the small, like-minded G7 an initial advantage over the larger, more diverse G20 (Kahler 1992). Third, the more members belong to other PSIs, the greater the global network and influence of and supportive inputs to the summit (Slaughter and LaForge 2021). This has expanded as more PSIs have emerged since the G7's inception in 1975 and the leaders-level G20's inception in 2008. This is broadly relevant to the leaders invited as guests to these summits. Which international organizations are invited to the summit is also relevant, especially if they are dedicated or relevant to climate change. Judgements about how much the leaders value a PSI club can be inferred initially from how often and regularly its summits take place as well as how often the leaders attend.

Connecting Causes to Performance

Most of these causes affect all dimensions of performance. However, as the dimension of delivery occurs following the summit, the relevant condition of these causes should extend to coincide with the compliance period. Compliance is both pushed by the summit that made the commitment and pulled by the forces on the road to the subsequent summit.

The match between these six causes and the six performance effects can arise both at the systemic PSI and global level and at the unit level among individual members. Among the performance dimensions, country- and member-specific data are available for the domestic political management components of attendance, communiqué compliments, and domestic attention and approval. Data are also available for delivery through compliance. Among the causes, information is available for shock-activated vulnerability, multilateral organization failure (given a country's control of a multilateral organization), predominant equalizing capabilities' GDP and emissions, the converging political characteristics of democracy and climate change performance, domestic political cohesion in all its components, and for the host of the club at the network hub. Performance likely increases sequentially over time, with conclusions about climate change leading to more consensus on principles and norms. This in turn leads to more commitments. Compliance requires commitments, which in turn require public conclusions. High compliance can boost domestic political management.

All the causes are analytically interrelated. When shock-activated vulnerability is high, multilateral organizational failure to cope is likely also high, leading a PSI with high global predominance and internally equal capability to fill the gap more actively. A PSI can do so more easily if members have highly converging political characteristics and if their leaders are backed by high domestic political cohesion and stand at the valued hub of a broader network of PSIs.

Application of the Model

Applying the model to the global governance of climate change in this study requires several extensions of the inherited systemic hub model. The five most prominent extensions are as follows.

The first is to emphasize results. This starts with the level and causes of members' compliance with their summit commitments and the low-cost, leaders-controlled accountability measures that may improve the implementation the leaders presumably want (Kirton and Larionova 2018b). It extends to how successful the summits have been, individually, cumulatively and collectively in solving the problem of relentlessly rising global heating, the emissions and concentrations, and the damage and deaths they cause. Here, the ultimate referent for evaluation changes over time. In 2015, Kirton and Kokotsis (2015) agreed it was appropriate to focus on how much one year's summit achieved relative to its predecessors and relative to the size of the problem as revealed by the current science then. By 2021, the ultimate referent shifted to the gap between the relentlessly rising problem and the growing summit performance, as the problem is expanding faster and further than summit performance has. The physical problem has increasingly overwhelmed the political solution.

The second extension is the impact of diversionary shocks. These begin with the great COVID-19 crowd-out in 2020 and into 2021. They highlight the complex cognitive dynamics of how shock-activated vulnerability led to individual and collective learning by key government leaders, especially after the intensified extreme weather events assaulting North America, Europe, Africa, Asia and elsewhere, and how much leaders connected climate change to human health in its COVID-19 and other forms.

The third extension is the role of individual leaders in and across these PSIs. Featured here are Obama in leading the production of the Paris Agreement for the United States, Trump then announcing the US withdrawal from it and Biden bringing the United States back in. Xi, Merkel and Macron are also important in replacing US leadership in global climate change governance. Indeed, Merkel is a key figure as the only leader to host two strongly successful summits and who managed Trump at Hamburg in 2017.

The fourth extension is to focus in finer detail on how the G7, G20 and UN worked alone or together on climate change, within the political cycle of a single year. Within a year, the G7 typically begins in late spring, the G20 follows in the summer or autumn, and the UN culminates the global governance sequence in November or December.

The fifth extension is to consider the distinctive value of summitry in global governance, not only from the summit-centred, plurilateral G7 and G20 and multilateral UN, but also from the ad hoc special climate summits arising and expanding since 2017. These special summits stand at the centre of the reconfiguration of global climate governance.

Ten improvements to the systemic hub model are offered in Chapter 9, which also summarizes the book's findings, suggests what lies ahead and offers possible areas for further research. There are grounds for hope that the critical, urgent problem of climate change can be solved in time.

Methods and Materials

The empirical application of the extended model in this book relies primarily on four methods. The first is input-output matching of the six causes and six effects, particularly on their levels and changes over the seven years from 2015 to 2021. Given the small number of years, even with a comparison of the two G7 and G20 summits annually, the resulting 14 sets of observations are too few for effective statistical analysis. However, such statistical methods have been employed for related research on the causes of compliance with G7 and G20 commitments on climate change (Kirton and Larionova 2018b; Rapson 2020a, b; Kirton and Wang 2021). These findings suggest the likely compliance-enhancing effects of ministerial meetings on the same subject, a high number of commitments on the same subject and the number of highly binding commitments. For the relevant materials, this study also relies on the extensive databases created by the G7 Research Group and the G20 Research Group on the six dimensions of performance, especially on 94 compliance assessments of 369 G7 climate commitments and 40 compliance assessments of 95 G20 climate commitments.

The second method is a quantitative content analysis of the communiqués issued in the leaders' name. This identifies the data for five of the six dimensions of performance, the catalysts and compressors of compliance contained in the commitments and the causal components of communiqué-recognized shock-activated vulnerabilities. Selective tests were conducted to confirm sufficient inter-coder reliability.

The third method is a textual analysis of the content and context of the communiqué passages to infer their meaning, especially as it connects with the known causes and consequences of climate change.

The fourth method is detailed process tracing to track the links between cause and effect. This process is based on published literature accounts as well as over 100 semi-structured, elite interviews with G7 and G20 participants.

2 Producing Paris, 2015

Introduction

In global climate governance, 2015 was dominated by high hopes and intense preparations for the landmark 21st Conference of the Parties (COP) to the United Nations Framework Convention on Climate Change (UNFCCC) in Paris at year's end — the first UN climate summit since the one in Copenhagen in 2009 that had largely failed. The preparations included both the formal UN process and the two plurilateral summits of the G7 at Elmau, Germany, on June 7–18, and the G20 at Antalya, Turkey, on November 15–16, just two weeks before the Paris meeting on November 30–December 13. Would these G7 and G20 summits lay a firm foundation to boost success at Paris? Or would they leave global climate governance to the UN alone, to focus instead on the many security, social and economic challenges they faced?

Angela Merkel's G7 Elmau Summit got off to a fast start, with a strong performance. But Turkey's Recep Tayyıp Erdoğan's G20 Antalya Summit, with only a small performance, did little to build on the G7's stronger beginning and broaden its consensus to embrace the major emerging economies. This left Paris to produce only a solid political performance, which failed to meet the escalating climate threat in the physical world.

This sequence of G7 strong performance, G20 small performance and UN solid performance was not caused by the rising climate shocks or worrying scientific evidence during 2015. It was driven by the hope that the fully inclusive UN, operating again at the summit level, could succeed where it had failed after the diversionary shock of the global financial crisis in 2008–2009. The rising capabilities, democratic convergence, domestic political support and climate-controlling convictions of all G7 leaders at Elmau were not shared by the G20's climate change–denying, far less democratic leaders at Antalya, led by Turkey as host, Saudi Arabia and Russia, and China's recently arrived Xi Jinping (Kirton 2016a). Above all, the UN's rigid procedures prevented the leaders, meeting in Paris after six years, from transcending the fixed political positions, interests and identities their ministers and governments had brought. Paris thus produced an agreement that continued to see countries such as China take advantage of the principle of common but differentiated responsibilities (CDR), with voluntary

DOI: 10.4324/9780429055485-2

cutback commitments not designed to be improved by leaders for many years. Even if fully complied with, the Paris commitments would not achieve the agreed goal of keeping the global temperature rise from pre-industrial levels below 2°C, and ideally 1.5°C.

G7 Elmau, June

Introduction

On June 7–8, 2015, G7 leaders met in Elmau under Merkel's experienced leadership. Climate change and sustainable development were prominent on the agenda, as the G7 prepared for two major UN summits on these subjects later that year. The first summit, in September, was to negotiate a sustainable development agenda to continue the unfinished work of the Millennium Development Goals (MDGs). The second summit, in December, was to negotiate a new legally binding agreement under the UNFCCC to respond to the growing climate threat. In the physical world, by 2015, the atmosphere's carbon dioxide levels had reached 400 parts per million (World Meteorological Organization [WMO] 2016). The years 2011–2015 were the hottest on record, with 2015 the hottest of all. Thus, all eyes were on the pre-UN summits, starting with the G7 at Elmau.

Debate

The G7 faced much pressure to perform well. Some argued that the Elmau Summit was "the last chance for the G7" to prove itself useful (Böhme 2015). The G7 needed to prepare for its "best chance to finally do something" at the UN in September where new goals for sustainable development — that would include climate change — would be adopted (Böhme 2015).

Most schools measured the summit's success by Merkel's announcement to decarbonize the global economy by the end of the century and her success, despite resistance, in leading the G7 to consensus here. Achim Steiner, executive director of the UN Environment Programme (UNEP), saw that as "a significant stepping stone toward a future of 100% renewable energy" (UNEP 2015). Hatiar (2015) welcomed Elmau's agreement to reduce carbon emissions to between 40% and 70% by 2050. Kirton (2015b) declared the summit a "strong success" on climate change. The European Climate Foundation, Avaaz and the ONE Campaign praised the G7 leaders' commitment to phase out "inefficient" fossil fuel subsidies, but World Vision, Wellcome Trust and 350.org were more reserved (Connolly 2015).

Argument

The Elmau Summit produced a strong performance on climate change, making it the most successful G7 summit from 2009 to 2020. Its 23 climate commitments had 80% compliance — the best between 2015 and 2020. Leaders dedicated 19% of the communiqué to climate change, the second highest portion in history, and affirmed democratic principles. But the ambition of their commitments varied and no money was mobilized for climate change. The outside development of global climate governance was high, demonstrating a clear consensus on the UNFCCC's facilitative role and the desire to make COP21 a success.

This strong success was somewhat driven by substantial climate shock-activated vulnerability. It was strongly propelled by fear of another multilateral organizational failure in Paris. Another contributor was stronger democratic political convergence and commonality among G7 members, with a de-democratizing Russia now suspended due to its annexation of Ukraine's Crimea in 2014. Significant domestic political cohesion of G7 members and, above all, Merkel's highly experienced and skilled leadership were significant causes.

The Elmau Summit did give full support to ensure the global community agreed at COP21. However, much was still missing to control the threat of relentlessly rising climate change, including credibly ending fossil fuel subsidies and killing killer coal immediately within the G7; ensuring the voices of small island developing states (SIDS), local communities, women and Indigenous peoples were heard; and strengthening links between climate change and its root causes by putting the environment and not economic growth first.

Plans and Preparations

The Elmau Summit was scheduled for June 7–8. Merkel chose Schloss Elmau for its secluded location in a nature reserve, against the backdrop of the Bavarian Alps, which aligned well with the summit's environmental agenda and the founding spirit of the G7 club — to foster free, unscripted discussion on the world's most pressing challenges.

Merkel was a highly experienced host. Elected in 2005, she hosted the 2007 G7 Heiligendamm Summit. As a trained scientist and former environment minister, she understood and respected the scientific concepts and evidence about climate change.

Participants

Joining Merkel were the UK's David Cameron, France's François Hollande, Italy's Matteo Renzi, Japan's Shinzo Abe, Canada's Stephen Harper, the US's Barack Obama and the two leaders of the European Union, Donald Tusk and Jean-Claude Juncker. The other invited leaders came from seven African countries and Iraq, and the heads of six core economic international organizations, plus the World Health Organization (WHO) and the African Union (AU).

Summit Theme and Agenda

Climate change and sustainable development were high on the agenda. On a visit to Tokyo on March 7–8, Merkel highlighted the priorities of climate change, COP21 and whether an ambitious binding climate agreement was possible in

2015. She noted that she intended to "demonstrate that the G7 states are willing to take on a leading role in fostering low-carbon development" that "does not mean renouncing prosperity" (Germany, Federal Government 2015).

Sherpa Process

Sherpas met in Frankfurt and Berlin in mid April. They heard presentations on climate change, COP21 and the post-2015 development agenda from academics, officials, trade representatives, the business community, civil society groups and union representatives.

Ministerial Meetings

Germany mounted an extensive set of ministerial meetings, but none for environment ministers. Ministers responsible for energy, foreign affairs and science and innovation addressed climate change. Finance ministers did not, despite a coalition of 120 business leaders urging them to provide greater business certainty by supporting a long-term global emissions reduction target (Patrick 2015).

Energy ministers mentioned greenhouse gas emissions four times in their communiqué. They acknowledged climate science and the need to move from traditional energy sources to low-carbon or no-carbon ones. They supported COP21 and committed to "do their part" to limit the global temperature increase to no more than 2°C above pre-industrial levels. They prioritized improving energy efficiency and employing technologies, such as carbon capture and storage for natural gas, without noting the climate impact of methane emissions from gas. They committed to support the development of renewable energy, particularly offshore wind energy, agreeing to improve coordination and transparency in global spending on clean energy. They also committed to eliminate inefficient fossil fuel subsidies.

Foreign ministers mentioned climate change nine times in their communiqué. They linked insecurity in the Sahel explicitly to climate change. Six paragraphs on climate change and security opened with "climate change is among the most serious challenges facing our world." Ministers noted that the impacts of climate change, including heat and changing precipitation patterns, "heighten the risk of instability and conflict." They recognized the science, adding that "deep cuts" in emissions were needed in line with the report of the Intergovernmental Panel on Climate Change (IPCC). They welcomed the study they had commissioned on climate and fragility risks and created a working group to evaluate its recommendations.

Science and innovation ministers referenced climate change three times. They related it to healthy oceans, including ensuring that oceans were included in the outcome of COP21.

On the Eve

In early May, the G7 was still searching for how to send a strong political signal, given that not all the world's major emitters would be at Elmau. There was a strong sense of European solidarity. The communiqué text was important to the United States with regard to the legal standing of any Paris outcome. The United States had secured satisfactory language on climate change at the G20's Brisbane Summit in 2014. Canada also had to consider its federal/provincial/territorial issues.

Germany drove the G7's climate change agenda, showing domestic leadership with its program to transition to clean energy. France offered strong support. Merkel and Hollande had pushed their G7 partners to announce their intended nationally determined contributions (INDCs) for COP21 to the UNFCCC secretariat before Elmau, which the United States, Canada and the European Union duly did. As host of the UN climate summit, Hollande had much at stake and thus sought support at Elmau. He pushed for more climate finance. French officials persistently reminded their G7 colleagues about climate change throughout the summit preparations.

Canada encouraged the G7 to see climate change from the broader perspective of a full UNFCCC negotiation. Both Canada and the United States wanted all major emitters to attend COP21 and commit to controlling their carbon under a new regime. This, along with climate finance, was also important for Merkel's coalition government and for the United Kingdom. There was some discussion on new climate financing mechanisms and mobilizing money. All G7 members had already committed to the Green Climate Fund, which had not yet started disbursing money. Canada considered the private sector key to bringing more countries on board.

At the Summit

The summit's third working session, on the morning of the second day, was dedicated to climate change and energy. This was followed by a session with African leaders and heads of international organizations to discuss sustainable development, although security topped African leaders' concerns.

Despite many points of agreement, G7 leaders' views differed on the strength of the language on climate change. Merkel faced pushback from both Canada and Japan for her initiative to have the G7 endorse a politically binding pledge to achieve net zero emissions (Blanchfield 2015). France offered the strongest support. Merkel secured consensus: the communiqué said "deep cuts in global greenhouse gas emissions are required with a decarbonisation of the global economy over the course of this century." This was achieved despite the climate change session being shortened to enable a discussion of recent terrorist shocks in Nigeria (Blanchfield 2015).

At the outreach session, G7 leaders met with South Africa's Jacob Zuma, Nigeria's Muhammadu Buhari, Senegal's Macky Sall, Tunisia's Béji Caïd Essebsi, Liberia's Ellen Johnson Sirleaf, Ethiopia's Hailemariam Desalegn and Iraq's Haider Al-Abadi. Some expressed concerns about Africa's disproportionately high vulnerabilities to the impacts of climate change and need for more climate financing from those historically responsible for emissions in order to adapt and build climate resilience (Kinkartz 2015). Buhari sought support from the G7 leaders to fight Boko Haram and support economic growth, which included environmental protection (Winsor 2015). Sirleaf included environmental issues in emphasizing the economy (Liberia, Executive Mansion 2015).

Also at the table were the heads of six major economic multilateral organizations: the UN's Ban Ki-moon, the International Monetary Fund's Christine Lagarde, the World Bank's Jim Yong Kim, Guy Ryder of the International Labour Organization, Roberto Azevêdo of the World Trade Organization and Angel Gurría of the Organisation for Economic Co-operation and Development (OECD). The G7 asked the OECD for support in tackling climate change. The WHO's Margaret Chan and AU's Nkosazana Dlamini-Zuma also attended. No international environmental organization was represented.

Notably absent were representatives of small island states at greatest risk of the consequences of climate change due to sea level rise and natural disasters. However, Vanuatu's Baldwin Lonsdale led representatives from Tuvalu, Kiribati, Fiji, the Solomon Islands and the Philippines to release a declaration on the sidelines of the G7 summit calling for climate justice (Greenpeace 2015). Cyclone Pam had hit these Pacific island states in March, affecting over 200,000 people.

Dimensions of Performance

At Elmau, the G7 leaders reversed the downward trend in climate deliberations and decisions since 2009, complied well with their decisions and gave the most guidance to outside international institutions to date (see Appendix C).

Domestic Political Management

Domestic political management was strong (see Appendix C). All G7 leaders attended the summit, including the second working session on climate change, development, health and women on June 8. All invited outreach partners also attended their sessions.

In the communiqué, one compliment on climate thanked Germany for publishing the Background Report on Long-Term Climate Finance.

Deliberation

In their public deliberations, the G7 dedicated 19% of the Elmau communiqué to climate change, ranking it second among all subjects (see Appendix C). This ended lows since 2009, ranging from 15% in 2014 to 4% in 2013.

The leaders offered full support for COP21, the Climate Risk Insurance Initiative, the Initiative for Renewable Energy in Africa, the Alliance on Resource Efficiency and the Marine Litter Action Plan.

In their private deliberations, leaders had rich exchanges, enhanced by the pastoral setting, full attendance, leaders-only dinner and the two-day event.

Direction Setting

In principled and normative direction setting, Elmau's 20 references to open democracy and 29 to individual liberty and human rights contained three links to climate change (see Appendix C). They used the democratic principle of transparency: to "enhance transparency and accountability ... to track progress towards achieving [emissions reduction] targets"; to ensure transparency in the implementation of climate finance goals; and to increase transparency in clean energy research and development. All three presented democracy as a cause of climate change control.

The G7 linked climate change with two other causes: economic growth, as leaders agreed to put climate change at the centre of their economic policies; and agriculture, as leaders acknowledged the Global Alliance for Climate Smart Agriculture, without committing to join it.

Decision Making

In their decision making, the leaders made 23 climate commitments (see Appendix C). This was the most since 2009, and about double the number in 2013 and 2014. Climate change thus ranked sixth overall.

Moreover, 65% of the climate commitments were highly binding, if not highly ambitious. This maintained recent levels, as 58% of the 12 climate commitments in 2013 were highly binding, as were 69% of the 13 climate commitments in 2014.

Elmau's commitments included supporting the UN in reaching a new legally binding agreement to stop emissions growth, transforming the energy economy by 2050 and mobilizing $100 billion annually by 2020 in climate finance. A new target-specific commitment was made — to increase disaster risk insurance coverage by 400 million people in the most disaster-prone countries by 2020, as part of the Climate Risk Insurance Initiative and by building on existing efforts in developing countries and regions.

The G7 also synergistically committed to put climate protection at the centre of its growth agenda. Leaders made a soft but significant commitment that underscored the need for cutting emissions deeply and decarbonizing the global economy. They endorsed the science, committing to the "upper end of the latest IPCC recommendation of 40 to 70% reductions by 2050 compared to 2010."

Delivery

Members' compliance with these 23 climate commitments was strong. The five assessed for compliance averaged 80% (see Appendix C).

The highest compliance, at 88%, came on putting climate protection at the centre of G7 members' economic growth agenda and on mobilizing $100 billion in annual climate finance. Next, at 82%, came supporting disaster-prone countries' efforts to build climate resilience, followed by 75% on developing national long-term and low-carbon strategies. The lowest, at 69%, was on the Climate Risk Insurance Initiative.

The three assessed commitments with synergies with other commitments had 80% compliance and the two with no such synergy had 82%.

This high compliance flowed partly from the broad nature of the climate commitments, making them relatively easy to comply with, and from familiarity through iteration. Of the two commitments with numerical targets, the one reiterated since 2009 (on climate finance) had strong compliance. The new commitment (on climate risk insurance) had the lowest.

Development of Global Governance

In developing global governance, G7 leaders made three references to three inside bodies: they referred to the G7 itself on climate change cooperation with Africa, the G7 energy ministers on clean energy and low-carbon technologies and the G7 Alliance on Resource Efficiency.

They also made 44 references to 19 outside institutions. COP21 led with eight references. The UN in general and multilateral development banks had four each. The UNFCCC (counted separately from the UN), the IPCC, the World Bank and UNEP had three each. Development finance institutions, the Green Climate Fund, the OECD and the AU had two each. The rest had one each.

The balance between 3 inside and 44 outside references, for a ratio of 1:15, showed Elmau was highly outward oriented, with the UN's COP21 at the core. Elmau focused strongly on making Paris a success.

The majority of Elmau's references to outside institutions showed the G7 supportively following the UN-led climate regime. A few had the G7 leading outside bodies in a new direction. These references included a call to the multilateral development banks to advance work on climate finance, a call to G7 members for "a further exchange in all relevant fora in view of COP 21," a request that their energy ministers advance climate issues, a request for the UNEP International Resource Panel to report on solutions for resource efficiency with supplementation from the OECD, and the establishment of initiatives on resource efficiency, climate insurance and renewable energy. On clean energy in Africa, the G7 stated it would develop a plan for further action with France as president of COP21 and with the AU.

Causes of Performance

Elmau's strong performance is adequately explained by the six causes highlighted in the systemic hub model (see Appendix A) (Kirton 2013). The most salient causes were the climate-conscious host, strong democratic convergence and fear of another UN failure to produce a legally binding climate agreement.

Shock-Activated Vulnerability

The G7 communiqué recognized no shocks from climate change (see Appendix E). The leaders recognized only one vulnerability, to vulnerable countries' need for resilience, identified as regionally located in Africa, Asia, Latin America and the Caribbean and as SIDS. The G7 committed to increase climate insurance coverage in these countries in response.

In the elite media salient to all G7 leaders, there were few reports of climate shocks or vulnerabilities on the front pages of the *Financial Times* and the *New York Times*. Between January and June, the *New York Times* front page reported only an American poll finding that Hispanics were more concerned than Whites about climate change, Hurricane Sandy in New York, an earthquake in Oklahoma, the severe drought in California, an earthquake in Nepal and flooding in Houston. There were far more non-climate shocks and vulnerabilities, led by political-security and health.

In contrast, physical shocks and vulnerabilities caused by or linked to climate change rose (see Appendix G). Heatwaves, droughts, fires and tropical storms were becoming the norm, including in G7 countries. In June, the historic drought in California increased the risk of fire and water stress. Still, non-G7 members were hit the hardest. Over 4,000 people died during heatwaves in G20 member India and neighbouring Pakistan. Rising sea levels continued to threaten SIDS.

Moreover, scientific reports showed the cumulative vulnerabilities from climate change. By year's end, they had confirmed that 2015 was the planet's hottest on record (NASA Earth Observatory 2016).

As 2015 began, business risk assessments put climate change in a modest place. The World Economic Forum's (2015) Global Risks Report released in January listed the most likely risks as interstate conflict first, extreme weather events second and failure of climate change adaptation only seventh. It ranked the most impactful risks as water crises first, spread of infectious disease second and failure of climate change adaptation only fifth. Thus, extreme weather or climate risks took 2 (20%) of the top 10 spots, with none in first. This was a decline from 30% in 2014, a tie with 20% in 2013 and a rise from 10% in 2012, but a sharp decline from 40% in 2011 when ecological risks first appeared.

In all, shock-activated vulnerabilities severely affected countries outside the G7 and were not highly visible in the elite media. The scientific reports were likely most salient to Merkel, the one leader with a scientific background, and particularly influential as an experienced G7 veteran and host.

Multilateral Organizational Failure

In 2015, multilateral organizational failure was substantial. It came largely from the searing memory of the UN's failure of Copenhagen in 2009 and how the G7, distracted then by the global financial crisis, could now help make Paris succeed in the financial crisis–free world of 2015 (Kirton and Kokotsis 2015). Obama (2020, 516) described Copenhagen as producing "an interim agreement that — even if it worked entirely as planned — would be at best a preliminary, halting step toward solving a possible planetary tragedy, a pail of water thrown on a raging fire."

Another spur came from the MDGs' failure to dedicate a goal to climate change. There was now a strong desire to include one in their successor Sustainable

Development Goals (SDGs), to be launched at a UN summit in September 2015. Another spur was the failure of the G20's 2014 Brisbane Summit to perform well on climate change. Thus, the G7 made multiple references at Elmau, and at three ministerial meetings, to achieve success at COP21 in Paris.

Predominant Equalizing Capabilities

In 2015, the G7's predominant equalizing capabilities were substantial, both overall and on climate change (see Appendix H).

In overall capability, measured by gross domestic product (GDP), as economic growth slowed globally, in nominal terms the G7 maintained the same global share of 46%, even though it no longer included Russia's GDP, as it had since 1998. This 46% was close enough to a majority to make the G7 globally consequential and able to commit credibly to provide the climate finance that would help Paris succeed.

Among G7 members, internal equalization held steady. From 2014 to 2015, the US growth rate increased from 2.5% to 2.9%, Japan's from 0.4% to 1.2%, France's from 1.0% to 1.1% and Italy's from 0 to 0.8%, while the other members declined.

The G7's steady economic growth came with a corresponding decline in emissions for the full year by an average of 2.2% (see Appendix I). This was less than in past years, with Russia no longer in the club. Emissions in the United Kingdom, the United States, Japan and Canada declined, while the others' rose.

Converging Characteristics

Converging and growing democratic characteristics also helped fuel Elmau's strong climate performance. It was the first fully prepared, scheduled summit without non-democratic Russia, a country heavily dependent on producing fossil fuels that had long resisted ambitious G7 climate action.

Nonetheless, there was a small G7 decline and divergence in 2015, relative to the hastily relocated Brussels Summit in 2014, the first without Russia since 1998. Among the 167 countries ranked for democracy from 2014 to 2015, within the G7, only Italy rose in the ranks (see Appendix J).

G7 members' climate performance also diverged and declined from 2014 to 2015 (see Appendix K). Among the 58 countries assessed, only Italy's rank improved.

Domestic Political Cohesion

The domestic political cohesion backing G7 leaders was substantial, and high for the host, propelling the G7's strong performance.

Merkel, personally committed to climate change control, had attended nine annual G7/8 summits (starting with Vladimir Putin's in 2006) and hosted at Heiligendamm in 2007. She led a traditional coalition government of the Christian

Democratic Union and Christian Social Union with the Social Democratic Party. She thus firmly controlled her legislature and had strong public approval ratings (although they dropped temporarily in May). Her next election was two years away. Germany was lauded as a climate leader, despite relying on coal-fired power. Its reputation was helped by a Green Party with substantial domestic support.

Obama, a lawyer, was highly experienced after his six G7 summits, including as host in 2012. However, his Democratic Party lacked control of Congress as the end of his second term approached in 2016. The opposition Republicans strongly opposed climate change control. Obama's public approval was modest. The Green Party in the United States had a tiny following.

In Japan, Abe had firm control of his legislature, and had recently won a coalition majority. A career politician from a prominent Japanese political family with a formal education in political science, he brought much experience to the table, although only a few years' summit experience. On climate change, a post-Fukushima Japan, dependent on coal power, and with the namesake Kyoto Protocol, primed him to support Merkel.

The United Kingdom's Cameron had surprisingly won a majority government in early May. A career politician, he entered politics with a degree in philosophy, politics and economics. He was committed to climate change control, but worried about losing votes to the right-wing, climate change–denying Independence Party (Payne and Pickard 2021). His five years of G7 summit experience, including hosting in 2013, put him in a strong position. But Scotland had overwhelmingly elected representatives of a separatist party and Cameron had promised to hold a referendum on a new deal for the United Kingdom within the European Union by 2017. Three months before Elmau, he signed a cross-party pledge to combat climate change regardless of the May election outcomes (Clark and Pickard 2015). This may have been spurred by a rise in popular support for the United Kingdom's Green Party, with the largest rise coming from young voters (Green Party 2014). Cameron strongly supported a new global climate agreement.

In France, Hollande's socialist party controlled the National Assembly. Yet, he stood at historic lows in the polls, despite a short-lived surge after the *Charlie Hebdo* terrorist murders. At the end of March, the conservative opposition party swept two thirds of the department elections. France's Green Party received less than 1% of the votes there. Hollande had been to three summits but hosted none. Hollande strongly supported a climate deal at COP21, held in his national capital.

Italy's Renzi presided over a complex coalition government struggling to secure his desired political and economic reforms. He had attended only one summit, the suddenly prepared Brussels one in 2014. The youngest Italian prime minister and the youngest G7 leader that year, he had a background in law and had worked in marketing. Italy's Green Party had little support. Yet, Renzi was set to co-host, with Hollande, a preparatory meeting ahead of COP21 only days after Elmau's end (Brocchieri 2015).

In Canada, Harper's Conservative Party held a majority government since 2011. In public approval and party popularity, he was tied with the opposition in the lead-up to the next election in October. Canada's Green Party held 5–8% of the popular vote. Harper, an economist, was a veteran of G7 summitry, having attended every summit that Merkel had. He had hosted the Muskoka Summit in 2010, which did not perform well on climate change and did not hold an environment ministerial meeting. In 2011, Harper withdrew Canada from the Kyoto Protocol.

The two EU leaders were attending their first G7 summit. As leader of Poland, a country heavily reliant on coal, Tusk had resisted climate action, but was now expected by some to be a pragmatic negotiator on climate change (Evans 2014). Juncker, from Luxembourg, placed EU preparation for COP21 third among his 10 priorities for his presidency (Politico 2014). The European Green Party had won 7.3% of the popular vote in the 2014 European Parliament election.

Club at the Hub

At Elmau, the G7's position as the leaders' valued club at the hub of an expanding network of global summit governance increased (see Appendix L). Its compact participation and interpersonal solidarity grew, as the former G8 — with a difficult, less committed Russian president — returned to the G7 club.

At the hub of the growing global network of plurilateral summit institutions, Germany was a member of only two others — the G20 and the North Atlantic Treaty Organization (NATO). Moreover, the suspension of Russia meant the G7 lost its link to the BRICS (Brazil, Russia, India, China and South Africa) and to the regional Shanghai Cooperation Organisation. But the United States, Japan and Canada connected the G7 to the Asia-Pacific Economic Cooperation (APEC) forum, the United States and Japan to the East Asian Summit, the United Kingdom and Canada to the Commonwealth, and France and Canada to the Francophonie. All countries but Japan connected the G7 to NATO. All were members of the G20.

Among the guests, South Africa connected the G7 to the BRICS and to the democratic India–Brazil–South Africa Dialogue Forum.

The presence of the six major economic multilateral organizations plus the WHO matched the outreach of recent G20 summits. However, no environmental organizations were invited, including the UNFCCC, UNEP or other relevant organizations such as the International Renewable Energy Agency or the Food and Agriculture Organization.

G20 Antalya, November

Introduction

After Elmau, there were hopes that the G20's 10th summit, in Antalya, Turkey, on November 15–16, would build on Elmau's strong performance (Kirton and Koch 2015).

Debate

At Antalya's end, the first school of thought saw success. Best et al. (2015) said sustainability "featured as a crucial issue" with an emphasis on climate change and success at COP21. They highlighted the G20's endorsement of several action plans related to sustainability, including for least developing countries (LDCs), and renewable energy deployment.

The second school saw failure elsewhere. Kulik (2015) and Kayagil (2015), focusing on gender, concluded that despite the new Women 20 (W20), the G20 paid little attention to gender and made no link between gender and climate change.

The third school saw failure specifically on climate change. Kirton (2015a) argued this was due to the G20 leaders' deference to the UNFCCC's "fundamentally flawed regime," and the diversion of terrorist and migration shocks.

Argument

The Antalya Summit produced a small performance on climate change. Led by the increasingly authoritarian Recep Tayyıp Erdoğan, the G20 deliberated far less on climate change than the highly democratic G7 did at Elmau. It made far fewer decisions to advance climate action, even if it complied well with the vague ones it did make. It backed the UN by expressing its support for a successful outcome at Paris.

This small performance flowed from Turkey and its European colleagues' preoccupation with the escalating diversionary shocks of terrorism and the crisis of refugees from the Middle East entering Europe. This made it easy for G20 leaders to defer to the UNFCCC and avoid trying to overcome the traditional deadlock over burden sharing between the industrialized and emerging economies in the G20. Climate action was further reduced by Erdoğan's weak conviction about the urgent need to counter climate change and his distraction by terrorist and refugee shocks.

Plans and Preparations

At the core of Antalya's crowded agenda was the G20's traditional objective of strong, sustainable, balanced and inclusive growth, which Turkey emphasized under the theme of inclusiveness, implementation and investment (Kirton 2015a).

On climate change, Turkey planned Antalya to support the UNFCCC and COP21. It wanted to promote access to electricity for the more than 1 billion people living without it, supporting the UN's 17 new SDGs launched in September, with one goal focused on energy and another on climate change. Ending inefficient fossil fuel subsidies and mobilizing climate finance were equally important.

At the World Economic Forum in January, Prime Minister Ahmet Davutoğlu (2015) said Turkey's presidency would focus on energy access, especially in sub-Saharan Africa. He said Turkey was committed to the UN climate process and emphasized the need for a "joint, strong message for the future of our planet."

For the G20 summit agenda, Turkey presented three pillars when it assumed the presidency on December 1, 2014: strengthening the global recovery, enhancing resilience and buttressing sustainability (G20 Turkey 2014). The third pillar would focus on the post-2015 development agenda, energy sustainability and climate finance. Under development, environmentally related considerations would include sustainable food systems and smallholder farming. Under energy, the focus would be on access, reducing the costs of renewable energy investment opportunities, deploying public and private resources for energy investments, phasing out inefficient fossil fuel subsidies, and promoting energy efficiency and energy market transparency. Turkey's agenda recognized the importance of COP21; on financing climate action, it would focus on collaboration and cooperation among the various climate funds and on the needs of LDCs.

Sherpa Process

To prepare the summit, the sherpas met four times, first on December 15–16, 2014. There, the United States wanted to reduce methane emissions by 40% by 2025, and pushed hard on eliminating fossil fuel subsidies, against much resistance. There was still no consensus on what an "inefficient" subsidy was, and the United States struggled to get Saudi Arabia and Australia to agree. There was anticipation that Antalya would not do much on climate change and would produce only one paragraph on the issue: the leaders would not negotiate COP21 at the G20.

There were three other meetings to prepare for the summit. China sent an official to the last two, who was reportedly quiet at both. The US sherpa also sent an official to the second meeting.

Ministerial Meetings

Turkey held 11 G20 ministerial meetings, but only three addressed climate change.

At their first meeting, on February 9–10, the finance ministers and central bank governors asked the Climate Finance Study Group (CFSG) to collaborate with the relevant international organizations and report back to the ministers' meeting in April. They expressed support for success at COP21. At their second meeting, on April 16–17, they continued to discuss climate finance, and asked the CFSG to finalize its work for the September meeting. They also asked the Financial Stability Board (FSB) "to convene public- and private-sector participants to review how the financial sector can take account of climate-related issues." On September 4–5, they dedicated one paragraph of their communiqué to climate finance, and welcomed the CFSG's report plus the work of the OECD and the Global Environment Fund. They repeated their support for COP21 and recognized the need to scale up climate finance for developing countries.

One of the 15 commitments made by the agriculture ministers on May 7–8 noted that the Paris Agreement acknowledged the impact of climate change on food security and expressed support for the UNFCCC. Another broader environment commitment recognized that food loss and waste was a global problem with "environmental and societal significance."

On October 1–2, the energy ministers made one commitment, of 20, that supported the work of the UNFCCC "to achieve a positive and balanced outcome at COP 21." Four commitments were on energy access, including sustainable energy. Four were on energy efficiency, including fossil fuel subsidies. Three were on clean and renewable energy. Erdoğan (2015) opened the meeting, recognizing "our struggle against climate change." He noted that "renewable energy and the improvement of energy efficiency should be our focus," including "lower costs for renewable energy sources and better integration of renewables into conventional systems."

At the Summit

India and Saudi Arabia were outliers at Antalya, resisting any robust text on climate change. France, hosting COP21 two weeks later, led the call for bolder action. Obama reportedly looked at the draft communiqué text and called for it to be more ambitious.

Turkey wanted to view the agenda mostly through the lens of development and the SDGs. On the development agenda, the United States wanted energy access in Africa.

Results

Antalya produced only a small performance on climate change. Despite two nights of intense, prolonged negotiations among the sherpas, G20 leaders added nothing new to support the struggling COP21. They passed the buck, stating that "the UNFCCC is the primary international intergovernmental body for negotiating climate change."

They did, however, finally acknowledge the need to keep the world's warming to under 2°C from pre-industrial levels. They also applauded the many climate change control measures submitted to the UN. But they did not acknowledge that the UN had already declared these measures inadequate, nor did they promise to help close the gap. Leaders, following their finance ministers, requested the FSB to continue its work on climate change risks to the financial sector (Carney 2021, 318). This linked climate to the G20's core foundational mission of global financial stability. But there was no link to the G20's other mission of making globalization work for all.

Dimensions of Performance

At Antalya, the leaders generated a small performance on climate change on almost all the major dimensions of G20 governance, including the number of commitments they made.

Domestic Political Management

In domestic political management, attendance at Antalya was low, with only 18 G20 leaders attending. France's Hollande stayed home after the terrorist attacks in Paris on November 13, two days before the summit. Thus, the host of COP21 was not present to secure support for the larger summit he would host in a few weeks. Argentina's Cristina Fernández de Kirchner was also absent. There were no communiqué compliments on climate change.

Deliberations

The leaders issued two documents. Climate change totalled 1,129 words (8%) in the documents, more than the previous two summits, but below three of the six summits from 2015 to 2020 (see Appendix D).

The G20 strongly backed the UN, saying their negotiators would prepare to ensure success at COP21, including on mitigation, adaptation, finance, technology, development and transfer, and transparency. Leaders stated that they were prepared to implement their INDCs.

Antalya's leaders fully deferred to the UNFCCC as "the primary international intergovernmental body for negotiating climate change." They advanced no new agreements or breakthroughs on issues deadlocked at the UN. They only endorsed a voluntary toolkit of options for renewable energy deployment. Although this was positive for participating countries regarding energy efficiency and vehicle emissions, it did not cover all participating countries. The G20's biggest contribution was to encourage the FSB to continue engaging with the public and private sectors on climate risks. Process, not product, was all Antalya prompted.

Direction Setting

In direction setting, Antalya's leaders did nothing. They made no affirmations in a climate context of the G20's distinctive foundational missions of financial stability or globalization for all, or on the G7's missions of democracy and human rights (see Appendix D).

Decision Making

Among their 133 commitments, leaders produced three on climate change. This was the least since the G20's earliest years in 2009 and 2010 (see Appendix D). The three commitments all related to COP21. One affirmed the leaders' "determination to adopt a protocol, another legal instrument or an agreed outcome with legal force under the UNFCCC." The second pledged to "underscore" the leaders' commitment "to reach an ambitious agreement … that reflects the principle of common but differentiated responsibilities and respective capabilities, in light of different national circumstances," and the third again committed "to work together for a successful outcome of the COP21." One commitment used highly binding language and two were low binding. One linked to both economic growth and sustainable development. The climate-related areas of energy and agriculture had three commitments each.

Delivery

The one climate commitment assessed for compliance averaged a high 85%. Yet, it merely upheld the leaders' determination to adopt an instrument with legal force that would apply to all the parties at COP21 and would also "support growth and sustainable development." Compliance with five more ambitious climate commitments made in 2014 was 76%.

Development of Global Governance

In developing global climate governance, Antalya's leaders made one reference to its own FSB and four to outside institutions.

Of the references to outside institutions, three went to the UNFCCC and one to COP21 itself. None contained a specific ask or call to action. One affirmed the UNFCCC's intergovernmental centrality for climate change negotiations. Two backed the UNFCCC and one backed COP21. The fourth acknowledged other countries' backing of the UNFCCC through their INDCs.

Causes of Performance

Antalya's small performance was driven by successive diversionary security shocks on the summit's eve, in the form of terrorist attacks in Turkey, France and Russia, and migrants streaming into Turkey and Germany. Also salient were Erdoğan's weak position on climate change, declining convergence on democracy and domestic climate performance, the absence of the core multilateral climate organizations, and Turkey's limited global summit network.

Shock-Activated Vulnerability

The most recognized shocks and vulnerabilities in Antalya's communiqué related to security and migration, with none on climate change (see Appendix F).

There were few media-highlighted climate shocks. On the *New York Times* front pages between June and November, there was just one story on COP21 and countries' pledges in the lead up. Other climate-related stories included the Pope's call to climate action and his speech in Washington on climate and poverty, the drought and wildfires in California, domestic climate change regulations, Japan's post-Fukushima restarting of its nuclear reactor, falling oil prices, the Volkswagen emissions scandal in America, earthquakes in Afghanistan and Pakistan, and Obama's rejection of the Keystone XL pipeline. In media coverage by major US television networks, climate change coverage declined by 5% in 2015 from 2014 (Seifter et al. 2016).

These shocks flowed directly from the physical world, which was dominated by security issues, even as the toll from climate change rose. By the end of 2015, the number of deaths from extreme weather events in G20 countries rose to 7,898 from 3,757 in 2014 (see Appendix G). The G7 countries had the biggest rise, to

3,736 from 556 deaths in 2014. Wildfires scorched millions of acres in the United States and Canada. In India and neighbouring Pakistan, heatwaves claimed over 4,000 lives.

Scientific reports confirmed the chronic, invisible, cumulating vulnerability. New heat records were set and carbon dioxide levels in the atmosphere soared (WMO 2016).

Multilateral Organizational Failure

The focus on terrorism and migration reinforced G20 leaders' hope that climate change could be left to Paris two weeks later. They may have also felt it was too late to change the outcome, given COP21's highly confining procedures. Their deference to the anticipated result of the UNFCCC's flawed regime and process, rather than strong support of it, and their own unwillingness to negotiate with each other in the G20 forum contributed to Antalya's small performance on climate change (Kirton and Kokotsis 2015).

Predominant Equalizing Capability

Changes in G20 members' predominant equalizing capabilities further reduced Antalya's performance (see Appendix H). They were led by the strong currencies and good economic growth in the G20's largest member, the United States, and in Turkey, while the gap was growing between a rising G7 and declining BRICS. Yet, the greatest growth came from climate-skeptical Turkey, India and Saudi Arabia.

From 2014 to 2015, the G20's aggregate emissions rose by an average of 0.42% (see Appendix I).

Neither the G20's decrease in GDP nor its increase in emissions spurred Antalya's performance. This suggests that economic recession had a lower causal impact on emissions than first assumed.

Converging Characteristics

The democratic characteristics and domestic environmental performance of G20 members diverged and declined, reducing Antalya's climate performance. The democracy divide grew thanks to a no longer democratizing Turkey and the presence of Russia, which had annexed Crimea in 2014. The fossil fuel preferences of autocratic Russia and Saudi Arabia strengthened.

Democracy declined in over half of the G20 members, including some traditionally democratic G7 members, although collectively the G20 rose to 35th from 47th in 2014 (see Appendix J). Yet, this did not propel strong G20 climate performance.

On domestic climate change performance, the G20's collective performance declined slightly, to 51 from 52 in 2014 (see Appendix K). By rank, of the 60 countries assessed, G20 members declined by 20 places. Among the G20's most powerful members and emitters, the United States, Japan, Germany and India declined far more than China alone rose.

Domestic Political Cohesion

Domestic political cohesion was mixed. Chairing his first G20 summit and attending his ninth, Erdoğan had recently secured a majority electoral mandate. Yet, he had low personal convictions about climate change. Political cohesion was also high for China's Xi and for Germany's Merkel, who was now focused on the shock of Syrian refugees.

Obama, ineligible to run for re-election in November 2016, came as a lame-duck US president, still unable to have the Republican-controlled Congress back his personal convictions on climate change. Canada's Trudeau, in contrast, came with a fresh majority mandate for his priority of climate change, but it was his first summit. Beyond the G7, the strongest climate change–denying leaders from Saudi Arabia and Russia had high domestic political support.

Club at the Hub

Antalya was poorly positioned as the valued club at the hub of a global network of plurilateral summit institutions on climate change (see Appendix L). The sense of the G20 operating as an interpersonal club came only at dinner, when the 18 attending participants were left alone to act flexibly as leaders, but on the scheduled subjects of terrorism and Syrian refugees.

Turkey was a member of neither the G7 nor the BRICS. It added only the Organisation of Islamic Cooperation and NATO, whose support for climate change control was weak.

However, Antalya started off an intense summit sequence that in quick succession took several of its members to the APEC summit in the Philippines, the Commonwealth Heads of Government meeting in Malta and then to Paris for the culmination of COP21 on December 13. Yet, the only G20 leaders to attend all four summits were the two newcomers — Canada's Trudeau and Australia's Malcolm Turnbull. Obama and Xi attended only three. It is possible that at Antalya, Xi did not support Obama, in order to wait for a deal at APEC, as he had the previous year when APEC took place just before the G20's Brisbane Summit.

None of Antalya's many guest leaders of countries and international organizations was strongly committed to climate change. Moreover, the core climate multilateral organizations of the UNFCCC, UNEP and the UN Convention on Biological Diversity were all absent.

UN Paris, December

Introduction

Antalya's small performance placed even greater pressure on COP21. The climate crisis was increasingly evident, with all 10 of the hottest years on record occurring since 1998 (Kahn 2014). Extreme weather conditions worsened, with ocean temperatures warming, sea levels rising and severe storms strengthening. A 30-year trend of melting polar ice caps could not be explained by natural variability (Stroeve et al. 2014). Over 1 billion people were in danger of being submerged by coastal surges.

The IPCC (2015) reported that warming would have catastrophic, irreversible consequences, for which the world was ill-prepared. The world's climate was changing faster than predicted and more rapidly than intergovernmental cooperation could cope with or control (Keohane 2015).

Under these alarming climate conditions, the UNFCCC convened for COP21 in Paris in December 2015. Governments sought a legally binding agreement with concrete pledges backed by essential finance. Hollande embraced France's critical role and responsibility in chairing this landmark meeting. All UN members attended, including the most polluting ones.

Debate

In the debate about the Paris Summit, most schools saw partial success.

The first school saw a promising new phase in the global response to climate change, in substantive and institutional ways (Chan 2016). Advances included enhancing actions to be taken before 2020 and offering the most comprehensive framework for engaging non-state actors (Chan et al. 2016). Government leaders of major powers were prioritizing climate change, significantly shifting positions and committing to act through the Paris Agreement they produced. They agreed to keep the pre-industrial global temperature increase below 2°C and ideally 1.5°C, and to submit more ambitious plans every five years, knowing their initial voluntary promises would not achieve their below 2°C goal.

The second school saw partial success in process, with decentralized policy coordination building confidence that could lead to bigger emission cuts, even if deep cooperation remained elusive (Keohane and Victor 2016). A broad informal coalition was formed to produce the Paris Agreement, with its several flaws and inadequacies (Savorskaya 2016). Emerging powers led by China spurred the core principle of CDR (Jinnah 2017). Some pointed to the greater opportunities to engage Indigenous issues, even if the basic state-centric convention would limit the ultimate results (Ford et al. 2016). Others added the educational opportunity that Paris brought (Noelle 2016).

The third school emphasized the partial failure in the process, as most government delegates and non-governmental actors attended only to network and build a community to facilitate a polycentric climate regime (Lövbrand et al. 2017).

The fourth school highlighted the partial failure in substance, as the agreement's highly ambitious target came with national pledges that together could not achieve the desired result. Moreover, there was little clarity about how to meet this target and what transnational support would fill the mitigation gap (Michaelowa and Michaelowa 2017). The agreement's flexibility enabled the inclusion of some principles of historical responsibility, necessity and capacity to define who would bear the costs, but the most complex dimensions of climate justice were not resolved (Santos 2017).

Argument

The Paris Summit produced a solid performance. It overcame the UN's failure at Copenhagen in 2009 and agreed on an appropriately ambitious collective target for global temperature rise. Due to the collective efforts of the LDCs and the Alliance of Small Island States, it included the aspirational and scientifically backed target of 1.5°C. It also included the world's largest global emitters in the collective political obligation to control net emissions. It secured a commitment from China, the biggest emitter, and several other large, rapidly growing emitters. It also reiterated the commitment of developed countries to provide $100 billion annually in climate finance for the developing world.

Yet, Paris failed to produce the bold, politically binding agreements on actions by individual members needed to achieve its targets. Even with its new, flexible, bottom-up approach, the voluntary INDCs were inadequate, even if fully implemented. Moreover, the prevailing CDR principle clearly divided the small group of developed countries obliged to control emissions ambitiously from the much larger group of developing countries obliged to do far less. Thus, emerging economies were largely free to develop in their own carbon-intensive, ecologically free-riding ways, even if the dichotomy between developed and developing countries was softened. There was no independent monitoring framework with adequate enforcement (Dwortzan 2017). Paris also failed to constrain the fossil fuel industry (Heinberg 2015). With leaders not due to meet again for another five years, lower-level officials were left to implement and strengthen the Paris Agreement.

These shortcomings flowed from the UN's organizational failure, which COP21's summit could not transcend. The annual COPs were battlegrounds among negotiators, delegates and other government representatives, who lacked the political authority to reach consensus on an adequate climate agreement. Country leaders arrived only at the very end, after most agreements had been reached. Paris's limited success also flowed from the lack of a boost from the G20's Antalya Summit, moderate equalizing capability among UN members, little common or converging democratic purpose, poor political cohesion and, above all, no constricted, controlled participation in a personally valued club attended by leaders every year.

On the Way to the Summit

Preparations for Paris began in 2014 at COP20 in Lima, Peru. Ministers produced the first broadly multilateral, UN-incubated deal in which all countries agreed to control their carbon emissions (Davenport 2014; McGrath 2014). They committed to submit national plans to reduce emissions by June 2015, indicating how much and how they would cut their emissions after 2020, with targets beyond their "current undertaking" (Davenport 2014; UNFCCC 2014). The UN would make these plans transparent. The UNFCCC (2014) secretariat would report on these INDCs, which would form the foundation for the regime to be agreed at COP21 (Davenport 2014).

But much was missing. With no country obliged to make significant cuts, the combined INDCs were expected to fall far short of the cutbacks needed to hit the 2°C target. Countries not submitting plans would not face sanctions. China, India and their Group of 77 (G77) allies secured a reaffirmation of the CDR principle. They eliminated a proposal requiring quantifiable data on how to meet emissions targets and vetoed any review mechanisms. The United States also opposed binding commitments (Clémençon 2016). The deal thus relied heavily on public and peer pressure. Still, India and Australia indicated they would submit their plans by June.

Attention thus turned to COP21. Preparatory meetings began in Geneva in February and continued in Bonn in June. Countries tried to shrink an 89-page draft text. Progress came on capacity building, education, public awareness and administrative, financial and institutional matters. Work also advanced on methodological guidance for reducing emissions from deforestation and forest degradation in developing countries, conservation, sustainable forest management and the enhancement of forest carbon stocks (*Earth Negotiations Bulletin* [ENB] 2015a). Bonn produced an agreement in principle and its streamlined text would form the basis of the Paris Agreement.

At the Summit

Convening in Paris from November 30 to December 13, COP21 had all G20 leaders attend except Saudi Arabia's King Salman bin Abdulaziz Al Saud and Argentina's Fernández de Kirchner.

There were 13 days of intense, all-night negotiations on complex, technical issues, including adaptation, loss and damage, ambition, response measures and differentiation. Negotiating positions differed markedly. The European Union pushed for climate targets for all countries "based on evolving global economic and national circumstances" (Appunn et al. 2015). France and China declared that the Paris Agreement should allow for revising the goals every five years, in addition to a five-year assessment of individual country progress on long-term goals (Geiling 2015). Merkel noted that the responsibility for climate action rested on many shoulders, including the emerging economies. She acknowledged that industrialized countries had contributed heavily to climate change, and that the developing world required financial support to mitigate and adapt to a changing global climate — a principle supported by the CDR concept. But she also stressed that "fair and binding rules for both industrialised and developing countries" were needed to secure global investment in low-emission development (Appunn et al. 2015).

China stressed maintaining the principles of CDR and respective capabilities. It supported differentiating between developing countries and developed countries with historical responsibility for high emissions and thus requiring absolute quantitative reduction targets. Developing countries, including China, would undertake mitigation based on support from the developed world.

The US refusal to make internationally binding targets largely reflected the legal limits of Obama's authority as well as what other countries were committing to (Purvis 2015). At US insistence, the 31-page Paris Agreement was crafted explicitly to exclude firm reduction targets and financing obligations from its legally binding components. It excluded any clause exposing the United States to liability and compensation claims for causing climate change. Other components, including the five-year reviews, would carry legal force, but would free Obama from having to submit the deal to Congress (Goldenberg 2015).

As one of the largest emitters, India officially opposed any long-term goal of decarbonization, carbon neutrality or the application of 100% renewable energy by 2050. It feared the added responsibility that could fall on emerging economies in the absence of any adequate or predictable technology transfer and finance commitments from developed countries (Vashist n.d.). India argued for a 10-year reporting cycle (Clémençon 2016).

As the world's largest oil producer, Saudi Arabia expressed concerns about low global oil prices. Although it hoped to reduce 130 million tons of carbon emissions by 2030, its INDC provided no threshold for measuring such reductions (Kingdom of Saudi Arabia 2015). Like India, Saudi Arabia also pushed for a 10-year reporting cycle (Clémençon 2016).

China and the G77 (including India) said they were "deeply concerned with the attempts to introduce economic conditions" (G77 2015). They stated that replacing developed countries' "core obligation" to provide financial support to developing countries with "arbitrarily identified economic conditions" violated the rules-based multilateral process.

The Umbrella Group of emerging and developed economies, including Australia, Canada, Japan and the United States, maintained that everyone should share the reductions needed to reach the 2°C target. It would not agree to a deal that excluded China and India. It also maintained that the stringency of reporting and accounting for emissions should be similar across all countries, as reductions should be determined on the basis of current, not historic, emission levels (Gupta and Mandal 2015).

The BASIC bloc of Brazil, South Africa, India and China, which had committed to act jointly at Copenhagen, stressed the need for the developed world to provide $100 billion annually to developing countries by 2020. On behalf of BASIC, China supported a transparent, party-driven process, but emphasized that the Paris Agreement should follow all UNFCCC principles and provisions, particularly on CDR (PTI 2015).

The LDCs called for safeguards for three overarching elements: the highest possible legal rigour under international law (i.e., a legal instrument with legally binding commitments); universal participation; and effective provisions to ensure adaptation and loss and damage from irreversible climate impacts (Yeung 2016). The poorest countries called for loss and damage as a stand-alone article and an enhanced system for transparency, action and support, with flexible terms for LDCs and SIDS (Abeysinghe et al. 2016).

Facing the prospect of severe land loss and mass migration, SIDS were in a very difficult position. For years, SIDS and African countries had been calling for the temperature rise not to exceed 1.5°C and continued to advocate this at COP21 (Clémençon 2016). Yet "the idea of even discussing loss and damage now or in the future was off limits," according to American officials who feared such discussions would "kill the COP" (Goldenberg 2015). Ultimately, the SIDS and LDCs prevailed.

Results

The Paris Summit overcame the UN's failure in Copenhagen by agreeing on an appropriate collective target to limit global temperature rise that included all the big emitters in the individual obligations, including China and India. Paris set high aspirational goals of limiting the increase to 2°C, and striving to keep temperatures at 1.5°C. These targets were far more ambitious than expected. Developed countries recommitted to providing $100 billion annually in climate finance. Paris accepted the G20 call for the FSB to establish the Task Force on Climate-related Financial Disclosures to recommend clear, common, trusted, high standards on climate risk reporting for private sector firms (Carney 2021, 318). Paris also advanced non-state actor engagement.

Yet, COP21 failed to produce a legally binding agreement with the credible targets and timetables needed to manage the consequences of a drastically changing climate. The Paris Agreement offered little legal weight and no means of enforcement. Each country was required to set its own target, which would become its nationally determined contribution, without legal repercussions for failing. Collectively, the INDCs would produce about 3.2°C above pre-industrial levels by 2100, thus falling far short of 2°C, let alone 1.5°C (Raftery et al. 2017).

Causes of Performance

The causes of Paris's solid performance are highlighted by the systemic hub model of global governance (Kirton 2013).

Shock-Activated Vulnerability

There were few climate shocks during the two weeks following the G20's Antalya Summit to make the key leaders arriving in Paris more aware of their countries' immediate vulnerabilities to climate change, and thus break the remaining deadlocks.

However, the deadly fires and heatwaves during much of 2015 had been on everyone's minds. The planet was more than 1°C warmer than the first recorded temperature in 1880 (NASA Earth Observatory 2016). The IPCC's findings were ever more clear and stark.

Multilateral Organizational Failure

Leaders remembered the great failure of Copenhagen, largely due to the status quo positions of emerging economies led by China. The COP21 process of creating an official document that all parties, without exception, had to agree on, sought to avoid such a situation. This was led by the G20 members of China, India and the United States, although LDCs and SIDS were influential too.

Still, the legacy of the 1997 Kyoto Protocol remained: developed countries were expected to produce large-scale, absolute emission reduction targets, and developing countries were to enhance mitigation efforts. Although emerging economies were now part of the Paris framework, the continuing separation between developed and developing countries prevented a universal, legally binding agreement that would keep temperature levels below 2°C. It also enabled most major emerging economies to continue developing in carbon-intensive ways (Stavins and Stowe 2010). Moreover, the developed economies also did not submit strong targets. This prevented the Paris Agreement from becoming fully legally binding, with consequences for non-compliance.

Predominant Equalizing Capability

Changes in relative capability both caused and constrained success. Since Copenhagen, China, India and other emerging economies had come to produce a much larger share of global GDP and emissions (see Appendices H and I). Developed countries and emerging economies alike recognized that they must agree to control their emissions if the UN was to reach an agreement to control the much greater climate problem of 2015 that they all helped produce. Moreover, emerging economies also recognized that they remained more vulnerable to the consequences of climate change than did the developed ones, which also had far more capabilities to adapt. China and its BRICS colleagues agreed to a somewhat weaker expression of the CDR principle, with the terms "developed" and "developing countries" in place of Kyoto's dichotomous Annex I and non-Annex I categories. Paris also specified the particular position of the SIDS (Pauw et al. 2019).

Still, China and the G77 reiterated the importance of financial contributions from developed countries for adaptation and mitigation and argued for quantitative targets based on historic emissions (Grieger 2015). The far less powerful SIDS knew that their acute vulnerabilities required much greater financial support from the far more economically powerful and responsible states. But the Paris Agreement merely stated the importance of averting, minimizing and addressing loss and damage. Above all, China and the BRICS refused to promise a proportional share of their reductions in their INDCs and ensured that provisions for independent monitoring and verification were very weak.

Converging Characteristics

Low-level, decreasing convergence on democratic characteristics underlay the UN's failure to reach an ambitious universal agreement (see Appendix J). The UN

Security Council, at the core of the UN system, had moved divisively in an anti-democratic direction with Russia's 2014 invasion of Ukraine. Russia appeared willing to weather the ensuing "temporary" sanctions, as the "acquisition of territory is permanent and strategic" (Volker 2014). Democracy was declining in the United States, Brazil, Japan and France as well, and populist leaders gained traction in Europe and the United States.

Domestic Political Cohesion

Low domestic political cohesion constrained success, particularly in the United States, the world's largest economy and second largest carbon emitter, with Obama's second term ending in a year and both houses of Congress controlled by Republicans strongly opposed to climate change control.

In sharp contrast, domestic political cohesion was high in the powerful countries opposed to ambitious climate change control. China's Xi was consolidating his power (Kirton 2016a). In Russia, Putin's political control and popularity soared after the annexation of Crimea. Political cohesion remained very high in Saudi Arabia too.

Club at the Hub

The UN's position as the leaders' valued club at the hub of a growing network of global summit governance was very low (see Appendix L). As host of the Paris conference, Hollande had little popularity at home and would soon leave the presidency. The UN had not held a leaders-level climate summit in six years. Few key leaders other than Obama had ever attended one.

Above all, Paris was far more a bottom-up, minister-produced process than a top-down, leader-dominated one. To be sure, more than 150 leaders came, the most to attend a UN event on any single day. But they were surrounded by over 36,000 participants, including over 23,000 government officials and 3,700 media representatives (ENB 2015b). This left very little opportunity for informal, spontaneous interactions and agreements among leaders.

Conclusion

The year 2015 marked a turning point in the global governance of climate change. It started strongly under Merkel's G7 leadership at Elmau. It then stumbled under the less democratic leadership of Erdoğan, whose G20 Antalya Summit merely pledged support for COP21 and the broader UNFCCC, while doing little to ensure their success.

The massively inclusive multilateral nature of the UNFCCC and Hollande's domestic distractions and political weakness as the Paris Summit host made it only a solid success. It set the 1.5°C target led by the now included LDCs and SIDS. It also reaffirmed the target of limiting the world's pre-industrial temperature rise to 2°C, relaxed the confining Kyoto Protocol categories of fully constrained developed countries and completely unconstrained developing countries, had China as the world's leading polluter agree to control emissions and had developed countries agree to provide major climate finance to developing countries (Pauw et al. 2019).

Yet, its voluntary, bottom-up approach with no mechanisms for implementation, monitoring, verification and timely upgrading meant that even with full compliance, the target would not be reached in time to avoid the climate crisis that scientists said would lead to much death, damage and disease. The Paris Agreement laid good groundwork, but its voluntary, aspirational incrementalism would not facilitate the transformative progress that reason called for and humanity required.

The varying success of the three summits was largely due to the differing shocks, institutional constraints, power balance between their democratic and non-democratic members, and the domestic support and political skill and will of their hosts, along with an across-the-board loyalty to the embedded fossil fuel economy by key members. Merkel's G7 Elmau Summit had few distracting shocks, full institutional flexibility and democratic commonality with Russia's Putin gone; she had strong political backing at home and high personal experience, expertise and conviction to drive strong success, relative to the G7's own past summits if not relative to the demands of the changing atmosphere and ecosystem. Erdoğan's G20 Antalya Summit had none of these propellers and thus made very small progress. Hollande's UN Paris Summit similarly suffered from diversionary shocks, and added more international institutional constraints, a minority of democratic members, climate-resistant leaders with high domestic political control and a summit that they neither valued nor controlled.

3   Relying on Paris, 2016

Introduction

After the landmark United Nations climate summit in Paris in December 2015, attention turned to ratifying the Paris Agreement, raising the promised climate finance and taking all the other actions necessary to make it work. As world leaders were not due to meet at another UN climate summit for another half decade, these tasks fell to the G7 in Ise-Shima, Japan, in May 2016, and the G20 in Hangzhou, China, in September 2016. Both these summits had far more compact, valued participation as clubs at the hub of global summit governance and the flexibility for leaders to do bold new things spontaneously. Both faced rising shocks with scientific evidence rapidly gaining force as global temperatures reached new highs. The G7 summit was hosted by Shinzo Abe, leader of a geographically small island country vulnerable to typhoons and sea level rise and long committed to climate change control. The G20 was hosted by Xi Jinping, president of a vast country that was the world's largest climate polluter and who was acquiring increasing personal control of the state.

Yet, in practice, relying on Paris meant relaxing after Paris. Japan's Ise-Shima Summit produced a substantial performance, but only half as many climate commitments as Germany's 2015 Elmau Summit, and it achieved lower compliance. China's Hangzhou Summit produced fewer commitments and less compliance than even Turkey's 2015 Antalya Summit. And the UN had to rely on ministers at its 22nd Conference of the Parties (COP) to the United Nations Framework Convention on Climate Change (UNFCCC) at Marrakech in November to try to implement and improve its own inadequate Paris regime.

The poor performance of all three global governance institutions shared some causes. The UN's failures at Paris included its messages that leaders could wait five years to deal with the substance of climate change and that the first task now was just to ratify the Paris Agreement. Many unrealistically hoped that the mere agreement on the 2°C target would inspire voluntary, individual, incremental improvements in time to meet the collective goal. A lame-duck US president Barack Obama was increasingly attacked by his climate change–denying Republican rival, Donald Trump, as the November election approached. This encouraged those G20 leaders who denied the impacts of climate change

DOI: 10.4324/9780429055485-3

to delay doing their part. They were joined by Xi, who now possessed prerogatives as G20 host.

G7 Ise-Shima, May

Introduction

Chaired by Abe, the G7's 42nd summit at Ise-Shima on May 26–27 was the sixth time Japan had hosted since its first in 1979 (Kirton and Kokotsis 2015).

Debate

The first school of thought about Ise-Shima's performance saw a solid success on climate change and energy, relative to previous G7 summits (Kirton 2016b). Yet, given "the urgency and magnitude of the climate change threat," it deserved "a grade of F for failing its own citizens and the world as a whole" (Kirton 2016b).

The second school put technology and development first. Kudo (2016) stated that to succeed on climate change and the UN's 17 Sustainable Development Goals (SDGs), the G7 needed to reduce emissions, and establish "a framework for providing financing and technical support, in coordination with the private sector, to emerging countries" to improve infrastructure and technologies for reducing emissions.

The third school saw Ise-Shima's emphasis on other subjects supplant progress on climate change. Kulik (2016) and Barnett (2016) noted Ise-Shima's advances on gender equality. Brown (2016) and Médecins sans Frontières (2016) highlighted health. Prince Michael of Liechtenstein (2016) declared failure over Iraq. *The Diplomat* (2016) added failure on economic growth.

Argument

The Ise-Shima Summit produced a substantial performance on climate change, but it was inadequate to meet the growing need and the great gap left by the Paris Agreement. Its deliberations exceeded those at Elmau in 2015 and covered more issues, but leaders focused only on ratifying the Paris Agreement. Compliance was high with its Paris-focused commitments but much lower on more substantive, specific ones. Its greatest advance was, for the first time, to specify the date of 2025 for ending inefficient fossil fuel subsidies. It also revived, after a seven-year absence, the G7 environment ministerial meeting, which acted effectively on the summit's eve.

Ise-Shima's substantial performance was driven by several forces. A deadly earthquake in Japan two months before the summit reinforced memories of the Fukushima tsunami disaster in 2011. The UN's launch of the Paris Agreement and the SDGs provided a focus, platform and many gaps for G7 leaders to fill. Members' convergence on democracy and climate performance grew. Above all,

Abe, at his fifth summit, inherited Japan's legacy of hosting successful G7 summits on climate change, now with the support of committed, experienced colleagues, led by Germany's Angela Merkel at her 10th summit and as the custodian of Elmau. Japan, along with Canada, highly valued the G7 club. Its global summit network grew with the Nuclear Safety Summit that year. But constraints came from the leaders' belief that climate action meant just implementing the Paris Agreement's immediate procedural requirements and from Obama's weak position at home.

Plans and Preparations

Japan's climate change priorities were to maintain post–Paris Agreement momentum with an emphasis on its implementation and to reduce emissions from major emitters. Energy priorities included promoting investments in upstream development, quality infrastructure, clean energy technology, gas market security and energy efficiency.

Pre-Summit Meetings

Abe visited several of his G7 colleagues, but only France's François Hollande — host of the Paris Summit — raised the issue of climate change.

Sherpa Process

Meeting on April 20–21, the sherpas agreed the G7 needed to lead on climate change, starting with ratification of the Paris Agreement and climate financing. Canada also sought action on short-lived pollutants, the Montreal Protocol on ozone-depleting substances, methane, heavy-duty vehicles, the Arctic and inefficient fossil fuel subsidies.

Ministerial Meetings

All seven pre-summit ministerial meetings referred to climate change in some way. On April 11, the foreign ministers' communiqué recognized the serious security and economic threat posed by climate change, and welcomed the Paris Agreement and COP21. Ministers committed to support the adoption of an amendment to the Montreal Protocol to phase out climate-warming hydrofluorocarbons and to provide funding for its full implementation. They endorsed the recommendations of the G7 Working Group on Climate Change and Fragility and committed to prevent climate fragility risks. US secretary of state John Kerry's personal commitment to climate change control drove the meeting's success.

On April 24, food and agriculture ministers produced a declaration that mentioned climate change 10 times. The preamble said climate change is threatening natural resources, farming systems, food security and livelihoods. Ministers supported the Paris Agreement and the Global Research Alliance on Agricultural

Greenhouse Gases and recognized the importance of the Global Alliance for Climate Smart Agriculture. They linked climate change to resilient infrastructure and illegal logging.

On April 30, ministers responsible for information and communications technologies (ICT) linked climate and digitalization. Climate change arose twice in relation to sustainable, inclusive development, including in a statement that a digitally connected world would help achieve the green information technology principles and goals of the Paris Agreement. Ministers committed to share experience with using ICT to contribute to the SDGs, especially regarding energy and climate change, resilience and disaster risk reduction, and sustainable transportation and logistics.

On May 1–2, energy ministers welcomed the Paris Agreement and committed to implement it and accelerate work to decarbonize the global economy. However, they continued to support methane-producing natural gas, stating that shale and other unconventional gas should be developed voluntarily within an "adequate regulatory and environmental framework."

On May 15, education ministers mentioned the general "challenge" of climate change.

On May 16, environment ministers made 66 commitments on climate change, many focused on sustainable material management.

On May 17, science and technology ministers linked climate and oceans, and noted the importance of innovation for low-carbon economies highlighted in the Paris Agreement.

At the Summit

Climate change was one of Japan's six summit priorities.

Canada's Justin Trudeau began laying the groundwork for the G7 summit he would host in 2018. Canada had succeeded in working through the environment ministers and sherpas to secure recognition in the leaders' communiqué of the disproportionate impact of climate change on women. It also pushed for references to black carbon, ozone-depleting substances, methane and climate finance.

The EU's Jean-Claude Juncker and Donald Tusk noted that it would be highly difficult for them to meet other leaders' calls for quick ratification of the Paris Agreement. With 28 members, it would take time to determine a single emissions reduction target. Germany allowed language in the communiqué that recognized nuclear energy as a zero-emissions energy source. All leaders agreed they needed to express support for the SDGs.

Results

G7 leaders recommitted to eliminate fossil fuel subsidies, as they had committed to do in the "medium term" every year since 2009 through the G20 and since 2012 at the G7. Now they specified a precise deadline of 2025 for the first time. Yet, a comprehensive definition for "inefficient" fossil fuel subsidies was not agreed upon, nor were the means for achieving the goal.

G7 leaders also pledged to ratify, accept or approve the Paris Agreement, but set no deadline to do so. They also did not formulate low emissions strategies, as they had promised to do by well before the 2020 deadline. On climate finance, they promised no new money, despite their previous promise to provide $100 billion annually by 2020. They recognized but made no commitments on climate risk insurance to follow up Germany's InsuResilience initiative.

Japan's G7 did revive, after a long seven-year absence, the G7 environment ministerial meeting. It focused on disaster waste management and a circular economy, rather than mitigation and adaptation, and resumed the cadence of meeting annually from 2017 to 2021.

Dimensions of Performance

Ise-Shima produced a substantial performance on climate change. It outperformed Elmau in deliberation and direction setting, but underperformed on commitments and compliance.

Domestic Political Management

Ise-Shima's domestic political management was significant. Attendance was complete. Yet, no communiqué compliments were paid to any member on climate change.

Deliberation

Deliberation was strong, with 3,802 words or 17% of the total on climate change (see Appendix C). This was much higher than the annual average of 721 or 5%. As a percentage, Ise-Shima ranked fourth among the 42 G7 summits to date.

The communiqué addressed the Paris Agreement, decarbonizing the global economy, investing in clean energy, engaging in environmental trade and investing in quality infrastructure to withstand climate-related disasters. It addressed the environment–migration crisis, climate–gender links and the environmental determinants of diseases, including antimicrobial resistance.

Leaders reiterated support for the InsuResilience initiative, welcomed the creation of the Carbon Market Platform, committed to carbon neutrality by 2020 in the aviation sector and to reducing short-lived climate pollutants and methane, linked climate and agriculture, and addressed resource efficiency, plastic litter, forests and green infrastructure.

Direction Setting

Direction setting was very strong. Leaders made four affirmations of the G7's two foundational missions of promoting democracy and promoting human rights in the context of climate change (see Appendix C). Only the 2009 summit had connected human rights to climate change, making Ise-Shima the most successful summit here.

The two affirmations of democracy centred on the transparent implementation of the nationally determined contributions (NDCs) under the Paris Agreement and of climate finance.

The other two affirmations were on women's rights. In a highly binding commitment, the G7 agreed to take a "gender-responsive approach" to climate change. They also recognized, in the context of development in Africa, that "security, development, climate change and gender equality are interdependent and instrumental."

Decision Making

Decision making was substantial. The 12 climate change commitments were the eighth highest, but only half the total at Elmau (see Appendix C). Most referred to the Paris Agreement and implementing rather than improving it.

In the preamble, leaders committed to implement the Paris Agreement "swiftly." They agreed that decarbonizing the global economy could facilitate economic growth, that environmental trade standards should be improved, that aviation should be carbon neutral from 2020, that $100 billion still needed to be raised annually by 2020, that they would participate in the Paris Agreement's five-year review, and that they would adopt an amendment to the Montreal Protocol and increase funding for its implementation.

Most of the commitments used strong, politically binding language. Nine were highly binding and three were low binding, for a high–low ratio of 75%. Synergies with other subjects were present too.

Delivery

Delivery of these climate decisions was substantial. Compliance with the three assessed climate change commitments was 73% (see Appendix C). Although lower than Elmau's 80%, it was close to the G7's 75% average on climate change. The two commitments on the Paris Agreement averaged 85%. The third one on the Montreal Protocol had 50%.

Development of Global Governance

Leaders' development of global governance was solid, if smaller than at Elmau (see Appendix C). They made no references to inside institutions, but 17 references to 10 outside ones. Eight went to the Paris Agreement and one each to COP21, the UNFCCC, multilateral development banks, the Lima-Paris Action Agenda, the International Civil Aviation Organization, the International Atomic Energy Agency, the Nuclear Energy Agency and the Association of Nuclear Operators.

Causes of Performance

Shock-Activated Vulnerability

Ise-Shima's substantial success was not driven by climate shock-activated vulnerabilities (see Appendix E). The communiqué recognized several diversionary health and regional security shocks as well as vulnerabilities, led by terrorism. On climate, it referred only to the 2011 Fukushima disaster and recognized nuclear energy's contribution to reducing emissions.

Front-page media coverage of climate shock-activated vulnerability in the months leading up to the Ise-Shima Summit was small. The *Financial Times* reported that over 90% of new electricity in the previous year was generated from renewables, the most since 1974 (Clark 2016a). The *New York Times* featured the Zika virus; in February an inside article linked that outbreak to the expanding range of disease-carrying mosquitoes caused by climate change (Gillis 2016).

In the physical world, the number of deaths from extreme weather events dropped sharply from 3,736 in 2015 to 693 in 2016 (see Appendix G). Environmental shocks were led by natural disasters from earthquakes. Japan suffered a 7.3 magnitude earthquake on April 16, two days after a 6.0 magnitude foreshock in the Kumamoto and Oita Prefectures. These killed 276 people and destroyed 43,000 homes (Kyodo News 2021). They reminded the Japanese of the much more severe 2011 earthquake and tsunami, from which they were still recovering and which had taken 22,000 lives (CNN 2021). Chronic cumulative ocean plastic pollution from debris and pollution as a result of the 2011 disaster, still making its way across the Pacific and reaching North America, created a shock for Japan (BBC News 2016). Japan thus emphasized sustainable materials management as an environmental issue at Ise-Shima. On climate change itself, Canada was shocked by unusual wildfires in British Columbia and Alberta, with its highly polluting oil sands, which forced 80,000 people to flee and caused significant economic damage.

Multilateral Organizational Failure

Multilateral organizational failure was significant but viewed as low. G7 leaders considered Paris a success, due to the new semi-legally binding global climate agreement. They committed to take the lead on early ratification and implementation. Yet, Paris's advances also narrowed the G7's decisional focus and lowered members' delivery of their broader climate commitments.

Another UN success, with substantial substance, was the SDGs, to which the G7 linked its own climate and gender statements. The Montreal Protocol was also seen as a success, leading the G7 to welcome its efforts and commit to support an amendment later in the year.

Predominant Equalizing Capabilities

Changes in relative capability constrained performance. The greatest global growth, in currency values and real gross domestic product (GDP), came from countries beyond the G7. Inside the G7, Japan, the G7's second most powerful member, had the slowest growth: 0.5% in 2016 down from 1.2% in 2015 (see Appendix H).

By the end of 2016, the G7's aggregate emissions declined by a very small average of 0.02% from the previous year (see Appendix I).

Converging Characteristics

Converging characteristics slightly boosted performance. Among 167 countries, the global democratic ranking of Canada, France and Japan rose in 2016 compared to 2015 (see Appendix J). Only the United States declined; the other members' rankings were unchanged.

G7 members' upward convergence on climate performance also grew. Of 58 countries ranked for climate change performance, G7 members collectively rose (see Appendix K). The United States jumped from 44th place to 34th. Nonetheless, this was not enough to lead to a significant decline in emissions.

Domestic Political Cohesion

Domestic political cohesion was substantial and strong. Japan's first G7 summit in 1979 had invented the global governance of climate change and was followed by reduced emissions for at least five years (Kirton and Kokotsis 2015). But this reduction was likely a result of energy diversification policies in response to the oil crisis rather than direct climate control policies. Japan's 2008 Hokkaido-Toyako Summit performed strongly, producing 54 climate change commitments, the most up to that time (see Appendix C). Inheriting this legacy at Ise-Shima, an experienced Abe faced formidable economic and security problems and a summer legislative election.

Obama, at his eighth and last summit, brought formidable experience but an opposition-controlled legislature and lame-duck status, which constrained US and others' compliance with Ise-Shima's climate commitments. The US Senate was controlled by a climate-resistant Republican Party whose presidential candidate Trump denied that climate change was occurring. His arrival as president in January 2017 caused lower compliance than the year before.

Merkel was attending her 10th consecutive summit, two of which she had hosted. She was set to build on the successes of her 2015 Elmau Summit, while coping with waves of migrants coming from the Middle East and Africa, and the security threat from Russia's Vladimir Putin in Ukraine and elsewhere.

The UK's David Cameron, at his sixth summit, brought his priorities of trade, tax, transparency and terrorism from his 2013 Lough Erne Summit, to show his voters who were facing a referendum why they needed the United Kingdom to remain fully engaged in the EU and the world.

France's Hollande, also at his sixth summit, focused on preventing the deadly terrorism that had struck Paris in November 2015 and on generating the jobs that he needed to stay in power in the 2017 presidential elections, but still promoted climate action.

Italy's Matteo Renzi, in office since 2014 and coming to his third summit, emphasized migration, food security and sustainable development to set up his strategy for the summit Italy would host in 2017.

Canada's Trudeau, coming to his first G7 summit, brought energy, a new global vision backed by a majority mandate and a concern with fiscal stimulus, climate change control and gender equality, with an eye on the G7 summit he would host in 2018.

Club at the Hub

At this second summit without Russia, G7 leaders valued their intimate club, led by Japan as host. Japan belonged to three other plurilateral summit institutions, one more than the previous year's host: the G20, the Asia-Pacific Economic Cooperation (APEC) leaders meeting (along with the United States and Canada) and the East Asian Summit (along with the United States since 2011). Canada's membership in the Commonwealth and Francophonie reinforced this Asia–Pacific emphasis, as did the United Kingdom and France to a lesser degree. The network hub was expanded with the Nuclear Safety Summit on March 31 and April 1, attended by the leaders of the United States, Japan, Germany, France, the United Kingdom, Italy and Canada, as well as China, India, Mexico, Korea and Argentina.

Further reinforcement came from the guest leaders: Sheikh Hasina of Bangladesh, Idriss Déby of Chad, Joko Widodo of Indonesia, Thongloun Sisoulith of Laos, Peter O'Neill of Papua New Guinea, Maithripala Sirisena of Sri Lanka and Nguyen Xuân Phúc of Vietnam. None was known to be committed to controlling climate change, but all led countries highly vulnerable to its effects. Déby chaired the African Union.

Abe also invited the heads of the Asian Development Bank, the World Bank, the International Monetary Fund, the Organisation for Economic Co-operation and Development and the UN. No environmental bodies were invited.

G20 Hangzhou, September

Introduction

On September 4–5, 2016, G20 leaders met in Hangzhou to help advance the Paris Agreement. As this was the last G20 that a climate-committed Obama would attend, some hoped he would join with China's Xi to have the world's two largest economies and greenhouse gas emitters lead in making Paris's promises a genuine ecological success (Kirton 2016d). These hopes were fuelled by the fact that the Hangzhou Summit took place soon after July, the hottest month in recorded history thus far, with every month of the previous year warmer than ever.

Debate

In the debate about Hangzhou's climate change performance, the first school saw China's global leadership. For Zhao (2016), "China and the US finally inked their commitment to global climate governance on the eve of the theatrical G20

Summit." He expected the Hangzhou Summit to issue a communiqué about climate change, due to China's rank as the largest developing country and its changing development strategy at home.

The second school saw significant success, as G20 members committed to and complied well with the relevant recommendations of the Business 20 (B20) and Think 20 engagement groups (International Chamber of Commerce 2016; Kirton and Warren 2017; Koch 2019). Koch (2019) noted that several G20 leaders, led by Xi, attended the B20 summit on Hangzhou's eve. The B20 and the International Chamber of Commerce G20 CEO Advisory Group made recommendations on energy sustainability. Their overall recommendations influenced the G20 summit's work.

The third school saw selective success on green finance and energy but not on fossil fuels subsidies. *Energy and Power* (2016) concluded that the leaders' recognition of green finance and clean energy production and their endorsement of the G20 Voluntary Collaboration Action Plan on Energy Access, the G20 Voluntary Action Plan on Renewable Energy and the G20 Energy Efficiency Leading Program were necessary for "environmentally sustainable growth." Yet, they set no timeline to phase out fossil fuel subsidies, despite a large group of insurance firms asking them to do so.

The fourth school saw substantial success but a need to do much more. Just before the summit, Kirton (2016e, 78–79) noted G20 leaders' progress on the "early entry into force of the 2015 Paris Agreement and more international cooperation" on climate change, but urged them to use Xi's "vision of an ecological civilization for China as their blueprint for the world." This would be possible due to China's traditional G20 leadership together with another partner, with Obama as the most likely one (Kirton 2016c). The *Cairns Post* (2016) wrote that Obama and Xi "sealed their nations' participation" in the Paris Agreement, and reported that Xi hoped other countries would follow suit and advance new technologies to help meet their targets.

The fifth school saw little progress on climate change or energy, and a "big show" short on substance (Byrne 2016). It underscored the US–China ratification of the Paris Agreement just before the summit as offering "the promise of a new model for G20 shared leadership" on climate change. However, the leaders' focus on the Syrian crisis, refugees, terrorism and migration displaced any such promise for climate action.

Argument

At Hangzhou, G20 leaders produced a small performance on climate change, despite the new act of Sino-American leadership on the summit's eve announcing their decision to ratify the Paris Agreement. The summit left little grounds to expect that Xi's eloquent ecology-first address to the B20 would radiate from China's G20. The summit made the lowest number of commitments on climate change, although they achieved 83% compliance. Hangzhou did little to develop global climate governance inside or outside the G20. Leaders did not advance ending fossil fuel subsidies, as agreed in 2009 and every year since then. There was no reference to killing coal, despite Xi's focus on his domestic actions in his opening address and the fact that Germany as G20 host in 2017 and Japan as G20 host in 2019 continued to operate, build or finance old and new coal-fired plants. In December 2015, Xi had put innovation first and sustainable development second on the G20 agenda (China's G20 Presidency 2015). By the end, innovation had slipped to second place and sustainable development to near the bottom of a long list.

Hangzhou's small performance was driven by diversionary financial and migration shocks, and by G20 leaders' complacency in leaving hard negotiations to the UN. Obama could not lead strongly, with his bold domestic climate agenda stopped by Congress, his presidency about to end, his party's fortunes in the 2016 election in doubt and Trump gaining strength. Xi, an engineer whose country's rapidly rising relative capability depended on carbon-intensive growth, was insufficiently committed for his summit to build on Paris and close the many gaps it had left.

Plans and Preparations

Priorities

China's plans for Hangzhou were first outlined in the final session of the 2015 Antalya Summit (Kirton 2016d). They acquired greater visibility on the G20 summit website in December 2015, when China formally assumed the G20 chair (China's G20 Presidency 2015). They sharpened in May 2016 when foreign minister Wang Yi (2016) outlined the 10 deliverables that the Hangzhou Summit was expected to produce. Xi (2016b, 3) himself outlined his priorities on August 30.

These lists showed that Hangzhou would continue China's G20 leadership evident since the G20's creation in 1999 (Kirton 2016a). Among China's priorities for Hangzhou, innovation usually came first and climate change last.

Xi put sustainable development in fourth place, by leading the implementation of the 2030 Agenda on Sustainable Development, a G20 Initiative on Supporting Industrialization in Africa and Least-Developed Countries, and the early entry into force of the Paris Agreement.

Pre-Summit Meetings

Xi visited the United States on March 31–April 1 for the Nuclear Security Summit. In a bilateral meeting with Obama, he discussed climate change and low-carbon development. He noted that the similarities between the two countries outweighed their differences.

Sherpa Process

Throughout the G20 sherpa process, the United States sought an enhanced environmental goods trade agreement and action on climate change and energy. It saw the United States and China as the world's two biggest economies, leading to help ensure the world met its climate goals.

By mid June, sherpas had agreed on an action plan for implementing the SDGs. China, which viewed the SDGs from a developing country perspective, resisted distinguishing between international assistance and G20 countries' national action plans.

Canada prioritized climate change, determined that the issue would not disappear as it had at Antalya before COP21. It strongly supported China's priorities on energy access, energy efficiency and clean energy. It supported the US lead on many of these issues, especially heavy-duty vehicles and methane emissions. It also continued to push on eliminating fossil fuel subsidies by 2025. Saudi Arabia and Russia remained the two most outspoken opponents.

China signalled that it sought to secure sufficient ratification of the Paris Agreement to bring it into force. However, G20 members were unlikely to agree to ratification by a fixed date. The EU required its members to vote to ratify it and had already indicated at the G7 Ise-Shima Summit that this would not happen before the end of 2016.

Ministerial Meetings

The pre-summit ministerial meetings did little on climate change.

On February 26–27, finance ministers and central bank governors welcomed the Paris Agreement, reaffirmed the commitment to phase out fossil fuel subsidies and had developed countries promise to provide financial resources, including through the Green Climate Fund. They welcomed a report from the Climate Finance Study Group (CFSG). They reiterated these points at their April 15 meeting and asked the CFSG to finalize its work on climate finance monitoring and transparency by their next meeting. At their final meeting on July 23–24, they welcomed the CFSG report that outlined voluntary options for scaling up green finance.

Meeting on June 2–3, agriculture ministers reaffirmed their "strong" support for the Paris Agreement and countries' implementation of their NDCs. They committed to support skills development and training for farmers to prepare for climate change.

On June 29, energy ministers focused on a clean energy transition, the Paris Agreement and the UNFCCC, a sustainable low-emission energy future, a clean energy future, inefficient fossil fuel subsidies and energy access. They continued to promote natural gas as a less emission-intensive fossil fuel, while ignoring its significant methane contribution. They made 25 commitments, dominated by clean energy and energy efficiency, and included a section on "cleaner fossil fuels."

On the Eve

Just before the summit started on September 4, it produced its first, and ultimately only, major achievement (Kirton 2016c). On September 2, China and the United States jointly announced that they would ratify the Paris Agreement by the end of the year. As these two countries together accounted for 38% of global greenhouse gas emissions, and the Paris Agreement required 55 countries representing 55% of global emissions to ratify it, this act of co-leadership did much to help.

G20 members had been slow to ratify the agreement. France started on June 15, half a year after the agreement had been reached in its capital city and about two months after the high-profile, summit-level signing ceremony at the UN in New York in April. Korea followed. Brazil moved ahead. But many players were missing — including Canada, whose popular prime minister had climate change control as his third highest priority at Hangzhou (Trudeau 2016b).

Meeting the day before the G20 summit, Xi and Obama discussed climate-smart agriculture in Africa and disaster response. They reaffirmed their commitment to support countries affected by the El Niño and La Niña events intensified by climate change. They also discussed wildlife trafficking, clean energy, sustainable development and oceans.

In his address to the B20, Xi (2016a) declared "green mountains and clear water are as good as gold and silver."

On September 5, on the sidelines of the summit, Xi met separately with Hollande and the United Kingdom's Theresa May. Hollande was pleased with the consensus on climate change and agreed to continue to jointly promote the environment. May sought cooperation on the economy, trade, investment, finance, security and law enforcement, to which Xi added clean energy.

The leaders of the BRICS (Brazil, Russia, India, China and South Africa) met on September 4 for their standard sideline summit. Their extensive media note referred only once to climate change. It made a weak commitment reaffirming the 2030 Agenda, the Addis Ababa Action Agenda on Financing for Sustainable Development and the Paris Agreement. It also expressed satisfaction with the first loans issued by the New Development Bank for renewable and green energy and a green bond denominated in renminbi.

At the Summit

Xi opened the summit with an address to the G20 members and guests. He acknowledged the G20's work on green finance.

On phasing out inefficient fossil fuel subsidies, the United States, Canada and Germany pushed to commit doing this by a fixed date, having just agreed at the G7 to do so by 2025. But as Russia, India and Saudi Arabia opposed this, no consensus was reached.

Obama read from a script that contained much on climate change and inclusiveness.

At the last session, leaders were due to discuss climate change and its links to energy. The G20 sought to define inefficient fossil fuel subsidies, but each member interpreted what that meant as it wished. Canada pushed hard on the Paris Agreement's entry into force. But China was not willing to push it.

Trudeau (2016a) said he had stressed that "our common growth must be clean," and encouraged his colleagues to reduce emissions and adopt clean technologies

so "everyone can make progress on the Paris Agreement before our next meeting." He said he had achieved his priorities on setting the 2025 deadline on fossil fuel subsidies, reducing methane and vehicle emissions and linking climate
change and gender.

Results

Hangzhou was only a small success on climate change. It did dedicate more words
to climate change and make more connections to outside institutions than Antalya
had. It loosely linked climate to inclusiveness for the first time and to democracy
for the first time in four years. Yet, it made only two commitments and compliance was slightly lower than Antalya's. Its only standout success was the pre-
summit announcement by China and the United States that they would ratify the
Paris Agreement by year's end, just before Obama retired as US president.

Dimensions of Performance

Domestic Political Management

Domestic political management was substantial (see Appendix D). All G20
leaders attended the Hangzhou Summit except Saudi Arabia's King Salman bin
Abdulaziz Al Saud, who sent Deputy Crown Prince Mohammad bin Salman.
There were no communiqué compliments on climate change.

Obama did not participate in the boat ride on the opening evening, after his
secret service agents were not permitted to accompany him. It was another slight
in a series that had begun on arrival, and made Obama's last G20 summit a domestic political management failure for him.

Deliberation

Public deliberation was strong. The outcome documents contained 1,175 words
on climate change, for 11% of the total (see Appendix D). Both these figures were
the second highest to date, exceeded only by Seoul in 2010.

The communiqué offered the "Hangzhou Consensus," which began with the
principle of intergenerational equity. It invoked the principle at the centre of ecological sustainability articulated in the 1987 Brundtland Commission report. The
sixth and final section on "Further Significant Global Challenges Affecting the
World Economy" noted the need to bring the Paris Agreement into legal force.

Private deliberations were few. The opening session for leaders alone was
interrupted by a ceremonial dinner for the leaders and their spouses, an artistic
performance plus a boat ride.

Direction Setting

Direction setting was solid. Leaders made one affirmation to the G20's mission of
making globalization work for all and one to the G7's mission of democracy (see

Appendix D). They committed to "foster an … inclusive world economy to usher in a new era of global growth and sustainable development, taking into account … the Paris Agreement." On democracy, the G20 referenced transparency in climate finance.

Ecological values were affirmed in 18 (13%) of the 139 commitments and integrated into almost all the policy areas of the communiqué, including all of the G20 summit's new issues.

Decision Making

However, Hangzhou's decision-making performance was very small, with only 2 of 139 commitments on climate change (see Appendix D).

The first climate commitment was a general, low binding promise "to address climate change." The second committed to ratify the Paris Agreement as soon as possible. It had a highly binding verb but depended on the differing and undefined "national procedures" of each member, rather than setting a common deadline date.

Four of the eight energy commitments had implications for climate change. Three were low binding and none linked to other subjects. They reiterated the commitment to phase out fossil fuel subsidies, encouraged G20 countries to participate in voluntary peer reviews, committed to build a low emissions energy future via energy sources and technologies, and promoted natural gas while minimizing environmental impacts.

Delivery

Delivery of Hangzhou's two climate commitments was strong at 83%, if lower than Antalya's 85%. The commitment to join the Paris Agreement had 93% and the one to "address climate change" along with sustainable development had 73%.

Enough large G20 members followed the US and China's immediate pre-summit promise to ratify the Paris Agreement to secure ratification by October 5 (EFE News Service 2016). By the end of December, only Russia and Turkey remained. Russia ratified the agreement in 2019.

Development of Global Governance

The institutional development of global governance was small. The leaders' communiqué made four references to inside institutions: two to the Green Finance Study Group, one to the CFSG and one to the G20 energy ministers. The G20 made five references to outside institutions in a climate context: two to the UN, and one each to the UNFCCC, the World Trade Organization and the International Civil Aviation Organization.

Causes of Performance

Shock-Activated Vulnerability

Hangzhou's small performance on climate change was shaped by diversionary shocks from the G20's core subject of finance, its recent concern with migration and North Korea, which had fired three ballistic missiles on the summit's first day. The communiqué recognized three shocks and three vulnerabilities on financial stability and two shocks on migration, but none on climate change (see Appendix F).

Front-page media shock-activated vulnerability was stronger but specialized. In the months leading up to Hangzhou, the *Financial Times* climate stories focused on the Paris Agreement rather than on new physical shocks. In November, while China still held the G20 presidency, the *Financial Times* highlighted divisions over "common but differentiated responsibilities" with a headline that India's Narendra Modi called for "rich nations" to fulfill their "moral imperative to lead the fight against global warming" (Clark 2016b).

In the physical world, Japan was still recovering from the Kumamoto earthquakes. North America had suffered from unprecedented wildfires. A month after Hangzhou, on October 4, Hurricane Matthew struck the Caribbean, killing 546 Haitians. Yet, in G20 countries there were fewer extreme weather events and deaths than in 2015, dropping from 7,898 to 4,350 deaths (see Appendix G).

Scientific reports subsequently confirmed that the years from 2015 to 2020 were the warmest since records began (*Financial Times* 2021a). Recorded global temperature rise averaged 0.2°C per decade since 1970, reaching a record high in 2016 during a warming El Niño (Hook 2021a). They were consistent with Hangzhou's strong compliance with its two climate commitments.

Multilateral Organizational Failure

Hangzhou's small climate performance was fuelled by the perceived success of the multilateral UN. The Paris Agreement provided an assumption of effective action. This left the Hangzhou leaders free to focus on the lengthy, legal, procedural issue of ratification, which would have no short-term impact on a steadily warming climate.

This attitude was seen six weeks later when, on October 15, Xi attended the BRICS summit in Goa, India. BRICS leaders welcomed the adoption of the Paris Agreement, called on industrialized countries to fulfil their climate financing obligations, promoted nuclear energy and natural gas and expressed their condolences to Haiti and the Caribbean for the damage wrought by Hurricane Matthew. They left climate change mitigation to others.

Predominant Equalizing Capabilities

The G20's predominant equalizing capabilities help explain Hangzhou's small performance. Although the G20's global predominance remained high, internal equality declined. In the first four months of 2016, the value of the US dollar had declined steadily compared to its peers from a level of 100 to 93, but then rose steadily to 98 in September and 103 by year's end (a level not matched again until March 2020). It later plunged to 92 by July 2017 and a low of 88 by February 2018, in ways consistent with Hangzhou's compliance performance (Szalay and Smith 2021).

GDP growth in 2016 was led by Turkey at 3.2%, Indonesia with 5.0%, China with 6.8% and India with 8.3%, countries all reluctant to act ambitiously on climate change (see Appendix H).

In specialized climate capabilities and vulnerabilities, G20 global predominance and internal equalization were high. The United States and China together produced almost 40% of global emissions, and the EU another 12% (EFE News Service 2016). This propelled the China–US bilateral agreement and G20 leadership in ratifying the Paris Agreement.

From 2015 to 2016, the G20's aggregate emissions rose at an average of 0.10% (see Appendix I). Only Indonesia, Canada, Argentina, the United Kingdom and France had declining emissions.

Converging Characteristics

Deepening political divergence fostered poor performance. The G20's collective democracy in 2016 declined from 2015 (see Appendix J). There was also downward divergence in their climate change performance (see Appendix K).

Domestic Political Cohesion

Domestic political cohesion was high in Xi's China but low in Obama's United States. Two months before the election, the lame-duck Democratic US president faced a Senate controlled by a climate-resistant Republican Party led by climate change–denying Trump. There was also high political cohesion in the other key climate-resistant members of Saudi Arabia, Russia and India.

Club at the Hub

Even lower was the G20's status as the leaders' valued club at the hub of a network of global summit governance (see Appendix L). Xi's summit had a formal approach rather than a flexible one. His slights of Obama at the arrival ceremony and the opening dinner left little personal trust or incentive to go beyond the preplanned bilateral agreement announced before the summit's start.

Moreover, China belonged to the BRICS, where Russia and India opposed ambitious climate change control and whose sideline summit made only one reference to the Paris Agreement. Climate change loomed larger at APEC, whose other ranking members beyond the United States, Japan and Canada included Australia, which resisted strong climate action. The network had expanded with the meeting of the Nuclear Safety Summit in Washington in the spring, where all

G7 leaders joined those from China, India, Mexico, Korea and Argentina. Here, Obama and Xi discussed climate change bilaterally.

UN Marrakech, November

Introduction

On November 7, 2016, representatives of 176 UN members assembled in Marrakech, Morocco, for COP22. The Paris Agreement had entered into force on November 4 — 30 days after 55 countries representing 55% of global emissions submitted their instruments of ratification, acceptance or accession.

Yet, by November 7, most focused on the United States as it prepared to elect a new president the following day. On the eve of the election, polls indicated a tie between Hillary Clinton, whose Democratic Party remained fully committed to implementing and improving the Paris Agreement, and the Republican Trump, who won.

Trump openly opposed the Paris Agreement and believed that environmental regulations passed by Obama impeded American business. He maintained that the agreement put the United States at a disadvantage when competing against large developing countries such as China, and had announced his intention to "cancel" it should he win (Samuelson 2016). He called climate change a hoax, invented by the Chinese, and promised to boost America's oil, gas and coal energy, while ending climate financing abroad and at home (Worland 2016).

Thus, by the end of the second day in Marrakech, COP delegates worried that US leadership in the hands of Trump would reverse any progress made since Paris. Morocco and COP22 co-host Fiji appealed to Trump's team to remain committed, but also signalled that they would march forward regardless.

Argument

COP22's performance was small. Eleven additional countries ratified the Paris Agreement, including Japan, the United Kingdom, Australia and Pakistan, bringing the total to 111 countries, responsible for 79% of global emissions.

Marrakech participants agreed to work on a rulebook that would help clear up numerous details left vague in the language at Paris, such as how countries would monitor their emissions reduction pledges and report progress. They reiterated the climate finance commitment of $100 billion from industrialized countries annually by 2020. Non-state actors played an increasing role, including joining state actors in announcing funding.

Yet, delegates focused on technical details, despite the current pledges being inadequate to stay below 2°C and bridge the estimated emissions gap of 12–14 gigatons. They set 2018 as the deadline for operationalizing the Paris Agreement and revisiting the so-called orphan issues not yet explicitly included in COP's agenda. Calls to ratchet up the NDCs were left unanswered. The announcement of $81 million for the Adaptation Fund, surpassing its fundraising target for 2016, helped, but did not come close to meeting the finance gap.

This small, largely procedural performance was caused in part by multilateral organizational failure due to the challenges of securing consensus among almost 200 countries with widely varying positions to which negotiators were unable or unwilling to adjust. There was no democratic or climate performance consensus that had unified G7 leaders at their summits. The election of Trump, which did not bode well for G7 unity, only compounded existing difficulties.

On the Way to the Meeting

COP22 was initially dubbed a gathering for action as the first major climate meeting after Paris, but Trump's election threw this into doubt.

Salaheddine Mezouar, Morocco's foreign minister and the conference president, urged Trump to join other countries in committing to actively limit emissions (DW 2016). Frank Bainimarama, prime minister of Fiji, an island state threatened by rising sea levels, invited Trump to Fiji to witness the effects of climate change for himself.

"Climate champions" Hakima El Haite, minister delegate to Morocco's minister of energy, mines, water and the environment, and Laurence Tubiana, France's ambassador for climate negotiations and special representative for COP22, declared:

> The Marrakech call is loud and clear: *nothing can stop global climate action* … At the same time, there is universal recognition that if we are to realise the goals of the Paris Agreement, *we must all go further and faster in delivering climate action before 2020*, enabled by adequate flows of finance, technology and capacity building.
>
> (High-Level Climate Champions 2016, 1)

At the Meeting

Marrakech's primarily technical agenda focused on financing, loss and damage (the Warsaw mechanism), technology transfer, accountability and transparency frameworks, common time frames for NDCs, education, training and public awareness (Sharma 2016, 14).

Delegates largely agreed that scaled-up finance flows, technology and capacity building, and adaptation mechanisms for long-term transformation were needed. They stressed that governments should partner with development banks, private finance institutions and other financial actors to mobilize finance at the needed scale (High-Level Climate Champions 2016, 2).

Outcomes

Following 11 days of negotiations, and without full support from the United States, the Marrakech Action Proclamation reaffirmed members' commitment to fully implement the Paris Agreement, calling for "the highest political commitment to combat climate change" as well as "strong solidarity with those countries most vulnerable to the impacts of climate change" (UN Climate Change 2016).

Ambition

At the heart of the negotiations lay ambitious goals to improve adaptation and mitigation strategies, funding mechanisms, transparency frameworks and technology transfer. Discussions included how to design a global stocktaking every five years for countries to assess progress and make adjustments (Deheza et al. 2016). COP22 set a deadline of 2018 to reach consensus on the implementation of rules, NDC work plans and a five-year framework on tracking loss and damage.

Members agreed to work collectively on a rulebook that set out guiding principles to address the numerous details left vague in the Paris Agreement. These details would address how countries would report on progress and how they would monitor their pledges to curb emissions (DW 2016).

Adaptation

Adaptation discussions focused principally on funding delays, highlighting a gap between the climate financing promises made and delivered. A road map established how developed countries would deliver $100 billion in climate finance annually. Negotiators worked carefully to optimize the linkages between accounting and reporting mechanisms. Some countries focused on direct access to climate finance, while others supported finance for adaptation measures. Developing countries also urged developed countries to scale up financial support in line with the needs and priorities identified on an individual country basis (Deheza et al. 2016).

Technology Transfer

On technology transfer, countries largely agreed to focus on innovation, implementation, capacity building, collaboration and stakeholder engagement. The strongest leadership came from the most climate vulnerable states. On the final day, the Climate Vulnerable Forum (2016), a group of 48 climate-vulnerable countries, committed to 100% renewable energy "as soon as possible and at the latest by 2030 and 2050." Achieving net carbon neutrality would shift the energy dependence away from fossil fuels and toward sustainable resources, economic growth, a just transition and disaster risk reduction (Climate Action Network 2016).

Transparency Framework

On transparency frameworks, flexibility was the focus. Some argued that developing and developed countries had different responsibilities based on national circumstances and capacity constraints. Identifying how to design an enhanced framework that acknowledged these needs while allowing common guidelines was key for all parties (Deheza et al. 2016).

COP22 established that by 2018, negotiations would focus on common guidelines and recommendations on transparent frameworks for improving tracking efforts through simplified NDC assessments. This facilitative dialogue would include technical papers published by the Intergovernmental Panel on Climate Change that focused on emission trajectories to keep global warming below 1.5°C compared to pre-industrial levels (Deheza et al. 2016).

Finance

On climate finance, delegates discussed delivery of the $100 billion pledge and financing for resilience and the Adaptation Fund. The High-Level Ministerial Dialogue on Climate Finance on November 16 offered an opportunity to make new commitments.

Discussions also covered accounting. Developed countries wanted to focus on reporting. But with accounting a prerequisite for reporting, Marrakech negotiators needed to work carefully to optimize the linkages between the accounting practices and reporting mechanisms (Mezouar 2016).

India highlighted the need to pinpoint sources within and outside the UNFCCC framework and called for the review of terms of reference for several key financial bodies. Views differed on a "needs assessment programme" for developing countries (*Earth Negotiations Bulletin* [ENB] 2016, 6). Some countries focused on access to climate finance, while others on the role of policies and enabling environments.

Global Climate Action

Outside the COP process, non-state and sub-national actors actively engaged with Global Climate Action, the platform for non-state actors, to highlight efforts across a wide range of sectors, from cities to forests to resilience.

Canada, Germany, Mexico and the United States became the first four countries to publish their climate action plans through to 2050. These plans outlined long-term approaches for reducing emissions to 2050 and beyond. Despite Trump's electoral victory, US delegates presented an optimistic 111-page report outlining multiple alternative pathways for deep decarbonization by 2050.

Results

Although COP22 generated some progress, experts expressed concerns that delegates were lost in technical details and appeared to forget that current pledges remained inadequate for remaining below 2°C. Calls to ratchet up NDCs were left largely unanswered.

By the end of the deliberations, COP22 delegates acknowledged the need to move forward with or without the United States, and remained committed to acting (Inside Climate News 2016). There were expressions of support for accountability and evidence-based decisions. Mezouar, profoundly disappointed by Trump's announced withdrawal from the Paris Agreement, later said "climate action remains undeniable and irreversible" and that collective efforts to fight climate change would accelerate (Xinhua 2017).

By the end of COP22, participants sought to push forward rather than wait to see what the United States would do (Inside Climate News 2016). This approach resonated with high-level Chinese, German and US administration officials. During the meeting's final hours, Jonathan Pershing, US special envoy for climate change, told reporters that at the beginning of the conference, "the election was the discussion and by the end of the meeting, the climate change issue was the discussion" (Inside Climate News 2016).

However, many observers, particularly those among non-governmental organizations such as ActionAid and the Climate Action Network, felt that the burden still fell on developing countries and great gaps remained between what was needed and what was provided at Marrakech (*Beyond* 2016). Oxfam's Isabel Kreisler deplored developed countries' "stubborn refusal" to provide adaptation financing and Greenpeace's Jennifer Morgan described COP22 as a "small step" (Greenpeace 2016; Oxfam 2016).

Causes of Performance

Despite some modest gains, some key determinants featured in the concert equality model of global governance caused Marrakech to produce only a small, largely procedural performance.

Multilateral organizational failure flowed from the diverse and conflicting positions and interests of the over 22,500 participants, including 15,800 officials (ENB 2016). The meeting included a high level of involvement of fossil fuel corporations including ExxonMobil, Chevron, Peabody, BP, Shell and RioTinto (Slezak 2016).

The lack of equalizing capability among governments contributed. Some observers noted the European Union's lack of a clear plan. France talked of becoming carbon neutral by 2050, but provided little detail on how it would achieve this goal (Robert 2016).

By contrast, the most motivated and ambitious initiative came from the world's most vulnerable countries in their announcement to switch to 100% renewable energy, emphasizing that shock-activated vulnerability was much more salient for those on the front lines of the climate crisis (Climate Vulnerable Forum 2016). Yet, despite 15 days of bilateral meetings, the 48-country bloc "failed to convince a single one of the other COP participants to join them" (Robert 2016).

Political cohesion among participating representatives was low. In the United States, Trump's election cast doubt and confusion, bringing into question what an effective multilateral climate agreement without the United States would look like (Karunungan 2016).

Conclusion

Following the landmark 2015 UN Paris summit, attention quickly turned in 2016 to ratifying the Paris Agreement and establishing the rulebook for making it work. With world leaders not due to meet at a UN climate conference until 2020, leader-level actions to implement and improve Paris turned to the G7 in Ise-Shima in May and the G20 in Hangzhou in September. The stage was thus set for Japan, an island state vulnerable to climate change, and China, the world's largest climate polluter, to lead the global governance of climate change.

Yet, the stage set by the G7 and G20 was less than strong. Progress was largely left to the UN's ministerial meeting at COP22 in Marrakech to advance progress on the Paris framework. The G7 Ise-Shima Summit produced a substantial performance, but only half as many climate commitments, which had lower compliance, than Elmau's the year before. The G20's Hangzhou Summit produced a small performance, with only two climate commitments, fewer than Antalya's three in 2015.

In Japan, the G7 set 2025 as the deadline to eliminate inefficient fossil fuel subsidies, but failed to detail how they would achieve this. It pledged to ratify the Paris Agreement, but chose no target date. It did not formulate any country strategies for reducing emissions by 2020 and did not mobilize new money. The substantial progress achieved was in part caused by the shock-activated vulnerability of destructive earthquakes in Japan, Abe's experienced leadership and increased democratic political cohesion. However, the low political control of Obama at home risked ushering in a climate change–denying US president who would lower the G7's democratic convergence.

This constraint followed leaders to the G20's Hangzhou Summit in September. Yet, at the larger G20, there was some hope that Obama and Xi would come to a bilateral agreement to advance climate action. On the sidelines of the summit, the two leaders promised to ratify the Paris Agreement by year's end. But even in the face of global record-breaking temperatures throughout 2016, the G20 produced only two commitments on climate change, the fewest since 2008. The long-promised commitment to ending inefficient fossil fuel subsidies again failed to advance, as did any serious commitment to eliminate coal. Hangzhou's small performance was driven by the G20's complacency in leaving sticky subjects to the UN, along with an innovation-focused host and Obama's low domestic political cohesion.

Attention then turned to COP22 in Marrakech. But with no foundation on which to build from the prior G7 or G20 summits and with COP delegates consumed by technical details, it could not move beyond the current pledges, already inadequate to meet the Paris Agreement goals. Calls to ratchet up the NDCs were thus left unanswered. The US presidential election at the beginning of COP22 brought in Trump, who attacked and promised to abandon the multilateral Paris Agreement. The process of operationalizing the Paris Agreement was initiated but no breakthroughs emerged.

The G7, G20 and COP22 collectively saw limited success that was insufficient to reverse climate change, in the short time science and sense said was left before a potentially irreversible climate catastrophe occurred. The failure of all three global governance institutions in 2016 thus shared several common causes: the juxtaposition between a weakening and democratically declining United States and a strengthening but also democratically declining China; an overestimation at the UN level in the ability of incremental technical progress, voluntary NDCs and the ambition of the major polluters to speed up implementation; and diversions that overshadowed sudden extreme weather events and the slower, less visible onset effects of climatic change.

The world now looked to 2017 with Italy at the G7 helm, Germany at the G20 helm and the UNFCCC in Bonn. The question was whether these plurilateral and multilateral forums would face up to or follow the still powerful but declining United States under the overt environmental antagonism of Trump.

4 Tackling Trump, 2017

Introduction

In 2017, the global summit governance of climate change was dominated by the test imposed by Donald Trump, who became US president on January 20 and brought the most aggressive anti-environmental approach of any G7 and most G20 leaders ever. At the G7 summit in Taormina, Italy, in May, G7 leaders energetically tried to convince Trump to reverse his intention to withdraw the United States from the Paris Agreement. They failed. Not long after, in July, Germany's Angela Merkel, hosting her first G20 summit, in Hamburg, tried another approach. She succeeded. Hamburg performed strongly on climate change, even with Trump there. This showed that a highly committed, experienced host could largely overcome the veto sought by its most powerful member, the United States, with Russia and Saudi Arabia's support. Hamburg's success, however, did not extend to the United Nations, at its annual ministerial-level Conference of the Parties (COP) to the United Nations Framework Convention on Climate Change (UNFCCC) at Bonn in November and its solid performance. However, the power of plurilateral summitry reinforced Hamburg with the significant performance of the new One Planet Summit, in December, when key leaders — without Trump — gathered again in Paris at the request of France's Emmanuel Macron, the World Bank's Jim Yong Kim and the UN's António Guterres. This was the first major step in reconfiguring global climate governance, as new plurilateral, climate-dedicated summits supplemented the G7, G20 and UN.

G7 Taormina, May

Introduction

The 43rd annual G7 summit, in Taormina on May 26–27, centred on the great drama of whether G7 leaders could convince their new colleague, Trump, to keep the United States in the Paris Agreement or advance climate action, with or without him, in other ways.

DOI: 10.4324/9780429055485-4

Debate

In the lead-up, Taormina's prospects inspired several schools of thought (Kirton 2017b). The first predicted little consensus on climate change, due to several pressing geopolitical and security preoccupations, disagreements between Trump and host Italy's Paolo Gentiloni on trade and climate change and poor US preparations for the summit (Hammond 2017; Kirton 2017b).

The second school predicted only a small success, due to the severe setbacks from Russia's 2014 annexation of Crimea, Britain's 2016 decision to leave the European Union and the advent of a unilateralist, protectionist and unpredictable Trump skeptical of climate change (Wouters and Van Kerchhoven 2017).

The third school saw a barely relevant G7 usurped by the G20. It argued that the G7 no longer represented the "global power base" and was only barely relevant because the United States, Germany and Japan were members (Reguly 2017). This implied that with China, India, Russia and Saudi Arabia outside, and a now climate-skeptical US president within, the G7 would do little on climate change.

The fourth school saw a seven-to-one outcome on climate change, with the six "G" countries, plus the European Union, stating their strong commitment to the Paris Agreement, with or without Trump's United States (Reuters Staff 2017a). It conjectured that, should the G7 issue an unprecedented statement on climate change without the United States, as Macron speculated it might, G7 members might act unanimously, without staying silent on Paris to appease the United States.

The sixth school, in the middle, saw the G7 sneaking climate change statements into the communiqué via less direct language, such as increasing renewable energy capacity or protecting the environment (Kirton 2017b).

After the summit, several new schools acknowledged Trump's isolation on the issue. All agreed that Taormina had failed on climate change.

A seventh school saw a "glimmer of hope" (Agence France-Presse 2017b). It predicted that despite Trump's refusal to include the United States in the Taormina commitment on implementing the Paris Agreement, he would choose to remain in, after considering the political and environmental repercussions of withdrawal. It highlighted the much-publicized comments by US secretary of defence James Mattis that Trump remained "wide open" on climate change (Miller 2017).

The eighth school accepted the seven-to-one outcome at Taormina and saw the emerging economies, including China and India, filling the gap left by the United States (Buncombe 2017). With the world's largest polluters on track to exceed their emissions reductions targets set under the Paris Agreement, a "green" lining for the planet could emerge.

The ninth school suggested the European Union would increase cooperation with other countries such as India to ensure global temperatures stayed well below 1.5°C (IANS 2017). A variant saw EU cooperation among its own members, noting Merkel's declaration that "Europeans must really take fate into our own hands" (Central Asia News 2017).

The 10th school saw that, if the United States withdrew, others might too. It assumed that the United States remained the world's most powerful country in economic capability and hard military strength, and other countries would rationally follow its lead, stalling international climate cooperation (Kirton 2017c).

Argument

Taormina produced a small performance on climate change. G7 leaders unsuccessfully tackled Trump to reverse his declared desire to withdraw from the Paris Agreement. Previous American presidents from the Republican Party had delayed ambitious G7 action on climate change, but never before had G7 members united against a United States so determined to go it alone (Kirton 2020e).

Taormina produced only one climate change commitment, the fewest since the 2002 summit, in the wake of the diversionary shock of the September 11 terrorist attacks on the United States (see Appendix C). The commitment excluded the United States — Taormina failed to tame Trump on climate change.

This failure flowed from the resistance of a newly elected, climate change–denying US president attending his first summit, backed by a fresh electoral mandate and strong legislative control. Diversionary security and migration shocks preoccupied other G7 leaders and spurred the climate-security link. An "America first" Trump had no attachment to the G7 as a valued interpersonal club, and thus resisted the passionate face-to-face pleas of his new G7 peers.

Plans and Preparations

On July 4, 2016, Italy's Matteo Renzi announced that Italy would host the next G7 summit in the southern coastal city of Taormina, Sicily, on May 26–27, 2017 (ANSA 2016). Taormina was chosen to spotlight key issues preoccupying Italy and the G7. Migrants flooding into Europe from across the Mediterranean had helped UK voters decide to leave the European Union and threatened to divide the European Union, of which Italy had been the most loyal pillar since the 1957 Treaty of Rome (Kirton 2016f).

The climate-related issue of energy was on the agenda, as Italy had strong connections to the Middle East and North Africa, which provided much of Europe's energy. Also on the agenda was implementing the 2030 Agenda for Sustainable Development and its 17 Sustainable Development Goals (SDGs), with Africa and Afghanistan on the front lines, and with climate change inside.

After Renzi resigned on December 4, 2016, Gentiloni chaired the G7. But the most important newcomer was Trump, at his first international outing. Macron was elected on May 7. Britain's Theresa May, prime minister since August 2016, now had a minority mandate. Merkel was a veteran of 11 G7 summits. Japan's Shinzo Abe had hosted the previous year's summit at Ise-Shima. Canada's Justin Trudeau, at his second G7 summit and preparing to host the following year, offered confident, youthful leadership. The European Union sent Jean-Claude Juncker and Donald Tusk, both facing the urgent question of whether the European Union would unify or unravel after the United Kingdom left (Kirton 2016f).

Italy's summit theme was "Building the Foundations of Renewed Trust" (G7 Italy n.d.). It was chosen to put citizens' needs first, in recognition of growing global anti-globalization sentiments (Oldani and Wouters 2019). The second of its three pillars, on economic, environmental and social sustainability, and reduced inequalities, included energy and climate change.

Pre-Summit Meetings

Trump visited the Middle East and Europe on his way to Taormina. Macron strongly urged him to remain in the Paris Agreement, saying that the international community still had "political responsibilities to assume" and the United States should not make a "hasty decision" (Waterfield and Kington 2017). At the Vatican, Pope Francis gave Trump his 2015 *Laudato Si': On Care for Our Common Home*, a 184-page letter that strongly supported the scientific consensus that the Earth's climate was changing due to human activity (Mathiesen 2017b).

Sherpa Process

Progress at the sherpa meetings was slowed by a lack of clarity from the United States about Trump's position, particularly on trade and climate change. At their first meeting, on January 26–27, sherpas discussed all issues, including climate change and energy. At the second meeting, on March 20–21, they focused on climate and energy, in light of the G7 energy ministers' meeting a few weeks later (G7 Italy 2017). At the third meeting on April 26–27 and the fourth on May 15–16, they began finalizing the communiqué.

Ministerial Meetings

All but three of the four pre-summit and nine post-summit ministerial meetings organized by Italy referred to climate change.

On March 30–31, culture ministers concluded that cultural heritage "is an important tool for … sustainable development." They expressed concern over the threat of increasing natural disasters to cultural heritage, such as museums, monuments, archaeological sites, archives and libraries.

On April 9–10, energy ministers had no communiqué, as the official US position was not yet determined. However, the chair's summary said all G7 members except the United States reiterated their commitment to the Paris Agreement and their leaders' 2016 Ise-Shima Summit commitment to phase out fossil fuel subsidies by 2025. Energy ministers also discussed the transition to a low-carbon economy, energy efficiency and low-carbon transportation.

On April 10–11, foreign ministers made 141 commitments, including one on climate change to improve resilience. Another commitment was on the conservation and sustainable use of marine biodiversity.

At the May 11–13 meeting of finance ministers and central bank governors, the United States prevented any reference to climate finance, including subsidies to the fossil fuel industry, from appearing in the communiqué.

After Taormina, G7 environment ministers met on June 11–12, just over a week after Trump's announcement of the US withdrawal from the Paris Agreement. Twelve (25%) of their 49 commitments directly referenced climate change, all without the United States. The other ministers emphasized their "strong commitment to the swift and effective implementation of the Paris Agreement," as their leaders had done in May. Their communiqué focused on climate change and included the 2030 Agenda, sustainable finance, resource efficiency, 3Rs (reduce, reuse, recycle), the circular economy and sustainable material management; marine litter; support for multilateral development banks (MDBs); environmental policies and employment; and Africa. A footnote stated: "We the United States do not join those sections of the communiqué on climate and MDBs, reflecting our recent announcement to withdraw and immediately cease implementation of the Paris Agreement and associated financial commitments."

On June 21–22, transport ministers stated that investment in infrastructure could "trigger" environmental benefits. They recognized that advanced technology could reduce vehicles' emissions. Yet, they made no politically binding commitments in this key sector, responsible for 14% of emissions according to the Intergovernmental Panel on Climate Change ([IPCC] 2015).

On September 25–26, industry ministers and ministers responsible for information and communications technologies (ICTs) recognized the positive impacts of the "Next Production Revolution" on the environment, and more generally on the SDGs.

On September 27–28, science ministers offered strong support for the sustainable use of the seas and oceans and explicitly recognized the importance of SDGs 13 (climate action) and 14 (life below water). The G7 Future of the Seas and Oceans Working Group recommended developing a global initiative for an enhanced, global, sustained sea and ocean observing and assessment system.

On October 14–15, agriculture ministers recognized the priority of safeguarding food production systems against the adverse impacts of climate change. Nonetheless, they ignored the sector's causal contribution to climate change, despite it being the second largest contributor of global greenhouse gases at 24% and the leading direct cause of tropical deforestation and ocean dead zones (IPCC Scheer and Moss 2012, 2015; Food and Agriculture Organization and United Nations Environment Programme [UNEP] 2020).

On November 5–6, health ministers discussed the "impact of the climate and environmental-related factors on health," including changing patterns in infectious diseases, extreme weather events, sea level rise, ocean acidification, biodiversity, soil pollution, water scarcity, food insecurity and malnutrition, and increased migration. They also looked forward to the COP in Bonn in November.

At the Summit

The two-day Taormina Summit started late in the morning of May 26. The third session that day, on the world economy and sustainable growth, included trade, climate change and energy. Trump listened intently as his fellow leaders argued credibly, often from the US perspective, for the United States to remain in the Paris Agreement. There was some direct disagreement, but not in ways that pushed Trump. No one lectured or criticized him.

On the second day, G7 leaders were joined by Tunisia's Beiji Caid Essebsi, Niger's Mahamadou Issoufou, Nigeria's Yemi Osinbajo, Kenya's Uhuru Kenyatta and Ethiopia's Hailemariam Desalegn. The other guests were the UN's Guterres, Angel Gurría of the Organisation for Economic Co-operation and Development (OECD), Christine Lagarde of the International Monetary Fund (IMF), the World Bank's Jim Yong Kim, Akinwumi Adesina of the African Development Bank, Alpha Condé of the African Union (AU) and also president of Guinea, and Mahamat Moussa Faki of the AU Commission. Only Adesina (2017) addressed climate change, by featuring renewable energy, particularly solar power.

Results

As host, Gentiloni said that the G7 leaders "were satisfied by how things went" but did not disguise any of the divisions that arose in their conversations (Irish and Balmer 2017). However, Merkel said "the entire discussion about climate was very difficult, if not to say very dissatisfying" (Agence France-Presse 2017a). Macron remained hopeful their arguments would have the desired effect and Trump would decide to keep the United States in the Paris Agreement (Irish 2017). The G6 plus the European Union reaffirmed their commitment to Paris, and did not unravel past progress, despite the resistance of the powerful United States. Yet, with the United States left out, the G7 largely failed on climate change and the G6 did nothing to ensure coordinated emissions reduction or resilience. On June 1, Trump announced that the United States would withdraw from the Paris Agreement (Shear 2017).

Dimensions of Performance

Taormina's small performance arose across all six dimensions of G7 summit per-formance, making it the lowest performing on climate change since 1990.

Domestic Political Management

In its domestic political management, Taormina's performance was small. All G7 leaders, as usual, attended. They stayed the entire time. Trump departed immedi-ately after to visit a nearby US military base. Neither of the two compliments in the communiqué concerned climate change (see Appendix C).

Deliberation

In its public deliberation, only 201 words, less than 3%, addressed climate change (see Appendix C). The main communiqué gave the climate and energy section 0.01%, referring to the growth and jobs that clean technology could bring. The paragraph on climate change mentioned the US review of its climate change policies and thus it was "not in a position to join consensus" on the issue.

Smart energy grids were mentioned in the separate G7 People-Centered Action Plan on Innovation, Skills and Labor. Mainstreaming gender equality into environmental policies was addressed in the G7 Roadmap for a Gender-Responsive Economic Environment.

Direction Setting

Direction setting was almost non-existent. None of the 158 references affirming the G7's core values of promoting democracy and human rights referred to climate change (see Appendix C). There was little consensus on the facts, causation and consequences of climate change, even among the G6.

Decision Making

Decision making was very small. There was only one commitment on climate change, less than 1% of the 180 made overall (see Appendix C). It had low binding text that reaffirmed G6 leaders' "strong commitment to swiftly implement the Paris Agreement" as stated in 2016. Nonetheless, it held the G6 together while the United States excluded itself. The United States joined the six energy commitments, including harnessing the economic opportunities offered by clean technology. Trade, with 11 commitments, made one broad link by committing the full G7 to improve "environmental standards throughout the global economy and its supply chains."

Delivery

G6 members' compliance with their one climate commitment was strong at 86%, up from the G7's 73% in 2016 (see Appendix C). This high compliance came from the commitment's generality, implicit recognition of the UNFCCC, explicit reference to international law and the US absence (Kirton and Larionova 2018a).

Development of Global Governance

The development of global governance was very small (see Appendix C). Inside the G7, leaders created one ad hoc ministerial-level body on culture and two official-level groups, but none related to climate change. Leaders made no climate reference to institutions outside the G7, other than two to the Paris Agreement that did not include the UNFCCC or COP21.

Causes of Performance

The overwhelming proximate causes of Taormina's small performance were Trump and related personal and political forces. The deeper drivers were diversionary security shocks and the low domestic political cohesion backing the inexperienced summit chair.

Shock-Activated Vulnerability

Shock-activated vulnerability arose strongly in security subjects that diverted attention and action away from climate change, to which no links were made.

The summit communiqués recognized 15 shocks, including seven on regional security and four on terrorism, but none on climate change (see Appendix E). Media-highlighted shocks also diverted attention from climate change. The *Financial Times* front pages in the first half of 2017 highlighted security and migration subjects.

At the time of the summit, the Paris Agreement did make the front page. The stories presented Europe, led by Macron and Merkel, pushing Trump to stick with the Paris Agreement at the summits of the G7 and the North Atlantic Treaty Organization (NATO). Immediately after Taormina, the front pages reported that Trump had clashed with the other G7 leaders on climate change as well as on trade, reporting that EU leaders described the summit as among the most challenging in years.

Canada's elite *Globe and Mail* also had front-page stories on Trump's refusal to sign the G7 commitment supporting the Paris Agreement. It added several vulnerabilities: climate risks for Canadian investors; Trump's efforts to boost coal-fired electricity; the domestic fight over pipeline expansion; Trump's weakening of fuel efficiency standards; and Canada's push to move toward zero emissions vehicles (Morrow 2017). It featured several diversionary subjects.

In the physical world, security and other shocks also crowded out those from climate change. The first set came from the terrorist attacks in France, Germany, the United Kingdom, Sweden and Russia. In the attack in Manchester, United Kingdom, just four days before the summit's start, 23 people died. The second set of shocks came from weapons of mass destruction. The third set came from migration, with more immigrants crossing the Mediterranean from Africa into Europe than the previous year. Although climate change was one root cause of this migration, the causal chain was complex, unclear and not made. The fourth set of shocks came from food and agriculture, led by famine afflicting several African states. The fifth set came from health, first from the Zika outbreak in Latin America and the Caribbean, now reaching the southern United States. The climate–health link was only made by the G7 health ministers meeting five months after the summit.

There were few energy and climate shocks leading up to the summit. Three months later, an unprecedented set of hurricanes hit British, French and EU territories in the Caribbean and the southern United States itself. By year end,

493 people had died from 44 extreme weather events in G7 countries, fewer than the 693 deaths from 47 events in 2016 (see Appendix G).

Multilateral Organizational Failure

In response to the very small climate shocks, there seemed to be considerable multilateral organizational success, beyond Trump's threatened withdrawal from the Paris Agreement. The United Nations Development Programme and the World Bank were implementing the SDGs. The UNFCCC secretariat had secured ratification of the Paris Agreement before Trump arrived. This inspired the other G7 members to offer strong support for the Paris platform. The failure on climate at Taormina flowed from Trump's distrust of UN multilateral organizations that could constrain the United States, but was not driven by any new UN failure to respond to new climate shocks. The IPCC continued to report on the most up-to-date science on climatic stability.

Predominant Equalizing Capabilities

The G7's declining globally predominant and increasing internally equalizing capabilities largely constrained performance (see Appendix H). The IMF (2017b) predicted that growth in gross domestic product (GDP) would remain around 2% for industrialized countries, but forecast China at 6.5% and India at 7.2%. In the first quarter of 2017, real GDP in OECD (2017a) members decelerated sharply to 0.4%, including in the United States, the United Kingdom, France and the European Union.

Although the G7 remained predominant in many specialized capabilities for controlling emissions, energy demand was growing from China, India, Russia, Saudi Arabia and other G20 members.

However, G7 members' internal equalization increased, driven by the recent decline in the value of the US dollar against the euro and yen and the renminbi beyond.

This internal equalization may have spurred the G6 to support the Paris Agreement and the United States to withdraw from it. According to the IPCC (2015, 46), economic and population growth were the most important drivers of increases in emissions from fossil fuels. So slower US and G7 economic growth could lower emissions and make leaders leave climate action to the future. Moreover, by year end, the aggregate emissions of G7 members declined on average by 0.80% from the previous year (see Appendix I).

Slower G7 growth also intensified domestic populist pressures that diverted the G7's focus toward economic issues, and away from climate change control and environmental protection, which Trump saw as harming economic growth.

Converging Characteristics

Divergence in members' political and ecological status also constrained G7 climate performance. The democratic character of G7 countries declined (see

Appendix J). Among the 57 countries ranked on climate change performance, the G7 members dropped and diverged (see Appendix K).

Domestic Political Cohesion

Low domestic political cohesion also reduced summit success.

Trump had a fresh electoral mandate and his party controlled both houses of Congress, but he had historically low approval ratings for a newly elected US president. He was deeply and broadly unpopular throughout the G7: he was only popular in Russia, recently suspended from the G8 after annexing Crimea. A Pew poll of 40,447 respondents in 37 countries from February 16 to May 8, 2017, found that the difference in confidence between Trump and Barack Obama in making the right choices in world affairs was 75 points lower for Trump at only 11% in Germany, 70 points lower in France at 14%, 57 points lower in the United Kingdom at 22% and 54 points lower in Japan at 24% (Wike et al. 2017). This drove Taormina's seven-to-one divide on climate change, as the G7's democratically elected leaders responded to what their publics believed.

Merkel, the G7's leading climate-committed leader, governed in coalition with the Social Democratic Party, which she would confront in a general election in September. Her approval ratings gave her a widening lead, confirmed by her Christian Democratic Party's victories in state elections just before the summit.

Macron was fresh from a decisive election victory as president but still had to fashion a majority in the National Assembly elections taking place soon after the summit — an outcome that, polls showed, his voters did not prefer.

Gentiloni led a caretaker coalition government, with his predecessor Renzi likely to return to fight the general election due in 2018.

May's governing Conservative Party seemed likely to win a decisive majority in the general election she had called for June 8. Yet, by May 22, her party's lead over the opposition Labour Party had dropped to only 9%.

Only Abe and Trudeau had high political control and capital, public support, and personal continuity, competence and conviction.

Continuity and the experience that comes with it were low, with new G7 leaders coming from Italy, the United States, the United Kingdom and France. Merkel was at her 12th summit, Abe at his sixth and Trudeau at his second. Seldom before had newcomers outnumbered veterans. Personal competence was limited by the absence of Trump's government experience. Only Merkel, Trudeau and Macron held strong personal convictions on the need for climate action, and only Merkel was competent in climate science.

Club at the Hub

The shared sense of the G7 as a valued club at the hub of a network of global governance was low. Trump did not personally value it at all. Among the guests at Taormina, there was no country leader who was a global champion of climate action and no head of a multilateral organization dedicated to climate change control. In the broad summit network, no G7 member, save Italy, participated in China's Belt and Road Forum on International Cooperation, which held its first summit in 2017 but did not support climate change control.

G20 Hamburg, July

Introduction

On July 7–8, 2017, G20 leaders gathered in Hamburg for their 12th summit. They too had to tackle Trump, now right after he had announced the US withdrawal from the Paris Agreement.

There was some chance the G20 could tame Trump enough for the G20 to succeed on climate action. It was the first G20 summit hosted by Germany, the world's fourth ranked economic power. In office since 2005, Merkel was the most experienced G20 leader and the only one to have attended every summit. Hamburg was the first G20 summit chaired by a former environment minister, a professional scientist and a woman, at a time when climate change, science and gender issues took centre stage. It came a few months before Merkel faced a general election. She stood opposed to a newly elected, internationally inexperienced, distracted US president opposed to climate change cooperation and much else. Thus, the looming collision threatened to make Hamburg the first failed G20 summit.

Debate

In the debate about the Hamburg Summit, the first school saw failure. Kaul (2017) viewed the G20 club as having become a "jack-of-all-trades" that should now focus on a few key policy areas, and suggested food security in Africa or climate change. Beaulieu (2017a) also saw failure, as the G20 did not link sustainable development and tourism to climate change.

The second school saw success. Kirton et al. (2017) expected scientist Merkel to substantially succeed, even if not enough to meet the goals of the Paris Agreement. At the end, Kirton (2017a, c) saw such success, even though a "domestically distracted" Trump, whose views were backed by Saudi Arabia and Russia, skipped part of the climate session. Snower (2017) also viewed the "G19" outcome on climate change as a "major achievement," as the G20 acknowledged the importance of climate change to most members and made climate commitments that "help strengthen global social norms to stop climate change."

Argument

The Hamburg Summit produced a strong performance on climate change. Its 22 climate commitments were the most at any G20 summit and complied with at an average of 71% (see Appendix D). They were reinforced by 58 commitments on the natural environment and 41 on energy. The G20 created far more official-level bodies to continue the work than any other summit and gave greater guidance to outside bodies too.

This strong performance arose despite Trump's presence, the G20's declining democratic members and its largely economic mission. Its strong performance, soon after Taormina's G7 small performance, came from Merkel's experienced, respected leadership and commitment to climate change control. She was aided by G20 leaders' growing experience with Trump's temperament, strengthened G20 internal institutional capacity through working groups and increased participation from international organizations.

Plans and Preparations

Host Priorities

Germany announced its plans for the Hamburg Summit on December 1, 2016, the day it formally assumed the G20 presidency (G20 Germany 2016). Its theme of "Shaping an Interconnected World" had three priority pillars. The second, improving sustainability, contained climate and energy, and the 2030 Agenda.

Sherpa Process

The short six-month cycle for preparing Hamburg, running from December 2016 to July 2017, led sherpas to focus on the value of the G20 and where it should focus its energy. The discussions on sustainability focused primarily on development with the 2030 Agenda and Africa, and on climate change, the environment, the Paris Agreement and clean energy.

As usual, climate change and energy were the toughest subjects, now even more controversial with the US opposing action. Reactions were visceral. Germany, with France's support, sought to follow up the Paris Agreement. Germany promoted the transition to a low-carbon economy, which was reasonable at the strategic level but difficult in its details. Russia and Saudi Arabia refused to deal with fossil fuel subsidies. Australia resisted carbon pricing. Some G20 members wondered about mobilizing private industry. Canada's proposal to reduce emissions from heavy vehicles was the lowest common denominator.

Germany specified 2025 as the deadline for ending fossil fuel subsidies in the draft communiqué for the G20 finance ministers meeting to be held in Baden Baden on March 18. G7 leaders had agreed to this at their 2016 Ise-Shima Summit, but it remained unacceptable to Saudi Arabia, Russia and India, as it had been since 2010.

Canada had climate change as its third summit priority. It argued that clean technology and climate change control were mutually reinforcing. It continued to push to end inefficient fossil fuel subsidies, limit emissions from methane and heavy-duty vehicles and promote carbon pricing. Progress depended substantially on the United States, so expectations were modest.

One month before the summit, the sherpas headed into extremely difficult negotiations. They worked with one-page "building blocks" that articulated high-level messages or priorities. For the first time at a G20 summit, several major issues, led by climate change, were left for the leaders themselves to resolve at the summit table.

Ministerial Meetings

The six pre-summit G20 ministerial meetings did little on climate change.

On January 22, agriculture ministers expressed full support for implementing the Paris Agreement and recognized that the climate change challenge must be met to give the world reliable, safe, nutritious and affordable food. Climate change was embedded throughout the declaration, especially the sections on sustainable development, the Paris Agreement, water and ICTs.

On March 18, finance ministers reaffirmed the commitment to phase out inefficient fossil fuel subsidies in the medium term.

On the Eve

On the summit's eve, Germany and the United States were on a collision course on climate change. Merkel, very engaged and ambitious, was the leading voice to counter the prevalent anti-globalization rhetoric. After her encounter with Trump at Taormina and comment that it was time for Europe "to take our fate into our own hands," she was not well disposed toward Trump, who was distracted by congressional hearings into his relationship with Russia and the firing of his FBI director (Klett 2017; Paravicini 2017). Trump and his policies were front and centre at Hamburg, and were a wild card there.

The dynamics and divisions at Taormina were likely to reappear at Hamburg in intensified form. The United States had quickly retreated from some of the consensus reached there. Ministers at the OECD's annual council meeting in June were unable to agree on a collective statement. A separate three-page chair's summary covered climate change (OECD Ministerial Council 2017a, b).

Climate and energy would be the subject of a separate discussion at the summit. Germany had prepared the very detailed G20 Action Plan on Climate and Energy for Growth early on, despite the United States saying it would have difficulty with it, even if the United States remained in the Paris Agreement. The entire action plan was threatened by the US withdrawal at the start of June.

Nearly all G20 members were arrayed against the United States on climate change. Among the many scenarios, there might be a separate chair's statement on climate change that would not bind the United States. Explicit references to implementing the Paris Agreement were out of reach and no consensus text was in sight, although there was common ground on investing in clean technology.

There could also be a green economy action plan. Argentina, host of the G20's 2018 summit, sought one focused on oceans. Germany was pushing a broader approach to include them.

At the Summit

The second session at Hamburg addressed sustainable development, climate and energy. Trump attended the first part and spoke on the subject before he and Vladimir Putin left for a bilateral meeting. Most leaders argued for the need to control climate change. The sherpas had to work hard behind the scenes to produce consensus communiqué text.

The session's highlights were on climate change and to a certain extent energy and accessibility. Trudeau spoke first, helping set the tone. He delivered a strong statement on climate change as the greatest global challenge, which leaders needed to act on collectively and recognize the advantage to leading on climate action. He insisted several times that a strong economy and a healthy planet were mutually reinforcing. He secured that message as the opening line in the communiqué passage on climate and energy.

Trump acknowledged that responsible stewardship and economic growth were complementary. He made a few pitches for fossil fuels, which he insisted would be clean. A few other leaders, led by Macron, expressed strong support for implementing the Paris Agreement. China and Russia led their BRICS colleagues of Brazil, India and Saudi Arabia in criticizing Trump's climate change decisions (Chazan and Wagstyl 2017).

Although the leaders discussed fossil fuel subsidies and carbon taxation, they focused on managing the 19-to-1 divide and reconciling a presidency that wanted a united G20 and consensus communiqué, given that only one country did not want to commit to implementing Paris. So, the sherpas negotiated a communiqué with a general statement on climate and energy, a paragraph directly related to the United States and a global recommitment by the other 19 members to implement Paris.

Discussions on carbon taxation continued, although several countries, including the United States, had reservations. There was no consensus on its value. In Australia, domestic resistance to a carbon tax had contributed to the resignations of both prime ministers Kevin Rudd and Julia Gillard in 2013. Canada continued to promote a carbon tax and tried to bring others on board.

Ultimately, Merkel would not accept a one-country statement as an appendix. The most constructive solution followed the example of the G7 summit. The three paragraphs at Hamburg contained the slimmest possible references to the Paris Agreement that all could accept.

Throughout the two-hour discussion, led by Macron and Trump, there was constant movement. They passed around a piece of paper, possibly proposed text, although the actual crafting of the communiqué was left to the sherpas. Macron and Trump did not leave the room for any separate bilateral negotiations.

There had been concern that the US withdrawal might trigger others to weaken their position. But none suggested reducing their commitment. The one small and surprising exception was Turkey's Recep Tayyıp Erdoğan, who said at his concluding press conference that the US withdrawal would "steer a course" for Turkey's parliament not to ratify Paris, because the withdrawal could jeopardize the promised funding for developing countries and the money France had promised Turkey (Clark and Wagstyl 2017). Turkey was among at least 40 countries that had not yet formally joined or ratified the agreement.

As the first day ended, Merkel was still urging compromises on the divisive issues of trade, climate change and migration, while noting that all disagreements would be resolved (Chazan and Wagstyl 2017).

Results

Despite these tensions, G20 leaders and their sherpas moved swiftly if painfully from confrontation and collision to a 19-to-1 division on climate change. On most issues, Trump's America adjusted just enough to make Hamburg perform strongly on climate change, relative to G20 summits before and since. Yet, relative to the growing global need, there was much more to do.

Dimensions of Performance

Domestic Political Management

In its domestic political management, Hamburg's performance was solid. All leaders attended, apart from Saudi Arabia's King Salman bin Abdulaziz Al Saud, who sent state minister Ibrahim Al-Assaf while he addressed tensions with Qatar. Trump attended the entire summit, but he and Putin skipped part of the climate change session to meet bilaterally. No communiqué compliments appeared (see Appendix D).

Deliberation

Public deliberation was very strong. Leaders dedicated 5,255 words to climate change, taking 15% of the documents (see Appendix D). This was the most words ever and the second-highest portion until 2019.

The preamble of the main declaration presented climate change as a challenge. Documents also referred to energy and climate, green finance and investments and climate-related financial disclosures, fossil fuel subsidies, the Paris Agreement and long-term 2050 pathways, the Kyoto Protocol, climate adaptation and resilience to disasters and extreme weather, climate risk insurance, climate and development, the Montreal Protocol, and climate and agriculture. In updating the G20 Action Plan on the 2030 Agenda, issued at Hangzhou in 2016, leaders noted that its "Sustainable Development Sectors" were "intrinsically linked to *people's* well-being, correlate with our *planet*, enable *prosperity* and foster *peace* and *partnership*."

In the private deliberations, in the session on climate and energy, Trump spoke before leaving for his meeting with Putin, unlike at Taormina, where he largely listened. Hamburg's discussions featured an intense debate between him and Macron, who, as had many of his French predecessors at summits, took a hard-line, antithetical approach to the US president. But that approach now centred on international economic openness, cooperation and strong collective climate change control with the Paris Agreement at its core.

Direction Setting

In its direction setting, Hamburg did not affirm the G20's foundational missions of globalization for all or global financial stability in regard to climate change (see Appendix D). However, on the G7's mission of human rights and democracy, leaders made one climate-related affirmation of each. The first loosely linked climate change with inequality and gender inequality in the same preamble paragraph. The second more directly linked to the democratic principle of transparency in climate financing through the Task Force on Climate-related Financial Disclosures (TCFD).

Several key environmental principles were affirmed, most notably renouncing overconsumption and overproduction, in a stand-alone document on marine litter. Hamburg also gave climate change priority placement in the main declaration, including it in the preamble, despite presenting it as a challenge equal to many others.

Decision Making

Hamburg's decision-making performance was very strong, with 22 climate commitments, the most made between 2008 and 2020 (see Appendix D). These commitments took 4% of the total, putting climate change 13th among 21 issues. The environment, with a strong emphasis on the marine environment, secured 58 commitments (11%), ranking second after development. Energy ranked third with 41 commitments and 8% of the total. Of the 22 climate commitments, 9 were highly binding and 13 were low binding, for a ratio of 39%.

Delivery

The delivery of these commitments was solid. The five assessed commitments averaged 71% compliance (see Appendix D). The commitment to overcome institutional and market barriers to green finance and private green investment was highest at 95%. The commitment to promote adaptation and resilience by cooperating within and beyond the G20 had 83%. The one to mitigate emissions through innovation, clean energy and energy efficiency had 80%. The commitment to support adaptation initiatives such as the National Adaptation Plan Global Network had 55%. The commitment to mobilize $100 billion annually by 2020 for climate finance had only 45%.

Development of Global Governance

In developing global governance institutions, Hamburg's climate performance was very strong (see Appendix D).

Inside the G20 there were 11 references to five bodies, far more than at any other G20 summit. The TCFD had four, the Green Finance Study Group three, the Environment and Sustainability Working Group two, and the Energy Efficiency Task Force and the International Financial Architecture Working Group one each.

Outside the G20, Hamburg made 26 references to nine different bodies, for the second highest number of references and the second highest spread. MDBs had eight references; the World Bank, the UNFCCC, the OECD and the International Energy Agency (IEA) three each; and the International Renewable Energy Agency (IRENA) two. The United Nations Environment Programme, COP and international organizations in general had one each. References focused more on the central economic and development bodies than on environmental and climate ones.

The G20 asked the IEA and IRENA to provide regular updates on energy sector transformation and to create an energy efficiency hub and supporting task force. The World Bank and MDBs were asked to identify further areas of cooperation on adaptation and resilience. The OECD, UNEP and the World Bank were invited to compile a list of public and private activities with the G20 for making finance flows consistent with the Paris Agreement. G20 leaders also asked MDBs to identify more opportunities for financing adaptation and mitigation and encouraging private sector finance to be aligned with the 2030 Agenda.

Causes of Performance

Hamburg's strong performance was driven primarily by Merkel's will and skill as host, backed by Macron's full support, in the face of multilateral organizational failure to prevent the US withdrawal from the Paris Agreement. Hamburg was nonetheless constrained by the anti-environmental passions of the inexperienced US president, who faced many other, more experienced leaders committed to climate action, such as Trudeau.

Shock-Activated Vulnerability

Shock-activated vulnerability spurred performance. There were spiking climate shocks in the physical world, amid the many diversionary shocks elsewhere (see Appendix F). Nonetheless, although the documents recognized many shocks and vulnerabilities, none were on climate change.

The elite North American newspapers in the first half of 2017 featured economic and security subjects. There were no new major issues or world events in the six weeks after the Taormina Summit. The dominating climate news focused on whether Trump would remain isolated on climate change.

Physical shocks from climate or the environment were significant, led within G20 members by earthquakes in Italy, extreme heatwaves in the United States, India and elsewhere, and forest fires in Portugal. There were deadly natural disasters from famine in Africa. Melting ice sheets in Antarctica and the Arctic produced severe vulnerabilities for all G20 members with major cities, populations, transportation facilities and military bases on their sea coasts. By the end of 2017, the G20 had recorded 131 extreme weather events and 4,125 deaths, slightly fewer than the 136 events and 4,350 deaths in 2016 (see Appendix G).

Multilateral Organizational Failure

Multilateral organizational failure strongly spurred Hamburg's strong climate change performance. It was led by the failure of the UNFCCC and the G7 to prevent Trump from announcing US withdrawal from the Paris Agreement. All other G20 leaders at Hamburg maintained their support for the agreement, despite some reluctance from Turkey and Russia. The G20 used the G7's all-except-one formula to govern climate change in other ways with Trump's United States on board.

Predominant Equalizing Capabilities

The G20's globally predominant and internally equalizing capabilities also supported Hamburg's strong performance on climate change. In overall capability, the G20 had 78% of the global economy, as good growth returned to the G7 members (see Appendix H). Internal equality increased, driven by the recent decline in the US dollar against the euro, yen and renminbi.

The G20's global predominance in emissions also increased (see Appendix I).

Converging Characteristics

Downward diverging political characteristics among G20 members, however, constrained Hamburg's success. The decline in democracy was most acute in Russia, Turkey and China (see Appendix J).

Members' climate change performance from 2016 to 2017 also diverged in a net downward way (see Appendix K).

Domestic Political Cohesion

High domestic political cohesion in most G20 members drove Hamburg's climate performance. Trump's deep, broad unpopularity among mass publics throughout G20 members made Merkel, facing an election soon, and most G20 leaders unwilling to defer to his anti-environmentalist convictions. The Pew poll found confidence in Trump in world affairs was low in Germany, France, the United Kingdom and Japan, and had dropped in Korea by 71 points to 17%, Australia by 55 points to 29%, Brazil by 49 points to 14%, India by 18 points to 40% and South Africa by 34 points to 39% (Wike et al. 2017).

Several of Merkel's colleagues had stronger, fresher electoral mandates than Trump, notably Macron, Korea's Moon Jae-in and Italy's Paolo Gentiloni. Hamburg was the second G20 summit for the United Kingdom's electorally weakened May, Argentina's Mauricio Macri and Brazil's Michel Temer. It was the third for Canada's popular Trudeau and the European Union's Tusk and Juncker. The veterans were India's Narendra Modi, Indonesia's Joko Widodo and Turkey's Erdoğan at their fourth summit, Japan's Abe and Mexico's Enrique Peña Nieto at their fifth, Russia's Putin at his sixth and South Africa's Jacob Zuma at his tenth.

Club at the Hub

Success was critically spurred by Merkel's skill in creating a summit that enhanced the G20's character as a club at the hub of an expanding network of global summit governance (see Appendix L). She constructed a summit schedule with an opening, free-wheeling, leaders-only general discussion on globalization, designed to match Trump's style and make the club an interpersonal as well as an instrumental one. It focused on the North Korean nuclear program, one of Trump's key concerns, and one where his G20 colleagues — led by China and Russia — could help. Having participated in an Arab Islamic American Summit as well as NATO and G7 summits in May, Trump was now more familiar with how such clubs worked. He participated in all Hamburg's summit sessions. The opening leaders-only retreat led to the planned advances on terrorism and more spontaneous ones on North Korea.

The network of global summit governance also grew. An important advance came with the first Belt and Road Forum Summit, created and hosted by Xi Jinping in Beijing on May 14–15. Leaders attended from Russia, Indonesia, Turkey and Argentina, helping make the G20 a more solid network hub.

UN Bonn, November

Introduction

The UN's COP23, taking place in Bonn on November 6–17, 2017, was the first COP after the formal announcement of the US withdrawal from the Paris Agreement. The spotlight shone on the US delegation, although Trump's announcement shifted policy, financial and institutional leadership to others, including China, the European Union, Germany, France, and even the United Kingdom and Canada. France initiated the new One Planet Summit in Paris on December 12, but did not invite Trump.

Debate

In the debate about the UN's performance at Bonn, the first school saw slow, uneven progress, with issues such as adaptation "slowed by more contentious items, such as mitigation" (*Earth Negotiations Bulletin* [ENB] 2017b). The second school, led by Maria Fernanda Espinosa, president of the Group of 77 plus China, saw progress on mechanisms to assess the "resources needed by poor countries to recover from climate disasters" (Phakathi 2017). The third school saw backsliding on emission cuts, due to the Trump administration's promotion of natural gas, nuclear energy and coal (Friedman 2017). The fourth school emphasized the clashes at COP, with developing countries, led by India, arguing that there was insufficient time to discuss the agenda leading up to the 2020 deadline for developed countries to meet their reduction targets (Sethi and Shrivastava 2017).

Argument

In the end, COP23 at Bonn produced a solid success. It fostered issue-specific coalitions and agreements among state and non-state actors. These included the Powering Past Coal Alliance co-launched by Canada and the United Kingdom; the America's Pledge coalition co-created by Michael Bloomberg and Jerry Brown, which convened US cities, states, businesses, citizens and universities committed to filling the gap Trump left; and the entry into force of the Kigali Amendment to the Montreal Protocol, due for ratification by Sweden and Trinidad and Tobago; some movement on the Talanoa Dialogue, designed to help countries enhance and implement their nationally determined contributions (NDCs); and alliances among local and regional leaders as well as mayors and urban councils.

These advances were inspired by the anticipated US withdrawal from the Paris Agreement and the trend since the 2014 Lima–Paris meeting of involving non-state actors, along with some momentum from the G20 Hamburg Summit.

On the Way to the Meeting

The UNFCCC subsidiary bodies met in Bonn on May 8–18, with several including members from the Umbrella Group, the Environmental Integrity Group of largely developed countries, the European Union, the G77 + China and the emerging economy bloc of Brazil, South Africa, India and China (BASIC). All were enthusiastic about negotiating technical and concrete issues. Two of the poorest blocs, the Least Developed Countries (LDCs) and the African Group, however, favoured less prescriptive outcomes, such as informal and reflection notes.

Emerging economies sought differentiation or flexibility in commitments, depending on national circumstances. China championed this position on behalf of BASIC, Bolivia for the Latin American and Caribbean bloc, and Saudi Arabia for the Arab Group.

Transparency, the technology mechanism and Article 6 on market and non-market cooperative mechanisms received much attention and a wide range of views. Guatemala called for using Article 6 as a source of finance. The European Union emphasized private and public sector involvement. The LDCs called for enhanced inclusiveness. The G77 + China asked for a clear definition of the role of non-state actors.

Negotiations also focused on the Paris rulebook. Progress was generally regarded as meaningful but slow (ENB 2017a). The US delegation engaged constructively, despite its unusually small size of seven negotiators and Trump's climate denialism (Evans and Timperley 2017). The Climate Vulnerable Forum called for a focus on 1.5°C rather than 2°C as the goal.

No consensus on the key topics of global stocktaking came. Discussions on Article 6 were "still in brainstorming mode" (Evans and Timperley 2017). There was also controversy on the non-inclusion of loss and damage on the agenda, although the agreement dedicated Article 8 to the issue.

Better progress came on the NDC registry and the initial design of the Facilitative Dialogue, a previously polarizing issue (Evans and Timperley 2017). Finance reporting moved forward. On the sidelines, there was a multistakeholder dialogue on the Local Communities and Indigenous Peoples Platform proposed at COP22.

Other high-level climate discussions continued throughout the year. Ministers of 35 countries attended the annual St Petersburg Climate Dialogue on May 22–23 in Berlin. The African Ministerial Conference on the Environment, on June 12–16 in Libreville, Gabon, reaffirmed members' support for the Paris Agreement.

By early August, US secretary of state Rex Tillerson confirmed America's intent to withdraw from the Paris Agreement, rather than renegotiate it. In August, the European Union and China, the likeliest candidates to replace US climate leadership, failed to deliver an expected joint statement on climate change and clean energy, although they joined Canada at a Ministerial on Climate Action in Montreal on September 16 (Canada, Environment and Climate Change Canada [ECCC] 2017c).

Later that month, at the UN General Assembly, Guterres and Fijian prime minister and COP23 president Frank Bainimarama emphasized the destructive impacts of climate change. By November, Nicaragua and Syria had committed to join the Paris Agreement (Maisonnave 2017a; Neslen 2017a).

At the Meeting

COP23 was chaired by Fiji, a small island developing country. The US withdrawal from the Paris Agreement raised concerns about the loss of $2 billion in promised American funding to the Green Climate Fund (Timperley and Pearce 2017). This was juxtaposed against the first volume of the US National Climate Assessment published on November 3 that acknowledged climate change and the attribution to climate change of extreme weather events (Wuebbles et al. 2017). Holding COP23 in Bonn increased attention on Germany's coal industry. In October, a leaked government report showed Germany would miss its 2020 climate target by a wide margin (Amelang 2017). However, as at the G20 in Hamburg, Germany was a competent host.

Some leadership came from the Powering Past Coal Alliance (Canada, ECCC 2017a). Institutional investors also increasingly valued sustainability, and private and public investors clean transportation, reforestation and sustainable cities (Caballero et al. 2017; Steer 2017). Here, some cautioned against greenwashing or falsely presenting an environmentally responsible public image.

Positions

Priorities at COP23 varied widely. Australia, representing the Umbrella Group, focused on transparency, where the United States was expected to work productively (Yeo and Mathiesen 2017). The United States emphasized its intention to continue engagement. Trump's representative George David Banks wanted to prioritize opposing differentiation and bifurcation. The European Union and the Environmental Integrity Group (represented by Switzerland) focused on implementation guidelines and encouraged partnerships between state and non-state actors. The European Union called for decision text drafts or elements for the entire rulebook. The Arab Group, represented by Saudi Arabia, called for an omnibus decision on the entire Paris Agreement.

Most other developing country blocs supported action and implementation. The Like-Minded Developing Countries (LMDCs) bloc emphasized finance, technology transfer and capacity building. The G77 + China, BASIC, the Alliance of Small Island States (AOSIS), the LDCs, the African Group and the Coalition for Rainforest Nations all wanted more climate finance. The LDCs expressed concern that support for the LDC Fund and the Adaptation Fund was waning. The G77 + China supported greater action on the Adaptation Fund. Bolivian chief negotiator Iván Zambrana commented that the operationalization of developed countries' finance commitments was reminiscent of development banks' focus on loans rather than transfers (Maisonnave 2017b).

A cross-cutting issue was designing the Talanoa Dialogue, focused on work leading up to 2020 and involving non-party stakeholders including cities, businesses, labour and faith groups. As incoming COP host, Fiji wanted it "to engage in an inclusive, participatory and transparent dialogue" (UNFCCC 2017). India emphasized that developing countries' need for finance and technology transfer not be downplayed. South Africa and the European Union committed to the dialogue as a mechanism to increase ambition and produce specific pledges. Some emerging economies opposed this, particularly if done in a compulsory way. The LMDCs considered the proposal unacceptably prescriptive (Li and Darby 2017).

Prospects for leadership by G20 members were initially encouraging. At the high-level segment of COP23, the United States was represented by Thomas Shannon, originally appointed by Obama (Mathiesen 2017c). Hope also grew for more progressive US participation in light of several gubernatorial elections in which environmental issues had been at the fore (Roberts et al. 2017). The European Parliament in October had called on participants to address conflicts of interest, echoing developing country calls at the May intersessional meeting. Xi said China had "taken a driving seat in international cooperation to respond to climate change" at the Communist Party congress in October (Li 2017b).

Negotiations

Negotiations at COP23 focused on advancing details across all aspects of the Paris Agreement Work Programme, particularly pre-2020 action, loss and damage, and finance.

Pre-2020 action attracted extensive debate. Developing countries worried that not enough parties had ratified the 2012 Doha Amendment to the Kyoto Protocol for it to enter into force and about failures in the annual provision of $100 billion in climate finance by 2020. The LMDCs called for an agenda item on pre-2020 commitments. In response, the COP presidency proposed that Morocco facilitate consultations during the second week of negotiations, but developing countries (including India but not China) criticized this (Li 2017b).

The European Union opposed adding any new agenda items to an already extensive list. This produced a deadlock by the end of the first week. Brazil's negotiator José Antônio Marcondes worried that this could cause a failure on the scale of Copenhagen (ENB 2017b). The European Union lead Elina Bardram maintained that a formal agenda item was not needed to discuss pre-2020 commitments, and that timelines for ratifying the Doha Amendment were impossible to provide due to domestic legislative processes (Li 2017a).

On November 15, developing countries agreed to disclose pre-2020 actions, including mitigation, finance and progress on ratifying the Doha Amendment, in exchange for not establishing a new agenda item on this issue. In addition, the final text mandated a climate finance ministerial meeting in 2019, with 2020 earmarked for reaching the $100 billion finance goal. The European Union announced its decision to ratify by the end of the year, with Germany and the United Kingdom doing so during the COP23 proceedings.

Loss and damage was discussed in the context of the Warsaw mechanism, and criticized for being overly technical. The LDCs and AOSIS called for it to become a standing agenda item. The G77 + China requested a place for it at the intersessional meetings. Developed countries opposed this, claiming it would become overly politicized. Germany remained one of a few developed countries defending insurance as a fast modality for mobilizing finance. The European Union and Canada sought to block these discussions. The resulting text established a task force on displacement and a commitment to meet in 2018, and acknowledged the "increasing frequency and severity of climate-related disasters" (AAP 2017). This met with strong objections from the United States and Australia (AAP 2017; Li 2017a). The text thus deferred the issue to 2018 (UNFCCC 2018, 20–21).

Finance debates centred on reporting financial contributions under Article 9.5 of the Paris Agreement. The G77 + China proposed that these discussions be connected to transparency and compliance. Developed countries largely objected. Australia and the European Union said it would involve questioning parties' individual contributions during negotiations. Developing countries' confidence diminished due to the limited attendance of European leaders, with only Merkel and Macron attending (Bauer 2017). Questions loomed over political-level financial promises, especially those made by France and others to fill the finance gap left by the Trump administration's withdrawal. Given the lack of consensus on finance, further discussions were deferred to 2018.

Throughout the negotiations, the US delegation engaged as usual (Siddique and Darby 2017). Nonetheless, the US withdrawal exacerbated vulnerable countries' concerns about inadequate ambition by the developed world (Mathiesen and Phakathi 2017). George David Banks described the White House as "reasonable" and anyone dismissing industries such as coal as "ideological and irrational" (Neslen 2017b). America's Pledge responded with its first report, which documented climate actions by entities constituting more than half of the US economy (Timperley 2017). However, the coalition did not replace the missing $2 billion from the Green Climate Fund left by the US withdrawal.

Some observers also worried that rather than fill the governance gap, countries such as China would follow the US approach (Mathiesen 2017a). The BASIC bloc pushed for bifurcation, with China dominating the G77 to secure compromises from developed countries. China's negotiator Xie Zhenhua said it was no easier to extract concessions from the United States than before (2017).

Results

Thus, progress on the Paris Agreement Work Programme was uneven, with information confusingly structured and with no agreed mandate to streamline the text. Concerns mounted over the deliberations focusing more on structure than substance, with the prioritization of a balanced outcome across topics contributing to the slow progress (Timperley 2017).

Developed countries were concerned that developing countries were adding more issues not explicitly mandated by the Paris Agreement, yet developing countries held that those issues constituted part of a package that was "consulted, not negotiated" at Paris (Mathiesen and Li 2017a). Although Xie Zhenhua and the LDC chair Gebru Jember Endalew were largely pleased with the outcomes, specific developing country concerns still included loss and damage, oceans, finance and NDC reporting (ENB 2017b).

COP23 produced the Talanoa Dialogue with a preparatory and a political phase. The result was a modest text that welcomed its launch in January 2018 (ENB 2017b). The presidency also launched the Ocean Pathway Partnership to increase consideration of oceans in UNFCCC processes.

Among the funds raised on the sidelines of COP23, more than $80 million each was allocated to the Adaptation Fund and the LDC Fund by Germany, Sweden, Italy, Belgium, Wallonia and Ireland. Macron promised that Europe would cover the IPCC's funding shortfall from the US withdrawal. The European Investment Bank and the Caribbean Development Bank established a $24 million fund for infrastructure reconstruction in the Caribbean, following devastating hurricanes earlier that year.

But little progress came on the NDCs (Phakathi 2017). The debate between bifurcated and universal requirements produced a deadlock. China and the LMDCs' push for bifurcation was opposed by the United States and the European Union, which characterized it as contrary to the Paris Agreement.

Disagreements also arose over the scope of information sharing, with vulnerable country delegates describing the talks as "re-litigating things that were resolved in Paris" (UNFCCC 2017, 3).

Other weak areas included adaptation, where progress was largely procedural; the Global Stocktake, where disagreements centred on the scope and the role of equity; and Article 6, over ensuring sustainability and environmental integrity and the operationalization of non-market mechanisms.

Commentators described the most promising COP23 achievements as arising outside the negotiations. These were the launch of the Canada–UK Powering Past Coal Alliance, America's Pledge report and the entry into force of the Kigali Amendment to the Montreal Protocol. Attention thus turned to the new One Planet Summit in Paris in December (Darby 2017).

One Planet Summit, Paris, December

Introduction

On December 12, 2017, two years after the formal adoption of the Paris Agreement and one month after COP23, Macron, Guterres and the World Bank's Kim convened the One Planet Summit in Paris. It assembled selected heads of state and government, international organizations, local authorities and civil society to motivate participants to act to achieve the Paris Agreement targets (Mathiesen and Li 2017b). It was the first of an expanding set of new plurilateral special climate summits that reconfigured the global governance of climate change by 2021. It produced a significant performance.

Climate finance remained Macron's primary concern. Without adequate financing, he believed that developing countries had few options for becoming greener economies or building capacity to counter damage from climate change (ENB 2017b). A Climate Finance Day was scheduled on December 11, the day before the summit, to highlight recent advances in the financial sector (World Meteorological Organization [WMO] 2017; Permanent Mission of France to the United Nations in New York n.d.-b). Any ideas or conclusions would be addressed the following day.

The One Planet Summit itself focused on effectively financing the global transition away from producing and consuming fossil fuels, and on announcing new financing in renewable energy, small-scale farming and support for women and forestry in developing countries (WMO 2017). It would also showcase existing initiatives that could be scaled up with the appropriate finance (Paris Europlace 2017).

At the Summit

Approximately 4,000 delegates participated, as did leaders from over 50 countries, and representatives from the public and private sectors and civil society (Worland 2017; WMO 2017). Macron invited all G7 and G20 leaders except Trump (WMO 2017). Macron had launched a funding proposal billed as "Make Our Planet Great Again," effectively enabling US scientists to live and work on climate-related research initiatives in France (Reuters Staff 2017b).

Summit participants addressed the availability of stable financing mechanisms, whose cost was estimated in the trillions (France, Ministry for Europe and Foreign Affairs n.d.-b). They recognized that the transformation to a greener economy with long-term impacts would require involving all actors, including the private sector, financial institutions and regulators. All levels of government would need to align guidelines and procedures effectively. Improved communication channels would ensure that cities, states and provinces had a solid framework for the execution of their commitments, allowing for increased transparency. Participants also acknowledged that public policies had to align with low-carbon solutions (France, Ministry for Europe and Foreign Affairs n.d.-b).

The summit produced several key announcements and a declaration containing 12 major commitments. The One Planet Coalition was formed to ensure these commitments were met (France, Ministry for Europe and Foreign Affairs n.d.-b).

Through its External Investment Plan, the European Union vowed to mobilize €44 billion worth of investment financing for Africa and countries neighbouring the European Union by 2020, with the expectation this would garner up to €9 billion in investments by 2020 (UN Climate Change 2017).

UNEP and the Banque Nationale de Paris Paribas launched the Sustainable Finance Facilities program to support small-scale renewable energy access, agroforestry, water access and responsible agriculture sustainability (UNEP 2017).

France, Sweden, the United Kingdom and more than 200 companies with a combined market capitalization of over $6.3 trillion pledged support for the TCFD (UN Climate Change 2017).

The Climate Action 100+ (2017) investor initiative assembled more than 450 global investors, with over $40 trillion in assets under management. It encouraged leading corporate emitters to strengthen their climate governance, curb emissions and improve the transparency of climate-related financial disclosures, thereby complying with standards set by the TCFD. Its results would be tracked by frequent progress reports.

A group of 89 French companies signed the French Business Climate Pledge, guaranteeing to invest generously in clean energy and emissions reduction strategies through carbon pricing and other science-based initiatives (Initiative of 89 French Companies 2017).

Canada and the World Bank partnered to assist several developing countries shift from fossil fuel dependence to renewable energy infrastructure, with minimum disruption to local communities (Canada, ECCC 2017b). The World Bank (2017) also announced a halt to funding upstream oil and gas projects after 2019, excluding projects consistent with countries' Paris Agreement commitments.

The OECD (2017b) launched, with France and Mexico, the Paris Collaborative on Green Budgeting to "drive the alignment of national budgetary processes with the Paris Agreement" toward achieving net zero emissions. It introduced a template for countries to present their budgets related to ecological transitions, clean taxes and green bonds. Canada, France, Mexico and Sweden announced their intent to become the first countries to incorporate this framework (One Planet Summit 2017).

China announced that by 2020, every Chinese listed company would be obliged to release information on its environmental impacts. Moreover, by committing to expand its green investment prospects, it encouraged banks to identify and capitalize on green initiatives, in addition to discouraging pollution-prone industries from continuing to offer pollution liability insurance (UN Climate Change 2017).

Results

The One Planet Summit produced a significant performance: it set directions, mobilized money, reinforced the Paris Agreement's commitments and built momentum for COP24 in 2018 (Worland 2017). Without the US president, it made promising commitments and raised funds for ongoing and newly promised green projects (WMO 2017). Patricia Espinosa said it "marked another extraordinary moment in the worldwide efforts to turn the promise of the Paris Agreement into a global reality" by trying to "reset the financial system" (UN Climate Change 2017).

Conclusion

The year 2017 was largely dominated by Trump's harsh anti-environmental convictions and actions. Taormina's poor performance flowed primarily from the US president attending his first global summit, buoyed by strong Republican congressional leadership at home. Its unprecedented declaration aligned all the other G7 members against Trump's United States on the Paris Agreement, but did little else.

Just two months later in Hamburg, G20 host Merkel confronted a domestically distracted Trump who had now officially announced the US withdrawal from the Paris Agreement. All other G20 members continued to support the Paris climate regime. Hamburg produced a record number of climate commitments, which secured solid compliance but were inadequate to meet the growing threat of climate change.

COP23 in Bonn in November, with Merkel and Macron present, produced a solid performance. More was done in December at Macron's innovation of the ad hoc, plurilateral One Planet Summit, which focused on climate finance, with Trump not invited. With its significant performance, it was the first major step in reconfiguring the global governance of climate change in response to Trump's unprecedented oppositional stance at the G7 and G20. The One Planet Summit, attended by over 30 leaders and heads of international organizations, reinforced commitments made under the Paris Agreement, mobilized badly needed climate finance and provided the collective high-level global leadership to forge a new path forward.

Global climate governance in 2017 was not propelled primarily by new shocks and vulnerabilities from climate change, which were overshadowed at the G7 and G20 summits by security and economic shocks and vulnerabilities. It was strongly spurred by multilateral organizational failure, led by those of existing climate and environment bodies struggling to overcome Trump's dislike of the Paris Agreement. Changes in relative capability shifted action from the G7 to the broader G20 with China there. Diverging and declining democratic characteristics and environmental performance constrained success at all 2017 summits and UN meetings. Low domestic political cohesion strongly constrained success at Taormina, but its strength at Hamburg, with Merkel as host, fuelled its strong performance.

The changing status of the G7 and G20 summits as the club at the hub of a global summit network drove this cadence. Trump's lack of attachment to summitry and multilateralism more generally reduced the G7 and G20's status as a personal club valued by leaders. But a climate-committed Merkel and Macron still valued both, and Xi had begun to do so too. And an important spoke was added by Macron with his innovative One Planet Summit.

The One Planet Summit started an expanding series of such plurilateral, climate-dedicated summits. A key element in this reconfiguration was effective global climate leadership from Europe. It came from Merkel at the Hamburg G20 as the custodian of her G7 Elmau Summit two years earlier, and from Macron at the One Planet Summit as the custodian of COP21's Paris Agreement also created two years earlier. Yet, whether that strong start in 2017 could overcome America's announced absence from the Paris Agreement and its consequences would depend on progress at the G7 and G20 summits over the next three years, with both an unrepentant Trump and a rapidly deteriorating climate.

5   G7 Leadership, 2018

Introduction

In 2018, leadership in the global summit governance of climate change returned to the G7, supported by the second One Planet Summit in the fall. Canada, America's highly integrated next-door neighbour, mounted the G7's Charlevoix Summit on June 8–9. It was hosted by Justin Trudeau, who was publicly as committed to controlling climate change as US president Donald Trump was to denying it. At Charlevoix, Trudeau stood by the core climate change agenda that his citizens and other G7 members valued, but focused on protecting the global oceans, which were a key carbon sink and a source of biodiversity under increasing threat. His strategy largely worked. Charlevoix produced a significant performance on climate change, even if it continued the G6-versus-G1 formula, and failed to get Trump and Japan to endorse Charlevoix's G7 Oceans Plastics Charter. Trump's impulsive immediate post-summit repudiation of the entire Charlevoix communiqué dominated media attention and deterred France's Emmanuel Macron from trying to advance climate action as host of the G7's Biarritz Summit the following year. But it had no impact on the high implementation of the many climate and other environmental commitments most G7 leaders made at Charlevoix. As with Germany's Angela Merkel at the G20 Hamburg Summit in 2017, a climate-committed G7 host delivered a successful summit.

The G20 summit at Buenos Aires on November 30–December 1 produced only a small performance. It was hosted by Argentina's Mauricio Macri, who had little personal commitment to climate action. His country depended financially on a US administration that controlled the International Monetary Fund (IMF) and thus the support that Argentina needed to survive its financial crisis that erupted in August. From the start, Macri did not even try to use his summit to counter climate change, focusing instead on an environmental agenda that added soil to Charlevoix's seas. G20 leaders claimed they should defer to the 24th Conference of the Parties (COP) to the United Nations Framework Convention on Climate Change (UNFCCC), which would meet immediately after the G20 summit on December 3–18.

With no G20 push, the COP24 in coal-dependent Katowice, Poland, also produced a small performance. It advanced on carbon markets but failed to align

DOI: 10.4324/9780429055485-5

members' climate pledges with the goals of the Paris Agreement. Although countries settled several sections of the Paris rulebook, they did not agree on the rules for voluntary market mechanisms. COP24 produced only incremental, procedural progress at best, thanks to divisive politics, and the rising populism and protectionism that also afflicted the G7 and G20. The host's inexperience and its sponsorship from the coal industry also constrained action.

But progress was kept alive by the substantial performance of the second One Planet Summit, on September 26 in New York. It forged public-private partnerships, reinforced the Paris Agreement commitments and mobilized climate finance. It delivered a road map, backed by concrete actions, for forging a reconfigured path forward. But it did not try to create major breakthroughs on the divisive topics traditionally negotiated under the UN umbrella.

G7 Charlevoix, June

Introduction

On June 8–9, 2018, G7 leaders and their many outreach guests met in Charlevoix, Quebec, amid continued divisions on key climate and environmental issues. One year earlier, Trump had announced that the United States would pull out of the Paris Agreement. The other G7 members remained united behind the Paris Agreement, declaring it irreversible. Trudeau, as host, sought to adapt to these political tensions through implementing an innovative format at the ministerial level, combining the meetings of environment, oceans and energy ministers over three days, three months after the leaders met, to advance several specific mandates from Charlevoix.

Debate

In assessing Charlevoix's performance, the first school forecast failure, due to the dark cloud cast by Trump. His first official visit to Canada came much later than most newly elected US presidents and he brought nationalism, populism, protectionism and unpredictability to a country whose distinctive national values were antithetical to his own. Many predicted a stormy summit, with Trump vetoing any serious advances on climate change (Kirton 2018g, h).

The second school saw self-inflicted Canadian caution. Fell (2018) noted that Canada did "reasonably well" by G7 standards, but the bar was too low. Mulroney (2018) saw Canadian foreign policy off track, due to Trudeau's impetuosity and "obsession with diaspora politics" and "exotic photo-ops."

The third school anticipated climate change being overshadowed by security shocks and ensuing multilateral organizational failure, due to the Russian nerve gas attack in the United Kingdom on March 4 and the failure of the United Nations Security Council (UNSC) and the North Atlantic Treaty Organization (NATO) to respond (Glavin 2018; Robertson 2018).

The fourth school saw an unwise choice of the G7 over the G20. The *Canadian Foreign Policy Journal* (2018, 11–12) argued: "Few international issues these days can be properly solved without G20 members like Russia, China, India, Brazil and South Africa."

The fifth school emphasized that intimate "relationships matter" (Carmichael 2018). Highlighting the summit's private deliberation and the value of face-to-face talks, it said that unlike the bigger, more diverse G20 summit, the G7 allowed members to complain about China and Russia, discuss common problems and learn about peers' successes and failures.

The sixth school saw substantial success (Bland 2018; Kirton 2018h). Kirton (2018g) forecast a difficult challenge on climate change, oceans and clean energy, but pioneering actions on plastics pollution, adding that Canadians strongly supported Trudeau on climate change (2018f).

The seventh school saw a disappointing performance on climate change. Banda (2018) argued that success should be measured against G7 members' domestic and collective emissions profiles. Here, Canada had high financial contributions to the fossil fuel sector and the largest per capita emissions in the G7 by far. Banda argued that the G7 "has lent critical political support to the international climate regime for decades," but Charlevoix missed several key commitments that built on this past work, particularly regarding climate risk and disclosure and the elimination of fossil fuel subsidies.

Argument

The Charlevoix Summit produced a significant performance on climate change. The "G6" (the six countries plus the European Union) stood by the Paris Agreement and a G5 — without the United States and Japan — emerged to tackle marine litter. Leaders produced 12 climate commitments, as they had at Ise-Shima in 2016. Compliance was a strong 84%, despite Trump renouncing Charlevoix's communiqué as he flew away, much higher than the 73% from Ise-Shima, which was the last Trump-free G7 summit, and much higher than the G7's 73% overall average. No member followed the United States in its climate stance. Yet, Charlevoix made no progress on fossil fuel subsidies and carbon taxes, despite strong popular support, including within the United States, and little in tempering the global warming that continued its relentless, deadly rise.

Charlevoix's significant performance was driven by Trudeau's skill, as a G7 veteran, in mobilizing his European and Japanese partners behind the Paris Agreement, and involving a coalition of coastal powers to protect the oceans beyond the G7, where Trump's views were less entrenched. The ministerial meetings in Halifax in September spurred compliance with Charlevoix commitments by creatively linking climate change, oceans and energy. Trudeau was backed by a majority government, strong support from the Canadian public and growing support for climate action from US voters next door. With climate shocks rising on US southern coasts and the UN failing on climate action, he restored G7 progress on climate change, even with the US government left out.

Plans and Preparations

The Charlevoix Summit was held on June 8–9, in Trudeau's home province of Quebec. The St Lawrence River offered the perfect visual backdrop for Canada's priority of clean oceans, with its regular visitors of beluga and blue whales in the protected Charlevoix Biosphere Reserve (Beaulieu 2017b).

Trudeau had recently developed Canada's first nationwide framework on climate change — the Pan-Canadian Framework on Climate Change — but was facing resistance for its imposition of a nationwide carbon price (Government of Canada 2016; Cowan 2018).

Participants

Attendees at Charlevoix were Macron, Trump, Theresa May, Merkel, Shinzo Abe, Giuseppe Conte and the European Union's Jean-Claude Juncker and Donald Tusk. The guests were Argentina's Mauricio Macri, Sheikh Hasina of Bangladesh, Haiti's Jovenel Moïse, Jamaica's Andrew Holness, Kenya's Uhuru Kenyatta, Hilda Heine of the Marshall Islands, Norway's Erna Solberg, Paul Kagame of Rwanda and chair of the African Union (AU), Senegal's Macky Sall, Danny Faure of the Seychelles, South Africa's Cyril Ramaphosa and Vietnam's Nguyên Xuân Phúc. The IMF's Christine Lagarde, Angel Gurría of the Organisation for Economic Co-operation and Development (OECD), the UN's António Guterres and Kristalina Georgieva of the World Bank also attended.

Preparations

Canada's summit team sought to encourage enthusiasm for its priorities at home. Canada intended to involve youth in particular (Kirton 2018g). Climate change was an important issue for youth and for Indigenous peoples. Meetings with the Youth 7 engagement group generated recommendations for G7 leaders on climate action (Mariani 2019). Ottawa was flooded with media calls about what could be expected if climate change continued.

Trudeau announced his five G7 themes via Facebook Live on December 14, 2017 (Canada, Prime Minister's Office 2017). Climate change, oceans and clean energy came fourth.

Pre-Summit Meetings

Trudeau attended the World Economic Forum in Davos in January 2018, where he held a private session on climate change and oceans. In late April, at the Summit of the Americas in Peru, he met Macri to discuss coordinating the G7 summit with the G20 Buenos Aires Summit in November. Trudeau then went to Paris to meet with Macron and the heads of the OECD, the Francophonie and the United Nations Educational, Scientific and Cultural Organization (UNESCO), and to London to meet May and the 53 leaders of the Commonwealth at their biennial Heads of Government Meeting from April 16 to 20.

Sherpa Process

Domestic consultations began in late November 2017. At the University of Toronto in early December, the town hall audience chose climate change as its top priority.

At the second sherpa meeting, in Victoria, British Columbia, on March 7–8, sherpas had a good discussion on the environment and oceans, which included exchanges with students and experts.

Climate change was a very broad file. Canada would focus on oceans, notably plastics and fisheries, and the resilience of coastal communities, especially small island developing states (SIDS) and coastal regions. Recent major hurricanes in Houston and Puerto Rico were a coastal resilience issue, whether one agreed that hurricanes were increasing because of climate change. The United States and other countries agreed to have a robust discussion of this topic. There would be a discussion of marine litter too.

Ministerial Meetings

On April 22–24, G7 foreign and security ministers met at the University of Toronto, chaired by foreign minister Chrystia Freeland. In their communiqué, they thanked the Working Group on Climate Change and Fragility for its work.

On May 31–June 2, just before the summit's start, G7 finance ministers, central bank governors and development ministers met in Whistler, British Columbia. They acknowledged the progress made by the InsuResilience initiative and InsuResilience Global Partnership for Climate and Disaster Risk Finance and Insurance Solutions. They asked international financial institutions to assess the role of disaster risk insurance coverage for vulnerable countries and financing mechanisms for managing gender-responsive disaster risk.

After the summit, on September 19–21, G7 environment, oceans and energy ministers met in Halifax (Kirton and Warren 2018b). The three-day sequence started with the environment ministers, chaired by Catherine McKenna, Canada's minister of environment and climate change and a veteran of the Paris Agreement negotiations. Environment ministers made 69 commitments, the most since environment ministers started meeting in 1992. Canada made financial commitments for renewable energy projects, launched initiatives on the circular economy with the private sector and committed to reduce plastic waste in its government operations by 75% by 2030. Compliance with three assessed commitments was 69% (Kirton et al. 2019; Warren 2019a).

At the Summit

At the summit, Trump was concerned with security, trade and health priorities and pursuing his "America First" policy (Trump 2018).

Macron (2018a) wanted the G7 to support the implementation of the Paris Agreement in the lead-up to COP24 in Katowice. He emphasized the need to commit to a low-carbon economy and to address biodiversity loss (Macron 2018a, 11). He welcomed Trudeau's focus on ocean conservation, and planned to have climate change and biodiversity as priorities at France's G7 in 2019.

May (2018) wanted to promote Britain's work on banning single-use plastics for ocean health, and said she was pleased that Trudeau was making marine litter a priority. Other members had economic, security and migration priorities.

At the table, G7 leaders discussed emissions reductions, a low-carbon society and economic growth and jobs (Japan, Ministry of Foreign Affairs 2018). They also discussed the harms of warming oceans to biodiversity and thus the need for implementing the Paris Agreement and a circular economy. Abe stressed Japan's technical capacity to advance decarbonization.

The outreach session focused on resilient coasts, ocean health and plastic waste, with "lively exchanges" on climate change and adaptation (Japan, Ministry of Foreign Affairs 2018). Abe noted the impact of plastic waste on human health and his intention to extend Ise-Shima's "3Rs" framework and environmental infrastructure to the G20 summit he would host in 2019.

Results

Charlevoix culminated in a consensus communiqué to which all members, except Trump on climate, agreed. It issued eight additional documents of so-called "commitments," released separately to highlight the importance the host attached to them.

Charlevoix's results on climate change were much greater than Taormina's in 2017. Performance was strong on the critical dimensions of decision making and delivery. However, Trump again created a seven-to-one divide on climate change. Another divide came on the G7 Oceans Plastics Charter, which both the United States and Japan refused to sign.

Domestic Political Management

All G7 leaders attended the summit and engaged seriously and collegially in almost all sessions. All members except Trump attended all sessions. At midday on the second day, Trump left to fly to his summit in Singapore with North Korea's Kim Jong-un on June 12, thus missing the climate change session. His early departure had precedents from other G7 leaders at past summits.

There were no communiqué compliments on climate change (see Appendix C). But the summit produced a very strong public opinion boost for its host (Kirton 2018b). An Ipsos poll showed Trudeau's approval ratings up 6% to the 50% level by June 18, even if the Conservative Party remained slightly ahead of his Liberal Party (Ibbitson 2018).

Moreover, an Ipsos poll of Canadians and Americans taken on June 13–14 found that 72% of Canadians and 52% of Americans approved of the way Trudeau had handled Trump's post-summit repudiation of the communiqué, and only 14% of Canadians and 37% of Americans approved of Trump's behaviour in the tweet Trump sent after his departure referring to Trudeau being "weak and dishonest" (Hopkins 2018). A majority of 57% of Canadians and 52% of Americans said Canada should not overreact to Trump's "political posturing."

Deliberation

Public deliberation was significant. G7 leaders dedicated 1,696 words (15%) of their nine outcome documents to climate change (see Appendix C). This was above the average of 13% from 2015 to 2017. It broke the "Trump shock" of 2% at Taormina.

Charlevoix referred to carbon pricing and ocean acidification for the first time. It emphasized coastal resilience to the impacts of climate change and recognized nature-based solutions, such as wetlands and mangroves, and women's equal participation in decision making for disaster risk reduction. G7 leaders also committed to collaborate with local and Indigenous partners on climate action. The Oceans Plastics Charter acknowledged the links between plastics and emissions.

In its private deliberations, Charlevoix's performance was strong. Over the summit's two days, the leaders engaged vigorously at almost all the sessions.

Direction Setting

In its direction setting, Charlevoix's performance was substantial. Leaders affirmed open democracy twice in the context of climate change. This was an increase from Taormina, which had none, and the third highest between 2015 and 2020. The G7 recognized the links between ocean acidification, plastic waste and emissions. This was the first time for such synergy.

Decision Making

Charlevoix's decision making was significant, with 12 climate change commitments for 3% of the total, far more than the one made at Taormina (see Appendix C). Five commitments explicitly linked to the Halifax environment ministers' meeting after the summit. Climate ranked seventh among all subjects. However, none of the climate commitments was highly binding, and the United States excluded itself from all.

Of the total commitments made, the environment, dominated by 42 on oceans, took 29% and ranked first, for the first time. Gender equality came second, which boosted climate action.

In mobilizing money, with five members pledging to prevent plastics pollution in oceans, Canada committed $100 million to the cause. The G7 also agreed to reinforce the resilience of coastal communities, with Canada pledging $162 million here. But there was no collective mobilization of climate finance (Kirton 2018b).

Delivery

Charlevoix's delivery of its decisions was strong. The five climate commitments assessed averaged 84% — due in part because they excluded the United States — higher than the 73% G7 average for climate change (see Appendix C). The commitment on climate finance had 100%; the commitments on the private sector and on multilevel governance had 93% each; the commitment on innovation had 85%; and the commitment on disaster risk insurance had 50%.

Development of Global Governance

The development of global governance was solid (see Appendix C). The Blueprint for Healthy Oceans, Seas and Resilient Coastal Communities referred to one G7 institution in the form of the ministerial meeting on the environment, oceans and energy.

The climate change passages referred to five institutions outside the G7. Most were to the UN system with one each to the UNFCCC, COP24, the Paris Agreement and the UN General Assembly (UNGA). One was to the One Planet Summit, which the G7 recognized for contributing to the "collective effort" to tackle climate change.

Causes of Performance

Shock-Activated Vulnerability

Charlevoix's significant performance was not caused by climate shocks or vulnerabilities recognized in its communiqués (see Appendix E). They only noted disaster risk prevention, reduction and response in general.

Media-highlighted shocks and vulnerabilities focused on plastics in the oceans. There was increased media attention to UN reports that the oceans would contain more plastics than fish by 2050. Powerful images of beached whales and other ocean animals trapped in plastic and with plastic in their stomachs inspired strong public outcry and changes in consumer behaviour (Kirton and Warren 2018b).

On the front pages of the *Financial Times*, climate change was pushed out by security and trade stories. In the *New York Times*, climate change was supplanted by security subjects. Canada's *Globe and Mail* ran energy, economic and security stories.

Yet, behind the headlines, the physical shocks from climate change were strong. Deaths in G7 countries (excluding the European Union as a whole) rose from 493 in 2017 to 857 in 2018 (see Appendix G). Even before Charlevoix, 2017 was the costliest year on record for natural disasters in the United States, reaching at least $306 billion in damages. Those disasters included wildfires in California, Hurricane Harvey and the blackout in Puerto Rico (Irfan and Resnick 2018). The fires were prolonged by the extreme heat that made 2017 the second hottest year on record, after 2016, and 2018 the fourth hottest year on record (Yale Environment 360 2018; NASA Global Climate Change: Vital Signs 2019).

Business risks from climate change were very strong. The World Economic Forum's 2018 Global Risks Report, released in January 2018, ranked the most likely risks as, first, extreme weather events; second, natural disasters; and fifth, failure of climate change mitigation and adaptation (World Economic Forum 2018). The five most impactful were, in order, weapons of mass destruction, extreme weather events, natural disasters, failure of climate change mitigation and adaptation, and water crises. This was the second consecutive year that extreme weather events or climate change had taken the first or second spot in both likelihood and impact.

Multilateral Organizational Failure

An influential cause was multilateral organizational failure, notably the inability of UN Climate Change to advance the Paris rulebook, including at a technical COP meeting just before the G7 environment ministerial meeting in September (Darby 2018b). Some progress was made on producing a legally binding instrument under the UN Convention on the Law of the Sea, but negotiations were pushed to 2019 and 2020, creating a gap on ocean conservation (Kirton and Warren 2018b). However, the G7 did not need to fill any governance gap on emissions reduction in the shipping sector, as over 100 members of the International Maritime Organization ([IMO] 2018) had adopted such a strategy in early April. In late 2018, after Charlevoix, UNGA launched the global Campaign Against Plastic Pollution (UN News 2018).

Predominant Equalizing Capabilities

Changes in G7 capabilities constrained Charlevoix's performance. G7 economic growth was predicted to be uneven in 2018 (CBC News 2017). It was expected to slow for Canada, Germany, the United Kingdom, Japan and Italy, but rise for the United States and France (IMF 2017a). From 2017 to 2018, all G7 members' economic growth slowed, except for Trump's United States (see Appendix H).

Canada as host was more motivated to act on the marine environment, given its geographic assets of bordering three major oceans and having large intact boreal forests, peatlands and freshwater. These specialized capabilities, where France and the United Kingdom (with their overseas territories) also ranked highly, helped spur Charlevoix's G5 performance on oceans.

By the end of the year, the G7's aggregate emissions would decline by 9.85% from 2017 (see Appendix I).

Converging Characteristics

Converging characteristics at a higher level on democracy and environmental performance supported Charlevoix's significant performance.

On their democratic rank among 167 countries, Canada led the G7, remaining sixth for a third year (see Appendix J). Only Italy declined.

On G7 members' climate change policy, divergence grew (see Appendix K). Canada's rank rose to 51st from 55th. Japan rose to 50th from 60th in 2017. Italy remained in 16th. But the others dropped, led by the US plunge to 56th from 43rd.

Domestic Political Cohesion

Domestic political cohesion was substantial, with very strong support for climate action in public opinion (Kirton 2018g).

In Canada, Trudeau was at his third G7 summit. His Liberals had a 14-seat majority and faced a general election in 2019. He had been elected in October 2015, due in part to his support for climate change and promise to reverse the previous Conservative government's poor international reputation on climate. Nanos polling showed that support for the Liberals had fallen from 41% at the start of 2018 to 36% by mid March, putting them only three points ahead of the Conservatives (Kirton 2018g). But by April 13, his substantial lead had returned to 41% against the Conservatives' 29% (Nanos 2018). Climate change ranked second among issues Canadians were concerned about, with extreme weather events in sixth (Kirton 2018g).

More broadly, 46% of Canadians were open to one another and the world, while only 30% felt economically and culturally insecure (Kirton 2018g; Levitz 2018). An Environics poll on Canadians' views on the world, taken from October 23 to November 26, 2017, revealed strong unity among Anglophones and Francophones (Environics Institute for Survey Research 2018). It also showed strong support for Canadians' traditional distinctive national values of environmentalism, multiculturalism, openness (in immigration and trade), antimilitarism (in the form of peacekeeping), globalism (of an Atlantic-centric kind) and international institutionalism of both a multilateral, formal bureaucratic and a plurilateral, informal, summit-centred sort (Kirton 2018a).

Trump was at his second G7 summit. His Republican Party controlled both houses of Congress. It faced mid-term elections on November 6 that previous polls and by-elections showed could end his control of Congress. In March, Trump abruptly fired his foreign minister, Rex Tillerson, following the resignation of Gary Cohn, head of his National Economic Council. Trump's approval ratings remained at unprecedented lows for a new president entering the second year in office (Kirton 2018g). Most Americans now supported policies to advance climate change, a level that had risen since 2014 (Kirton and Warren 2018b). They strongly supported carbon taxes on fossil fuel companies.

Merkel was the most experienced leader in the G7 and at home. But she had struggled to assemble a coalition government with the Social Democratic Party (SPD) and Christian Democratic Union/Christian Social Union (CDU/CSU), as both parties had suffered losses in the general election in September 2017 (Kirton 2018g). A Civey poll taken on March 20–27 found support for the governing CDU/CSU at 31% (down 2% from a month earlier), its SPD coalition partner at 17% and the Greens at 12% (Civey 2018; *Speigel* 2018). The governing coalition thus had a minority of 48%. Merkel was a scientist and former environment minister whose coalition was 33 percentage points more popular than the Alternative for Germany, which was just 3% ahead of the Greens.

Abe, another summit veteran, governed in coalition with the New Komeito Party and thus controlled both legislative chambers (Kirton 2018g). But by March 17, his support had plunged nine points to 39% in a survey taken from March 9 to 12 by Jiji News, due to a scandal that kept his finance minister Taro Aso from attending the G20 finance ministers' meeting in Argentina (Takenaka and Lies 2018). For the first time since October, Abe's support was below those who did not support him.

May had a faction-ridden, feuding, minority government in the United Kingdom, having lost her majority in the fall of 2017 (Kirton 2018g). However, a YouGov poll on March 26–27 on who would make a better prime minister gave her 38% against Labour Party opposition leader Jeremy Corbyn at 27% (YouGov 2018). Most voters supported strong climate action, with 80% also wanting urgent action on single-use plastics (Kirton and Warren 2018b).

At his second G7 summit, Macron had a legislative majority and was incoming host for 2019. His popularity going into the Charlevoix Summit was stable. However, a month before the environment ministerial meeting, environment minister Nicolas Hulot resigned because Macron's government was not taking strong enough action on climate change and environmental protection. Hulot's popularity rose, and Macron's declined (Kirton and Warren 2018b; Martens 2018).

Conte was attending his first summit, having been appointed Italy's prime minister on June 1. After the March 4 elections ended the Democratic Party government led by Paulo Gentiloni, Conte's Five Star Movement and the Northern League Party struggled to create a coalition (Kirton 2018g).

More broadly, a Pew Research Center poll conducted across 38 countries in 2017 found the issue of greatest concern among global publics was ISIS at 62%, followed closely by climate change at 61% (Poushter and Manevich 2017).

Club at the Hub

The G7 summit remained a club at the hub of a growing network of global summit governance among all its leaders, except Trump (see Appendix L). Trudeau as host connected the G7 to the Francophonie and Commonwealth and thus half the countries in the world. Charlevoix's guests connected the North American and European G7 members to Latin America and Africa, including least developed countries (LDCs), such as French-speaking Haiti and the Seychelles. The presence of Haiti and Jamaica connected the G7 to the Caribbean Community (CARICOM); Rwanda and South Africa to the AU, with Rwanda's Kagame the 2018 chair; and South Africa to the BRICS (Brazil, Russia, India, China and South Africa). Vietnam added a stronger connection to the Association of Southeast Asian Nations, along with Japan's connection. Thus, the G7 remained a highly globally connected club.

G20 Buenos Aires, November

Introduction

The 13th G20 summit took place in Buenos Aires on November 30–December 1. It was the first G20 summit hosted by Argentina, and the first by a G20 member outside the G7, BRICS and the MIKTA group of Mexico, Indonesia, Korea, Turkey and Australia (Kirton 2018c). It was the first hosted in South America. It was hosted by Mauricio Macri, first elected in 2015. In late 2017, Argentina set out a long list of summit priorities that included climate change, environmental sustainability and clean energy. Climate change thus had a small place at the start of Argentina's year. It had an even smaller one at the end.

Debate

Assessments of Buenos Aires's performance largely overlooked climate change but agreed that Trump's attacks on the global multilateral order would create divisions difficult to overcome.

The first school saw weakening performance on climate change. On seeing a leaked draft of the communiqué, Mathiesen (2018a) argued that the G20's resolve to stand up for the Paris Agreement against critical voices was weakening. The draft did not give full support to the agreement or mention the report released by the Intergovernmental Panel on Climate Change (IPCC) one month before the summit. By acknowledging "varied" energy choices and "different possible national pathways," it implicitly permitted the use of coal.

The second school saw a struggle to produce any communiqué, noting that a few weeks earlier there was none from the Asia-Pacific Economic Cooperation (APEC) summit for the first time (Champion et al. 2018; *Nikkei Asia* 2018). El-Erian (2018) thought any communiqué would be bland. Emorine (2018) argued that given the geopolitical tensions, a joint communiqué would be "an achievement in and of itself."

The third school saw poor performance on climate change. On the release of the final communiqué, Kirton (2018i) noted surprising, meaningful but still inadequate progress.

Argument

Buenos Aires produced a small performance on climate change. It offered only 532 words and three commitments (see Appendix D). It featured the G20's great divide, with all members except Trump's United States agreeing to two paragraphs and two commitments on climate change, and the final paragraph for the United States alone reaffirming its intention to withdraw from the Paris Agreement. This text was negotiated and agreed by all G20 members, to reduce the discordance it contained.

This small performance arose from the diversionary shocks of a financial crisis erupting in September in Argentina, which supplanted substantial climate shocks. Multilateral organizational failures continued, although the G7 Charlevoix Summit in June, with its significant advances on climate change, reduced the pressure on an economically distracted Argentina to act here. A substantial constraint came from rising US economic capability and its declining share of greenhouse gas emissions within the G20, whose emissions continued to rise. Divergence grew as the most powerful United States joined Russia, Saudi Arabia and now Brazil in opposing climate action. Trump's dismissive personal convictions about climate change persisted, despite the Republicans' loss of control of the House of Representatives. However, the key causes were the low climate change convictions and weak domestic political position and international power of Macri, who was consumed by Argentina's 2018 financial crisis that made him need the goodwill of Trump and the IMF, which the United States ultimately controlled.

Plans and Preparations

Host Priorities

A year before its summit, Argentina set its theme as "Building Consensus for Fair and Sustainable Development." Its three priorities omitted climate change and the environment but included food security from more inclusive, efficient and sustainable agricultural productivity.

The environmental focus was thus on the land. Macri's priority divided the labour between the G20 and Trudeau's G7, which had emphasized climate change and oceans. Argentina, a major agriculture exporter, sought more responsible use of fossil fuels by transitioning to cleaner technology and emphasizing science as key to greater productivity and the effective use of finite resources, especially as the global population would reach 9 billion people by 2050. Its G20 agenda would address the climate's impacts on soil, rather than on the conditions and causes of climate change itself.

Sherpa Process

A month before the summit, Macri's sherpa Pedro Villagra Delgado said the negotiations on the Paris Agreement were "the most complicated" part of drafting the communiqué's text on climate change, due to Trump's objections to strong language (Mathiesen 2018a).

Ministerial Meetings

Argentina relied heavily on a bottom-up strategy for its preparations. Leaders would largely endorse the work and many commitments made by the G20's ministerial meetings for 10 subjects. Environment was not on the list.

On June 15 in Bariloche, energy ministers made 18 commitments. The first was to work toward low emissions through increased innovation on sustainable and cleaner energy systems. Ministers promoted the use of methane-producing natural gas, but reaffirmed the commitment to phase out inefficient fossil fuel subsidies.

On July 28 in Buenos Aires, agriculture ministers made eight commitments on climate change and five on the environment. They supported the UNFCCC and nationally determined contributions (NDCs), long-term strategies for lowering emissions, efforts to reduce the impact of climate change on agricultural jobs, adaptation and mitigation, and the water-related risks to agriculture caused by climate change, including a link to the Sustainable Development Goal on clean water.

On the Eve

Buenos Aires was not expected to do much on climate change, as the UN's COP24 in Katowice was scheduled immediately after it. Nonetheless, some leaders pushed for more action. Macron emphasized the dual crises of climate change and biodiversity loss. May focused on plastics in the oceans. But most members prioritized populism, protectionism, security and the threats to the established global multilateral order. Moreover, given strong resistance from the world's biggest economy, the United States, and the proximity of COP24, G20 leaders felt restrained on climate change.

Going into the summit, and aware that Argentina was seeking a different direction from Germany's in 2017, Canada asked for communiqué references to continuing to explore financing for a transition to a cleaner future. Trudeau wanted the G20 summit to align his G7 summit's climate, economic and investment priorities, as he had discussed with Macri. This would include the Canada–Argentina voluntary peer review on the G20's commitment to phase out fossil fuel subsidies. As a result, the communiqué was expected to carry over some of Charlevoix's G7 priorities, including infrastructure resilience to climate threats and energy transitions, even though it would only pay lip service to climate change.

At the Summit

At the table, the leaders could improve, as well as approve, their sherpas' draft communiqué at their morning working session on sustainable development, climate sustainability and climate change and their working lunch on Argentina's priorities of infrastructure and food security. The morning session reviewed progress on the UN's 2030 Agenda on Sustainable Development. The lunch session included cleaner, more flexible and more transparent energy systems and agricultural productivity. None of the leaders' bilateral meetings addressed climate change.

In preparing the communiqué, Argentina had signalled it would draw heavily on the G20 ministers' documents. The firm US stance on the Paris Agreement and climate change meant the other leaders would seek middle ground. They chose energy efficiency and energy transitions. Trudeau was invited to speak early on climate change. Going into the summit, he felt the G20 should act on its commitments under the Paris Agreement but did not expect that the United States would join in. As G7 host, Trudeau sought to explain that Canada was investing in clean technology, drawing on science and seeking a cleaner energy mix.

The contentious atmosphere of the 2017 Hamburg Summit, with its 19 + 1 outcome, was expected to reappear. Argentina's summit sought a short communiqué with little room for climate change, especially as economic growth had become fundamental to its approach.

Sherpas worked through the first night of the summit on the communiqué. By the next afternoon, they had a draft for the leaders (Stauffer and Squires 2018). Officials said the wording on climate change would show "no backtracking" (Stauffer and Squires 2018). The draft leaked a week earlier, however, was already weaker on climate change than the Hamburg communiqué (Mathiesen 2018a).

Dimensions of Performance

In the end, the Buenos Aires Summit produced a small performance.

Domestic Political Management

Domestic political management was small. The timing did not work well for Mexico's leaders: Andrés Manuel López Obrador was to be sworn in as president on December 1, so outgoing president and summit veteran Enrique Peña Nieto attended the summit's first day and returned home for the presidential handover. Merkel missed the first day of the summit, due to a mechanical malfunction that grounded her aircraft at home overnight (Chazan 2018). Apart from Macri, all of the leaders ducked in and out of sessions. Trump put treasury secretary Steven Mnuchin in the chair during the finance session. There were no communiqué compliments on climate change (see Appendix D).

Deliberation

The short passage on climate change had only 532 words, or 6% of the communiqué, which was the fewest words since 2014 (see Appendix D). The communiqué only noted the IPCC's (2018) Special Report on Global Warming of 1.5°C, which had been approved in October, because Saudi Arabia did not want to welcome or endorse the latest climate science. The text addressed long-term low greenhouse gas development, adaptation, biodiversity and the impact of extreme weather events on agriculture. On COP24, it simply stated that the leaders looked "forward to successful outcomes" and to the Talanoa Dialogue.

Direction Setting

None of the 23 affirmations of global financial stability and the 53 of globalization for all referred to climate change or the environment (see Appendix D).

Decision Making

G20 leaders made only three climate change commitments, or 2% of the total of 128. This was the third lowest number, after Hangzhou in 2016 and Washington in 2008 (see Appendix D). The "G19," without the United States, reaffirmed its commitment to fully implementing the Paris Agreement. It also committed, vaguely, "to continue to tackle climate change, while promoting sustainable development and economic growth." It promised more specifically to support developing countries, with an emphasis on helping SIDS in the Caribbean adapt to and build resilience to extreme weather events. All three commitments were low binding.

Delivery

Compliance with the two assessed climate commitments averaged 79% (see Appendix D). This was higher than the G20's 71% compliance with climate commitments made at Hamburg. The general promise to implement the Paris Agreement averaged 87% and the commitment to support SIDS had 70%.

Development of Global Governance

Leaders made no references to G20 institutions (see Appendix D). They made only three references to outside institutions — the fifth lowest — to the UNFCCC, COP24 and the IPCC.

Causes of Performance

Shock-Activated Vulnerability

The Buenos Aires Summit's small performance was driven partly by the diversionary shock of a financial crisis in host Argentina, which overwhelmed the mounting shocks and scientific evidence on climate change.

The Buenos Aires communiqué did not recognize any shock-activated vulnerabilities on climate change (see Appendix F). It made one reference to the vulnerability of the financial system and debt vulnerabilities in low-income countries.

There were also few climate shocks or vulnerabilities noted in the elite media in the months leading up to the summit. The front pages of the *Financial Times* never included stories on climate change.

On its front pages, the *New York Times* featured the 7.5 magnitude earthquake that struck Indonesia on September 28, which, by October 7, had killed 1,649 people. Front-page coverage included oil prices, Trump's controversial leadership pick for the US Environmental Protection Agency and his promotion of fracking on public lands. In Canada, the *Globe and Mail*'s front page had no environmental stories.

During the summit, the front page of the *Buenos Aires Times*, Argentina's only English-language newspaper, covered the G20 as a challenging leadership role for the country, but not climate change. *La Nación*'s front page featured Macri's efforts to manage expectations of what his G20 summit could achieve given national economic circumstances and the recession.

More broadly, for all of 2018, coverage of climate change on US networks, including ABC, CBS, NBC and Fox, took only 112 minutes (Macdonald 2021).

Yet, in the physical world, climate shocks were strong. In mid September 2018, in the United States and the Caribbean — which includes overseas territories of the United Kingdom, France and the Netherlands — Tropical Storm Florence became a category 4 hurricane that killed at least 17 people and left nearly 1 million without electricity in North and South Carolina (Kirton and Warren 2018a). Many Puerto Ricans still dealt with the aftermath of Hurricane Maria. In July 2018, torrential rain in Japan killed 200 people, the most since the 2011 earthquake and tsunami. Then, on September 4, Typhoon Jebi, the deadliest in 25 years, left six dead in Japan and disabled the airport serving Osaka, Kyoto and Kobe (Lewis 2018; Rich and Dooley 2021). A heatwave and heavy flooding near Hiroshima followed. On September 6, 44 people died and millions lost power after an earthquake on Hokkaido triggered landslides (*Japan Times* 2020; Rich and Dooley 2021). In mid July, China was hit by Typhoon Mangkhut, which also struck the Philippines, Taiwan and Hong Kong, and left dozens dead (Kirton and Warren 2018a).

By the end of 2018 in G20 countries, excluding the European Union as a whole, the number of deaths from extreme weather events was 8,441, more than double the 4,125 in 2017 (see Appendix G).

Scientific shocks on climate change were also strong. The IPCC's report and the second volume of the US National Climate Assessment underscored the seriousness and urgency of the threat (IPCC 2018; Reidmiller et al. 2018). In 2018, global emissions surged to a new peak, rising 2.7% for their fastest pace in seven years. Concentrations rose to 401 parts per million, helping make 2018 the planet's fourth hottest year ever.

Yet, these climate shocks were completely overtaken by the financial crisis erupting in Argentina in September, two months before the summit, and the first in over half a decade. The G20's first distinctive mission was to promote financial stability, rooted in its creation in response to the financial crises arising from Asia in 1997, the United States in 2008 and Europe in 2010–2012. In 2018, the crisis created worries about G20 member Turkey, G7 member Italy, BRICS members and other countries such as Pakistan. On September 26, the IMF gave Argentina a $57 billion rescue package, the biggest in IMF history.

Other shocks such as famine in Yemen and elsewhere directly matched Argentina's initial priority of food security, which easily connected with its new need for agricultural exports to cope with its financial crisis.

Multilateral Organizational Failure

Multilateral organizational failure continued, with Trump's announced withdrawal from the Paris Agreement. Although the United States still remained in the UNFCCC, many felt the United States would impede progress at the COP24 in coal-dependent Poland.

Still, it was easy for G20 leaders to leave the divisive issue of climate change to the UNFCCC, and to the World Bank, which had secured a $13 billion capital increase in April 2018, and to the IMF, with its bailout for a bankrupt Argentina.

Predominant Equalizing Capability

Of greater causal salience was declining economic equality within the G20 (see Appendix H). This was due to rising gross domestic product (GDP) and emissions in the United States as it became a leading producer and exporter of energy due to fracking. In addition, the renminbi weakened, and China's GDP slid. The annual percentage change in the G20's emissions declined from +1.18% in 2017 to +0.66% in 2018, reducing the incentive for ambitious G20 climate action (see Appendix I).

Converging Characteristics

Downward divergence among G20 members lowered Buenos Aires's performance. Democracy marginally declined and the United States shifted to join the de-democratizing, climate-resistant coalition of Saudi Arabia, Russia and Brazil.

On democracy, from 2017 to 2018, among 167 countries, the United States rose one rank to 25th and China, Japan, South Africa, India, Argentina and Indonesia also rose (see Appendix J). It declined in six members, including Russia. It remained at very low levels in Saudi Arabia and Turkey.

Environmental performance declined and diverged. Among 61 countries, although the risers outnumbered the decliners, the climate change–denying United States plunged to 56th from 43rd and Argentina, now financially dependent on the United States and the IMF, fell from 46th to 36th (see Appendix K).

Domestic Political Cohesion

The poor domestic political cohesion of key G20 leaders also drove Buenos Aires's small performance. Macri as host had limited G20 summit experience, and no BRICS or G7 summit experience (beyond his guest appearance at Charlevoix); he also had no political or professional experience or expertise on climate change, and no personal commitment to its control. His legislative and public standing plummeted as the financial crisis arrived.

In October 2018, in more powerful, neighbouring Brazil, presidential elections replaced Michel Temer with the climate change–denying Jair Bolsonaro, who would assume office on January 1, 2019. Temer thus came as a lame duck and the first president in Brazilian history to be charged with crimes while in office. Mexico's López Obrador, who took office on December 1, did not attend.

Trump's Republicans had lost control of the House of Representatives in the mid-term elections on November 6, two weeks before the summit. His approval ratings remained well below the 50% mark.

Canada's Trudeau maintained his parliamentary majority and popularity, but faced re-election in 2019.

At their first or second summit were Korea's Moon Jae-in, the United Kingdom's May, France's Macron, South Africa's Cyril Ramaphosa and Italy's Conte. The G20 veterans, backed by high domestic support, were climate-committed Merkel from Germany, Abe from Japan, who would host in 2019, and the European Union's Tusk and Juncker. However, the more numerous, but far less climate-concerned domestically powerful leaders were Russia's Vladimir Putin, China's Xi Jinping, India's Narendra Modi, Turkey's Recep Tayyıp Erdoğan, Indonesia's Joko Widodo, Saudi Arabia's King Salman bin Abdulaziz Al Saud (who sent Crown Prince Mohammed bin Salman in his place) and Australia's Scott Morrison.

Club at the Hub

With this balance, and the focus at Buenos Aires on the meeting between Trump and Xi among other bilateral meetings, the G20 had low value as a club at the hub of a growing network of global summit governance. Macri, a relative newcomer, was leader of a country that did not participate in any of the G20's connected clubs of the G7, BRICS (whose summit was hosted by coal-fired South Africa in 2018), APEC (whose leaders met just before Buenos Aires), the European Union, the OECD or the IBSA grouping of India, Brazil and South Africa (see Appendix L). Argentina did belong to the Summit of the Americas, now a fragile institution.

Nor were prominent, powerful climate advocates plentiful among the summit's invited guest leaders from Spain, the Netherlands, Singapore, Rwanda, Jamaica (chair of CARICOM), Senegal and Chile. The invited multilateral Financial Stability Board, International Labour Organization, OECD, WTO and the UN, along with the IMF and the World Bank as members, included no environmental organizations. The existing plurilateral summit network was weakened by Trump's absence at the APEC summit and the lack of a Belt and Road Forum in 2018.

UN Katowice, December

Introduction

The UN COP24 was held from December 3 to 18, 2018, in Katowice, Poland — a country that produced enough coal to supply almost 80% of its own electricity (Ruiz

2018). Negotiators understood that the Paris Agreement was critical in binding all countries to the common goal of ambitiously controlling climate change. But they also knew that details remained to be negotiated on procedures and implementation mechanisms. With COP24 as the deadline for this task, Katowice became key to transitioning to a low-carbon, climate-resilient world (Banda 2018).

Debate

The first school of thought saw Katowice "offering hope" because countries were still working together, and proving that "global warming treaties can survive the era of the anti-climate strongman," referring to Trump and Bolsonaro (Heffron 2018). Despite Trump's announced intention to withdraw the United States from the Paris Agreement, Katowice proved "that while multilateralism has been damaged, it is not dead" (*Guardian* 2018).

The second school saw Katowice fail to prove that countries such as the United States, Saudi Arabia, Russia, Australia and Brazil "were taking this seriously" (Christian Aid 2018; Oroschakoff and Tamma 2018). COP24 revealed "a fundamental lack of understanding by some countries" of the current crisis, and the failure of "all countries to commit to raising climate ambition before 2020" (AFP 2018).

The third school concluded that the standoff over Article 6 showed that financial interests still undermined environmental integrity, despite the "indisputable evidence of the climate crisis," with a few countries obstructing "the most essential accounting requirements" (Carbon Market Watch 2018).

Argument

COP24's performance was small. It largely failed to align members' climate pledges with the goals of the Paris Agreement. Despite settling several sections of the Paris rulebook, countries did not agree on the rules for voluntary market mechanisms in Article 6 or the draft decision texts proposed by the Polish presidency (Evans and Timperley 2018).

On the Way to the Meeting

There were two intersessional meetings on the road to Katowice. At the first, in Bonn, Germany, from April 30 to May 10, Patricia Espinosa, UNFCCC executive secretary, set out three goals for 2018: building on the pre-2020 agenda of climate action, completing the Paris Agreement Work Programme and increasing the ambition of the NDCs (Darby et al. 2018).

Negotiators anticipated that climate finance and differentiation in the Paris rulebook would be key sticking points (Darby et al. 2018). In the opening plenary, Saudi Arabia opposed the IMO's target of halving shipping emissions by 2050. SIDS called for a strong position on loss and damage, with payment by key carbon emitters (Isaac et al. 2018). Seven European countries called on the United

States to adopt emission reduction targets in line with the Paris Agreement. The Netherlands and Sweden called for the European Union's NDCs to be more ambitious by 2020 (Waskow et al. 2018). Several other countries called for carbon neutrality in the second half of the century.

These stark divisions caused Bonn to fail. Espinosa stressed the need to "improve the pace of progress to achieve a good outcome in Katowice" (Doyle 2018). A second intersessional meeting was announced on May 7 because deadlocks on finance and uniform rules made it impossible to negotiate a single text. Some parties used the Bonn negotiations to address geopolitical issues, with the United States, Australia, Canada and Ukraine calling for the complete withdrawal of Russia from Crimea.

There was still the second meeting in Bangkok from September 4 to 8 where progress might be possible. There was a strong sense of urgency about the amount of work to do in such a short time (Darby 2018c). New Zealand's Jo Tyndall and others emphasized the importance of "balance" between issues, narrowing down options and providing an agreed basis for negotiations (Darby 2018a). Espinosa described progress as "uneven" on finance reporting, transparency and the differentiation in rules (UN Climate Change 2018).

There was no progress on the NDCs. Accounting rules, the degree of differentiation and the use of common time frames all remained contentious. Blocs of emerging economies called for developed countries to be bound by common time frames. China pushed for bifurcated rules. Little attention was paid to increasing ambition, with the Polish COP24 presidency, the Group of 77 (G77), Fiji and several Latin American delegates calling for more substantial outcomes. As the conference concluded, concerns centred on Poland's overall readiness to host a successful COP24 in December (Wyns 2018).

At the Meeting

At Katowice, fears quickly arose that a strong outcome would not be reached on the Paris rulebook, particularly on whether a common approach should govern all parties. The principle of common but differentiated responsibilities (CDR) for developing and developed countries from the original framework convention and the less clear-cut notion of "different national circumstances" from the Paris Agreement remained contentious. Other elements expected to be highly divisive were Article 6 on market mechanisms, Article 9 on finance, Article 13 on transparency, Article 14 on the global implementation stocktake and Article 15 on compliance mechanisms (Sauer 2018).

UK energy minister Claire Perry (2018) suggested that the Paris rulebook create a "level playing field" to "ensure scrutiny of progress toward targets and increase sharing of best practice." She declared the United Kingdom's exit from the European Union would not affect its "ambitious climate action."

Australia played an apparently constructive role, but was criticized for planning to carry over carbon credits from the Kyoto Protocol to meet its Paris Agreement targets (Doherty 2018; Gorrey 2018).

The United States sought to advance its bipartisan consensus that China must be held to the same rules as the United States. It thus opposed bifurcated requirements for developing and developed countries (Mathiesen 2018b). The United States limited consideration of loss and damage, human rights and equity, with Trump tweeting during the negotiations that the Paris Agreement was "fatally flawed" (*Climate Home News* 2018). Along with Kuwait, Russia and Saudi Arabia, the United States opposed "welcoming" the IPCC report and its recommendation to limit the global average temperature increase to 1.5°C (Climate Action Network 2018).

The G77 and China upheld the principle of CDR and respective capabilities and emphasized the role of finance (*Earth Negotiations Bulletin* [ENB] 2018). They sought a "balance between action and support," which meant opposing increased NDC ambition before 2020 and refocusing on adaptation.

Argentina and Brazil called for ambition and a balance between mitigation, adaptation and response (ENB 2018). Some Brazilian states contemplated forming a coalition to uphold Brazil's NDC in the event of national backsliding (Malo 2018).

The African Group emphasized finance and differentiation (Sethi 2018). The LDC bloc prioritized effective rules as well as loss and damage (ENB 2018).

Results

Despite such differences, Katowice produced a more comprehensive outcome than many expected. Mathiesen et al. (2018) said it struck "a delicate balance" among the most vulnerable countries, developed countries most responsible for global warming and the emerging economies wary of a big burden. Parties agreed on uniform guidelines with self-determined flexibility, blurring the distinction between developing and developed countries. They also agreed to begin a process in 2020 to determine a post-2025 long-term climate finance goal to exceed the current goal of $100 billion annually.

However, more significant failures came on Article 6, due to disputes over adjustments to prevent double counting as well as the replacement of the Kyoto Protocol's system of carbon credits. Moreover, although the IPCC report highlighted the criticality of enhanced pre-2020 mitigation action, NDCs were not adequately strengthened and the pathway to reducing emissions below a 1.5°C increase fell short. Despite support from LDCs and several European countries, there was no clear call to increase ambition prior to 2020, due to opposition from China, India, Japan, the United States and others (ENB 2018).

COP24 could not create a compliance mechanism that would impose penalties or sanctions. It did establish a compliance committee, which had limited power to do much more than consider issues related to a party's failure to communicate its NDCs or provisions related to finance mechanisms (ENB 2018).

Katowice thus produced no comprehensive, ambitious or operational rules for the Paris Agreement that would enable parties to both implement their NDCs and transparently track progress toward their goals.

One Planet Summit, September

With the slow progress at the UN in 2018, the reconfiguration of global climate governance continued with the second, summit-level, plurilateral One Planet Summit, held in New York on September 26. Held alongside the General Assembly meetings, it was now fused with the whole-of-global-governance UN. It produced a substantial performance.

It attracted over 4,000 participants, including economists and political, business and civil society leaders. It had two objectives: to report on implementing the 12 commitments made at the first One Planet Summit in Paris the year before and to accelerate their implementation through additional climate finance (Permanent Mission of France to the United Nations in New York n.d.-b). It also sought to strengthen public-private climate action by encouraging industry leaders to account for climate risks in their long-term business strategies and the role of regional and local authorities in implementing the commitments of the Paris Agreement.

Participants reported on 30 initiatives to accelerate implementation (One Planet Summit 2018a). The summit released a report on progress made on the 12 commitments (One Planet Summit 2018b). It outlined its accountability approach that detailed the concrete actions, initiatives and implementation strategy for delivering tangible results and preparing next steps.

The summit announced additional collaboration and investment initiatives. One was the Climate Finance Partnership among philanthropies, governments and private investors committed to jointly investing in climate infrastructure in emerging markets (One Planet Summit 2018a). Investments in low-carbon economies increased. The European Commission proposed dedicating €320 billion (25% of its 2021–2027 budget) to sustainable infrastructure investments. The World Bank's $1 billion commitment to accelerating battery storage in developing countries was expected to leverage an additional $4 billion in concessional funding and private sector investments. Michael Bloomberg, the UN secretary general's special envoy for climate action, said he would help convene the Wall Street Network on Sustainable Finance, aimed at encouraging sustainable finance innovation across US capital markets.

Conclusion

Global summit leadership on climate action returned to the G7 in 2018, supported by the second One Planet Summit. However, adequately ambitious action remained constrained by the divisive politics and rising populism fuelled by Trump. With poor results from the G20 and COP24, progress was sustained by the G7 and reinforced by the second One Planet Summit, which marked a further step in reconfiguring the global governance of climate change.

Trudeau's G7 Charlevoix Summit produced a significant performance. It delivered important — if low binding — climate and environmental commitments that its "G6" members complied with well. But Charlevoix failed to convince Trump to join them or to endorse its Oceans Plastics Charter. Instead, after departing from Charlevoix, Trump tweeted his repudiation of the entire Charlevoix communiqué.

Mauricio Macri's G20 Buenos Aires Summit delivered a small performance, among the lowest ever. Distracted by Argentina's latest financial crisis, he failed to advance climate action there. The great divide spread, with the United States against the "G19" on the summit's three climate commitments, and America's intent to withdraw from the Paris Agreement recognized in the communiqué.

With no push from the G20, the UN's COP24 in Katowice two weeks later also did little to propel progress. It settled some of the Paris rulebook, but not the rules for voluntary markets and implementation mechanisms nor increased climate finance. The second One Planet Summit in September did help narrow the critical climate finance gap. Held during UNGA in New York in September, it forged new public-private partnerships, reinforced commitments made under the Paris Agreement and spurred some of the needed climate finance. It also delivered a road map, with actions for a path forward. It thus institutionalized One Planet summitry in an innovative way, reinforcing the reconfiguration of the global governance of climate change at a critical time. But it was a supplement to the UN, rather than a substitute, as it did not seek to break the many remaining log jams.

6   G20 Leadership, 2019

Introduction

In 2019, global leadership of climate change governance shifted back to the G20, when the Osaka Summit hosted by Japan's Shinzo Abe on June 28–29 produced a significant success. The G7 summit, seven weeks later, in Biarritz, France, on August 24–26, added only a small performance. The third One Planet Summit in Nairobi, Kenya, on March 14, and a special climate summit at the United Nations in New York in September, also added small advances. The 25th Conference of the Parties (COP) to the United Nations Framework Convention on Climate Change (UNFCCC) on December 1–12 in Madrid had a small performance too.

This shift to G20 leadership and the small performance of all the other bodies were partly caused by changing capabilities in G20 members, where the globally predominant source of the climate crisis and the capabilities required to respond grew, especially in China and India. Moreover, Abe, an experienced leader who valued the G7 club, finally agreed at Osaka to the strong pro-climate push of his colleagues from France, Germany and Canada to produce more progress than he initially sought. In contrast, at the G7's Biarritz Summit, France's Emmanuel Macron was deterred by the US's Donald Trump, and did not even try to advance climate action. The special UN climate summit added some momentum, even as the ministerial-level COP did little yet again.

G20 Osaka, June

Introduction

Abe chaired the 14th G20 summit in Osaka, Japan, on June 28–29. It was the first G20 summit hosted by Japan, the third most economically powerful country in the G20 and the world. It took place after an unusually short interval of only seven months since the previous G20 summit and after campaigning had begun for the 2020 US presidential election. Abe was a veteran of six G20 summits and had hosted the G7's productive Ise-Shima Summit in 2016.

DOI: 10.4324/9780429055485-6

Debate

In assessing Osaka's prospects and performance, the first school of thought predicted an ecological success. Krupp (2019) argued that in addressing the current climate catastrophe, highlighted by soaring methane emissions, China was playing an innovative role and the G20 was the institution required to respond.

The second school forecast substantial success on climate change and oceans, due to ecological and terrorist shocks, the many innovative supporting G20 ministerial meetings and the summit experience of its host (Kirton 2019d).

After the summit, the third school saw failure from a powerless G20, which had produced a declaration that once again excluded the United States on climate change. On June 30, Martin (2019) concluded that as host, Abe "staked his chairmanship largely on moving the ball forward on climate change," but Osaka underlined the G20's "ineffectiveness, with the final declaration merely repeating language used last year." For Berger et al. (2019), the continued 19 + 1 "constellation" indicated the G20 needs to deal with a lack of cohesion and to bring the United States back in on climate change.

The fourth school saw a low bar met. The *Financial Times* (2019) editorial on July 1, noted that despite the "G19" commitment to the Paris Agreement, which included Turkey and Brazil, international cooperation was "still functioning — even if only just."

The fifth school detected some progress. Goodman (2019) saw value in the agenda on marine plastic waste and the G20 Principles for Quality Infrastructure, due to Japan's ability to broker an agreement with China. However, the communiqué made "vague commitments with no clear plans for implementation." Giles and Harding (2019) similarly saw progress, due to Abe's desire for consensus and the many pre-summit ministerial meetings that tested language the United States and China would accept.

The sixth school saw relative success. Luckhurst (2019) argued the G20 made awkward compromises on climate with partial progress on marine plastics, quality infrastructure and the Sustainable Development Goals (SDGs), despite distractions from bilateral meetings and Trump's post-summit visit to North Korea.

The seventh school saw great success, due to Abe's skilful chairing. Miyake (2019) highlighted the achievements on marine pollution through the Osaka Blue Ocean Vision.

Argument

The Osaka Summit produced a significant performance on climate change. All members but the United States affirmed their support for the Paris Agreement, and all backed COP25 and reaffirmed the unfulfilled commitment to phase out fossil fuel subsidies. Osaka innovatively introduced the instrument of nature-based solutions and the contribution of Indigenous people. It made 13 climate commitments, complied with them at 72%, and guided G20 and global institutions well.

Yet, Osaka made no advances in reducing emissions or reinforcing sinks to immediately stop the relentless rise in emissions and their deadly, destructive effects.

Osaka's success flowed from several forces. Although visible shocks from terrorism and trade diverted attention from the major wildfires in North America and Australia, heatwaves across Europe and Russia and an earthquake in Japan, new science from the Intergovernmental Panel on Climate Change (IPCC) pushed climate performance. The UNFCCC secretariat had again failed to keep global emissions from rising to new heights. G20 members remained the globally predominant source of emissions, and of the capabilities to respond. Its biggest emerging economies, with an increasingly climate-conscious China and India, grew more than the established G7 ones. Although G20 members' democracy declined and diverged slightly, their ecological performance rose. Domestic political cohesion was high in Japan, India and China, and low in the United States. The G20 summit hub of a global network of summit governance was strengthened by the newly institutionalized, expanding and more ecologically aware Belt and Road Forum. Above all, a consensus-oriented Abe as host, who downplayed climate change control at the start, adjusted when his determined G7 colleagues from France, Germany and Canada threatened to ruin the summit if he did not act on climate change.

Plans and Preparations

The Osaka Summit assembled veterans Angela Merkel of Germany, host in 2017; China's Xi Jinping, host in 2016; Canada's Justin Trudeau; India's Narendra Modi; Turkey's Recep Tayyıp Erdoğan, host in 2015; Russia's Vladimir Putin, host in 2013; Argentina's Mauricio Macri, host in 2018; and the European Union's Jean-Claude Juncker and Donald Tusk. Saudi Arabia's Crown Prince Mohammed bin Salman represented his father, King Salman bin Abdulaziz Al Saud, who would host in 2020. The newer leaders were Korea's Moon Jae-in, the United Kingdom's Theresa May, South Africa's Cyril Ramaphosa, Italy's Giuseppe Conte, Australia's Scott Morrison, Macron and Trump. Osaka was the G20 debut for climate-skeptic Jair Bolsonaro of Brazil. Mexico's new president Andrés Manuel López Obrador did not attend.

At Osaka, these leaders addressed Japan's declared priorities that included climate change, ocean plastic waste, resilience against natural disasters and the SDGs (Kirton 2018e; Abe 2019b; Kirton and Warren 2020b).

Priorities

Abe first publicly outlined his seven priorities during the final session of the Buenos Aires Summit (Kirton 2018d). Climate change was covered in the fifth. It was presented as the need to create a circular economy that would enhance environmental protection and economic growth and include private sector investment. Priorities also included marine plastic pollution, biodiversity, energy and the SDGs. Nothing highlighted climate change itself.

Abe (2019a) elaborated his priorities further at the World Economic Forum on January 23, 2019. Climate change now came second. Noting the recent IPCC report on the impacts of global warming of 1.5°C above pre-industrial levels, he now sought net zero carbon emissions by 2050 through technology such as artificial photosynthesis, methanation, carbon capture and sequestration, and hydrogen, and through environmental, social and corporate governance. He also sought a global commitment on marine plastic litter to reduce microplastics and toxic contamination on the ocean floor.

Guests

In addition to the usual guest leaders of the Netherlands, Spain and Singapore, Abe invited the leaders of Vietnam, Thailand (representing the presidency of the Association of Southeast Asian Nations [ASEAN]), Egypt (representing the presidency of the African Union [AU]), Chile (representing the presidency of the Asia-Pacific Economic Cooperation [APEC] forum) and Senegal (representing the New Partnership for Africa's Development [NEPAD]). He also invited the heads of the UN, World Trade Organization (WTO), International Labour Organization (ILO), Financial Stability Board, Organisation for Economic Co-operation and Development (OECD), Asian Development Bank and the World Health Organization, plus the International Monetary Fund (IMF) and the World Bank as G20 members, but no environmental bodies.

Sherpa Process

At their first meeting in January, sherpas addressed climate change and resilience against natural disasters, with Japan highlighting technology in controlling climate change through artificial photosynthesis, methanation, carbon capture and sequestration, and hydrogen production. Climate-related financial disclosure was also discussed.

The United States still professed to participate fully in the COP process but not as a party to the Paris Agreement. Osaka would focus on the energy transition, as all G20 members agreed on energy efficiency and moving toward a greener, cleaner economy. One way was through the "responsible use" of fossil fuels. The US administration heard from business, mayors and governors who were pushing such action.

Japan had its legacy of the UNFCCC's Kyoto Protocol. Domestically, China and India needed to be seen to be cleaning up their environment. So, climate change control would be prominent at the summit. It was unlikely anyone would gang up against Trump, who was supported by Saudi Arabia, Russia and Brazil, but all would reiterate their respective views.

Sherpas debated what energy sources were "renewable" and "sustainable." Some emphasized hydroelectricity and nuclear power. After the shutdown sparked by the Fukushima disaster in 2011, Japan had returned to using some nuclear power; Germany remained opposed to it. Some G20 members liked the alignment of carbon capture and storage with the concept of clean fossil fuels.

On fossil fuel subsidies, the United States and China had been the first to submit to the peer review process anchored in the OECD, in 2016, followed by Germany and Mexico in 2017. Canada and Argentina announced they would be third, followed closely by Italy and Indonesia. All agreed that fossil fuels would be used for decades, particularly as no infrastructure existed to convert vehicles to electric cars. Japan planned to bring the energy and environment ministers together just before the Osaka Summit.

As an archipelagic country, Japan added oceans to the agenda, in particular marine pollution, including ships' ballast, and plastic pollution in most forms, an issue on which Japan had done much work.

Japan would leave climate change mostly to the G7 Biarritz Summit in August, where Macron was trying to champion it.

Ministerial Meetings

One of the 15 commitments made by G20 agriculture ministers in Niigata on May 11–12 was on climate change.

Energy and environment ministers met in Karuizawa on June 15–16. This was the first time that G20 environment ministers met. Japan wanted to treat climate change through clean energy produced by technology, an approach that aligned well with Trump's United States. Japan had just announced its long-term strategy to reduce emissions, which emphasized decarbonizing, reducing the use of coal and nuclear power, recycling carbon dioxide into fuel and building materials, storing carbon dioxide underground and in oceans and using hydrogen energy (Jiji Press 2019). It would also publicly disclose how firms were controlling emissions.

The ministerial meeting also focused on marine litter, an issue of concern to Asia and the United States (Eilperin and Dennis 2019; *Japan Times* 2019). Japan, stung by criticism of its failure to join the Oceans Plastics Charter at the 2018 Charlevoix Summit, had drafted the Strategy for Plastics Resources Circulation in March to prevent microplastics from entering the ocean by 2020.

Japan also proposed the world's first international framework to compile data on plastic waste (Suzuki 2019). It offered technical assistance to ASEAN members for waste disposal. It would create a monitoring post for plastic garbage in the Pacific Ocean.

Ministers agreed to the first step of data gathering and to choose and implement pilot projects. Their leaders, without the United States, might also agree on a plastic product ban, similar to Japan's recently announced ban on plastic shopping bags.

Ministers also agreed to create the first international framework for members to voluntarily reduce their plastic pollution in the oceans, building on the 2017 G20 Action Plan on Marine Litter (Obayashi 2019). Implementation would start in the fall of 2019 when Japan would host the meeting of the G20 Resource Efficiency Dialogue.

The energy ministers alone made 39 commitments, the environment ministers made 28 and the two combined made another 12. On climate change, the

United States refused to promise to reduce emissions in accordance with its Paris Agreement pledges.

On the Eve

At the last sherpa meeting, in Osaka on June 25, there were substantial practical advances on the "global commons" of climate, oceans and the environment, but deadlock on climate change specifically. The United States resisted communiqué references to "global warming" and the Paris Agreement. Macron and his sherpa team insisted the communiqué affirm the willing members' intentions to implement the agreement and take strong action.

Korea sought a commitment to make early down payments on replenishing the Green Climate Fund. Others doubted the fund's effectiveness. Some wanted the money targeted to global health.

The United States again stressed clean energy, backed by Japan.

China supported climate language, as implied by its announcement on the sidelines of the Osaka Summit to reach peak emissions and increase its use of renewable energy by 20% by 2030 (Stanway 2019).

Sherpas could resolve some outstanding issues including ozone-depleting hydrofluorocarbons (HFCs). The United States did not agree on an international ban and did not sign the Kigali Protocol agreed to in 2016. Because France wanted a reference to HFCs, the sherpas came up with text on the cooling sector without specifically mentioning HFCs.

The United States also opposed any mention or discussion of the SDGs.

Discussions on the concept of the circular economy raised the possibility of producing a work plan for the following year, but it was unclear if this would fall to the G20 or the UN Environment Programme.

Ultimately, the big divisive issues on climate change remained for leaders themselves to resolve.

At the Summit

Abe and his team sought consensus wherever possible, including on climate change (BBC Monitoring European 2019). They proposed omitting any references, watering down the text or mixing it with energy to appease Trump. Macron, Trudeau and the Europeans refused. The United States saw no point in repeating the 19 + 1 text from Buenos Aires.

But the Europeans and Canadians said with no reference to the Paris Agreement or climate change, there would be no agreement on anything. The United States threatened to walk away from everything if there were a reference. This impasse continued as the summit began. Macron, Merkel and Trudeau insisted that the only conceivable outcome was the Buenos Aires solution.

Once Abe realized that a full consensus was not possible, a compromise came. At the end of the first day, Trump agreed to the 19 + 1 solution instead of a consensual but less ambitious text (BBC Monitoring European 2019). Negotiations continued through to the final minute of the summit. In the end, the 19 members' paragraph reaffirmed the Paris Agreement without the United States. There were rumours of other leaders wavering — of Turkey seeking financing, Brazil's new government resisting Paris and Saudi Arabia wanting more support for fossil fuels and less talk about weaning itself from its primary income source. But the arrangement held. With little time left, the long US-only paragraph was produced by the Americans without negotiation. Although some G20 members felt it lacked credibility, it did not damage the overall result.

Results

Nineteen G20 members thus reaffirmed their intention to implement the Paris Agreement and improve their commitments under it. The United States repeated its intention to withdraw from the Agreement and extolled its own accomplishments in clean energy technology and reducing emissions. The full G20 repeated its commitment to phase out inefficient fossil fuel subsidies in the medium term.

Broader progress came on sustainable development, across many SDGs, beyond SDG 13 on climate change. Osaka launched the world's first G20-led global regime to curb plastic and other waste in the world's oceans and launched the Osaka Blue Ocean Vision to reduce additional marine plastic litter pollution to zero by 2050. Biodiversity was largely left to Biarritz, where Macron had made it a priority. On resilience against natural disasters, G20 leaders reinforced the risk insurance facilities that were created to support vulnerable small island developing states and African states.

Dimensions of Performance

The Osaka Summit's significant performance on climate change arose across most dimensions of performance.

Domestic Political Management

Leaders' attendance was substantial as all but Mexico's López Obrador and Saudi Arabia's King Salman attended. No communiqué compliments were issued on any subject (see Appendix D).

Media attention was mixed. On June 26, the *Financial Times* reported that the draft omitted "the phrases 'global warming' and 'decarbonisation'" in "an attempt to placate the US" (Hook 2019b). Yet, a day later, the *Asahi Shimbun* reported Macron's refusal to back down and that demonstrators in Kobe and Osaka were demanding Japan lead on climate change, decarbonization and clean energy (Aose and Tomioka 2019; *Asahi Shimbun* 2019a). On June 29, *Nikkei News* (2019) reported that despite concern that some countries would follow the United States, the issue of environmental protection was gaining traction in public opinion. A day later, however, *Sankei News* (2019) called the declaration a "walk on a tightrope," referring to the Paris Agreement.

Media approval was neutral to mildly negative. The *Asahi*'s (2019b) editorial on June 30, titled "As the G20 Summit Ends, the Limits of Abe's Diplomacy Are Revealed," criticized Abe for emphasizing summit success and diplomacy and ignoring difficult issues, ahead of the approaching House of Councillors election. However, the impact on public opinion was generally positive, as one poll showed approval ratings rebounding for Abe's cabinet (Miharu 2020).

Deliberation

In Osaka's public deliberations, performance was strong. Climate change had 2,034 words, the second most after Hamburg's 5,225 in 2017 (see Appendix D). As a portion of the communiqué, Osaka stood first, with 31%, double Hamburg's 15%.

The communiqué recognized the "urgent need" to act on "complex and pressing global issues and challenges" such as climate change. This would require participation at all levels of government, as well as consideration of clean technology and approaches, smart cities, ecosystem-based approaches and traditional and Indigenous knowledge. Leaders again expressed their full backing for the UNFCCC.

Direction Setting

There was one affirmation each in a climate change context of the G20's distinctive foundational principles of financial stability and globalization that benefits all (see Appendix D).

The communiqué's preamble referred to "tackling economic, social and environmental challenges" in a list where the economy came first. This balance was supported by two affirmations of sustainable development.

However, the communiqué also affirmed principles that were antithetical or inadequate to controlling climate change. Leaders acknowledged "the role of all energy sources and technologies in the energy mix" — a category that implicitly embraced wood, peat and coal. They emphasized technology as a solution, especially technologies not yet proven in the market. They affirmed the traditional principles of "common but differentiated responsibilities" as part of the UNFCCC regime.

Decision Making

Osaka's decision-making performance was strong. Leaders made 13 (9%) of their 143 commitments on climate change, the second highest after Hamburg in 2017 (see Appendix D). Nine bound all members, two bound all but the United States and two bound the United States alone.

Five climate commitments focused on clean technology, including one that bound the United States alone. Two dealt with the Paris Agreement, one where the United States reaffirmed its intention to withdraw and the other where all other

G20 members said they would stay. Yet, innovatively, one commitment encouraged the use of nature-based solutions and another the role of Indigenous people. And one had all members promise to make COP25 a success.

Eleven commitments used highly binding language. Two used low binding language, for a high–low ratio of 85%.

Seven commitments had synergies with five other subjects — the second highest spread thus far, following the six at Hamburg in 2017. They were led by the environment and development with four each, followed by the economy with two and infrastructure and employment each with one. Thus, the economy and development were prioritized with climate change.

The energy commitments included the unfulfilled one on phasing out fossil fuel subsidies, with no hints about how the G20 would finally do so.

Delivery

Compliance with the five assessed commitments was solid at 72% (see Appendix D). The commitment linked to sustainable development and the mobilization of public and private financing secured 93%. That on innovation for low emissions development had 85%, that on clean technologies and smart cities had 70%, that on implementing the Paris Agreement had 63% and that on nationally determined contributions (NDCs) had 50%.

Development of Global Governance

Osaka's institutional development of global climate governance was significant (see Appendix D). Its documents made 3 references to three institutions inside the G20, close to the G20 norm, and 10 references to nine outside bodies, well above average. The resulting ratio of 3 inside to 10 outside continued the G20's focus of looking beyond itself on climate change.

The three references to inside institutions were to G20 energy ministers, past G20 summits and the Resource Efficiency Dialogue. The 10 references to outside institutions were led by the UNFCCC with three, followed by one each to the World Bank, the IMF, the WTO and, for the first time, the UN High Level Political Forum on Financing for Development, the African Development Bank (AfDB) and the Climate Action Summit.

Causes of Performance

Shock-Activated Vulnerability

Osaka's significant performance was propelled by substantial climate and ecological shocks, but constrained by stronger energy, economic and security ones. The communiqué only partly recognized one climate shock, in the form of "recent extreme weather events and disasters" (see Appendix F). It also referenced diversionary energy supply and terrorist shocks.

There was a similar balance in the elite media. The *Financial Times* front page featured the earthquake that hit California on July 7, whose effects were made worse by climate change, and another earthquake in California just days before. It also reported climate change rising up the political agenda, Japan's resumption of whaling after three decades, the importance of nature and fresh air for health, the rapid deforestation in the Brazilian Amazon and Japan's efforts to dilute the G20 climate pledge in order to win US trade favours.

Although the *New York Times* front page had stories on climate change, other subjects dominated. Canada's *Globe and Mail* featured the circular economy and China's support of it, and floods in Quebec forcing 6,000 people to evacuate.

The front page of the English-language *Japan News*, produced by the *Yomiuri Shimbun*, featured the G20 summit, its priority of marine plastics, whaling in Japan and an earthquake in Japan a week before the summit in which 26 people were injured.

In the United States, the television nightly news reported on climate change 68% more in 2019 than it had in 2018 (Macdonald 2020). However, this still only made up 0.7% of the news reports.

Strong, steadily spreading ecological shocks from extreme weather events encouraged summit action on climate change, although the number of deaths from extreme weather events in G20 countries declined from 8,441 in 2018 to 5,696 in 2019 (see Appendix G). But the United States had historically high and damaging spring floods in the Midwest farm belt. In June, Japan set high heat records, Canada's western provinces suffered escalating wildfires and there were unusual heatwaves in Russia, India, Poland and France. Germans remembered their highest average temperature and driest summer in 2018.

Adding more visibility were the soaring and spreading mass protests and strikes over climate change. Inspired by Greta Thunberg, student strikes spread across Europe and beyond. The Extinction Rebellion movement partially shut down the centre of London and the global financial services located there. The BBC's Blue Planet II documentary series with Sir David Attenborough, first broadcast in 2017, had grown in global popularity, and by 2019 "the Attenborough effect" was creating a compelling argument about the damage from plastics to the oceans and marine life (McCarthy and Sánchez 2019).

In making natural disasters a priority for Osaka, Japan was driven by the typhoon that flooded and closed the Osaka airport in 2018 and the earthquake and tsunami that devastated Fukushima and created a nuclear radiation threat in 2011.

Yet, stronger diversionary shocks arose in energy prices and supply, in health from the deadly Ebola outbreak in the Democratic Republic of Congo, in several security and terrorist threats and in the US–China trade war. They surpassed the alarming scientific findings highlighted in the IPCC (2018) report that predicted emissions would pass a tipping point in 2030 that would leave little time to prevent the possible extinction of human life, and the report by the Intergovernmental Science-Policy Platform on Biodiversity and Ecosystem Services ([IPBES] 2019) showing severe, unprecedented biodiversity loss. Osaka recognized these bodies' "important work" but not the clear results they showed.

Multilateral Organizational Failure

Multilateral organizational failure was significant. The UNFCCC secretariat's advances were inadequate to meet the growing need for climate action, as it stuck to its Paris timetable and members' struggled to overcome their divisions. The UN environment system had no body focused on marine litter, especially in its plastic form, which inspired Osaka to fill this gap. The UN's scheduled climate summit and SDG summits during the General Assembly (UNGA) in September offered some hope that ambitious climate action could be left to them.

Predominant, Equalizing Capability

The predominant equalizing capability of G20 members overall and on climate change was high. Their overwhelming share of gross domestic product and the internal equalization of all relevant climate capabilities were fed by the steady currency values and high economic growth in China and India, relative to those in the G7 members led by the United States and Japan (see Appendix H).

Yet, rising connectivity and vulnerability demanded action. G20 emissions rose in 2019, at an average of +0.67% by the end of the year (see Appendix I). They declined in only Germany, Japan, the European Union, the United Kingdom, the United States, Italy, Russia, France and Australia.

Converging Characteristics

A constraint came from the moderate downward convergence among G20 members in their democratic characteristics, led by closure in China, Russia, Turkey and Saudi Arabia (see Appendix J). Ranked out of 167 countries, the G20's average ranking from the year before dropped four spots to 47, led by China's plunge.

A further constraint came from the G20's climate change performance, which converged among all members except the United States and Brazil. The G20 stayed at an unimpressive rank of 37th (see Appendix K).

Political Cohesion

High domestic political cohesion in Japan, China, India, Germany, France and Canada helped spur significant performance, while lower levels in Trump's United States reduced this constraint.

Abe's experience as a veteran of six G20 summits helped him broker the consensus on climate. His coalition government controlled his legislature with a two-thirds majority in both the upper and lower legislative chambers. It would be re-elected in the July 21 election, although climate change was not on the election agenda.

Support for Abe's cabinet had slipped to 47.6% in mid June from 50.5% in mid May, but rebounded (Kyodo News 2019a; Miharu 2020). In a poll conducted by Japan's Ministry of Foreign Affairs in March, 49% of respondents chose ocean plastic waste as their top concern, followed by climate change and energy at 48% (Johnston 2019b).

Trump, at his third G20 summit, had been weakened by the loss of his Republican Party's majority in Congress in the mid-term elections in November 2018.

Xi, a G20 veteran, had complete executive and legislative control but faced street protests in Hong Kong. As an engineer, he favoured Japan's technological approach to climate control.

Merkel had the most G20 summit experience, and had hosted the 2017 Hamburg Summit that had performed strongly on climate change. She led a coalition government with a secure majority in both legislative chambers. However, her party's first-place position had sagged to a narrow lead in the recent elections for the European Parliament, and dropped just below the rapidly rising Green Party (Bittner 2019). By mid June, the Greens had risen to 26%, just ahead of Merkel's party at 25%, driven in part by the climate shocks of Germany's hot and arid summer in 2018.

Macron's personal commitment was reinforced by the rising Green Party in French and European polls and elections.

In Canada, an Ekos poll released on June 17 showed Trudeau's Liberals only two points behind the opposition Conservatives (Ekos Politics 2019). For the election scheduled for October 21, the Liberals could well depend on the Green Party. A Nanos poll covering the week to June 21 showed the Liberals and Conservatives tied at 33% (Nanos 2019).

Fresh from a majority victory in his general election, India's Modi had high political control.

In May 2019, Australia's Morrison secured a majority government in a Liberal–National coalition.

Argentina prepared for elections in the fall of 2019, with Macri doing poorly in the summer primaries.

Bolsonaro's approval rating in July was only 33%.

In Italy, Conte's hold on the government was shaky and his popularity hovered around 47%.

In Indonesia, Joko Widodo won an election in April by about 55% and a 10% lead over his rival, although his campaign was taken to the Constitutional Court to determine if the elections were fair.

In Korea, Moon's popularity was declining, with 45% disapproving of his administration's performance, in the middle of his five-year term.

In Russia, Putin's popularity dropped all the way to 31.7% ahead of the G20 summit.

Ramaphosa won an in-party referendum vote in May in South Africa.

In April, Erdoğan was on shaky ground, losing 7 of Turkey's 12 main cities in local elections.

In the United Kingdom, polls showed that Britons were largely unsatisfied with the way the leadership was handling the Brexit issue, which was top of mind.

On July 16, the European Commission narrowly elected Ursula von der Leyen, who would take office in December. She pledged to make climate change and the environment top priorities.

Club at the Hub

The G20 strengthened as a club at the hub of an expanding network of global summit governance. Osaka's very short cycle created very high engagement and leaders spoke to one another more.

The second Belt and Road Forum in Beijing on April 26 added a Chinese spoke to the hub (see Appendix L). In his opening address to it, Xi (2019) promised "open, green and clean cooperation" and to "make green investment and provide green financing to protect the Earth which we all call home." The Belt and Road Initiative now included G7 members Italy, the United Kingdom and half the EU members.

The Osaka Summit came first in a sequence of other global summits, including the G7 Biarritz Summit in August, four UN summits in New York in September (including one on climate change) and the BRICS (Brazil, Russia, India, China and South Africa) summit in Brazil in November. They offered both the opportunity to build on the G20 summit before Japan's presidency ended and an incentive to leave climate action to them.

G7 Biarritz, August

Introduction

The G7 Biarritz Summit, on August 24–26 on France's Atlantic coast, faced a climate crisis at a critical stage. The liberal multilateral order was under assault from growing populism, protectionism and provincialism, spurring doubts about democracy itself. These forces were embodied by Trump, who loomed large throughout Macron's year as G7 host.

Debate

In the assessments of Biarritz's prospects, no school saw a prominent place for climate change. The first school saw differences dominate. Goodman (2019) forecast differences over trade, climate change and tax.

The second school highlighted divisions and distrust between Macron, Merkel, Trudeau and Abe on the one hand and Trump, Conte and Boris Johnson on the other. These divisions led Macron to exclude climate change from the agenda and any formal summit communiqué (NHK 2019; Smee 2019). Others highlighted the diplomatic quagmire due to Johnson and Brexit (Boffey 2019).

The third school foresaw a solid success, even on climate change. Kirton (2019c) saw this coming from the extreme heat, floods and wildfires throughout the G7 and beyond, despite Trump.

After the summit, Biarritz was largely seen as a failure, due to Trump. The fourth school saw few achievements, due to Trump's America-first approach and Macron's willingness to defer to him (Palacio 2019). A variant saw large failures, including the retreat on climate change, due to G7 leaders' desire not to offend Trump and the rising importance of the G20 (*Yomiuri Shimbun* 2019).

The fifth school saw positive momentum without significant structural change, due to Macron's strategy on Trump. Tiberghien (2019) noted the concluding statement, the positive energy between the two leaders and new openings on climate change and biodiversity.

The sixth school saw leaders' trust restored. Postel-Vinay (2020, 125) concluded that France was "determined to avoid a disappointing conclusion that it lowered expectations," which re-established "a sense of common interest."

Argument

The Biarritz Summit produced a small performance that failed to meet the compounding global threat of climate change (Kirton 2019a). Its only visible achievement was acting against the fires ravaging the Amazon rainforests. On climate change, it produced 892 words but no commitments for the first time since 1988; there were only four commitments on other environmental issues.

This small performance was caused by the shocking images of the Amazon fires and the rationale that climate action was otherwise better left to the special UN climate summit in September, following Macron's One Planet Summit in the spring. It was also caused by Macron's decision to avoid confronting Trump after his repudiation of the 2018 Charlevoix communiqué. Macron, who had declared climate change his second priority for Biarritz in September 2018, also faced yellow vest protests after he raised gas taxes and reduced fossil fuel subsidies and speed limits, with no support for the working class and poor. These receded in November, just as Trump moved into his presidential re-election campaign.

Plans and Preparations

The Biarritz Summit was scheduled in August, to allow the G20 to go first at Japan's request. France chose the coastal town of Biarritz, with panoramic views of the Atlantic Ocean and the Pyrenees Mountains, beaches and forests, a picturesque background for the G7's environmental agenda (Filliâtre 2019).

Macron originally planned to hold his One Planet Summit, focused this time on biodiversity, at the same time as Biarritz. However, he decided to shift it to a regional event in Africa in March.

Participants

Macron, at his third G7 summit, welcomed Merkel to her 14th summit, Abe to his eighth and Trudeau to his fourth. Trump was at his third. The European Union sent veteran Tusk, but Juncker missed his final summit for health reasons. Conte came to his second and Johnson to his first, preoccupied with the United Kingdom's scheduled October 31 withdrawal from the European Union.

Macron invited Egypt's Abdel Fattah El-Sisi as AU president, Senegal's Macky Sall representing NEPAD and Chile's Sebastián Piñera as APEC president, as well as the G20's Morrison, Modi and Ramaphosa, Spain's Pedro Sánchez, along with Burkina Faso's Roch Marc Christian Kaboré and Rwanda's Paul Kagame. Moussa Faki, chair of the AU Commission, was invited too.

Invited leaders of international organizations were from the UN, OECD, World Bank, AU Commission, AfDB, IMF, WTO, ILO and UN Women.

Theme

In presenting his initial vision for the summit to UNGA in September 2018, Macron (2018b) promised to support the Paris Agreement, implement the Kigali Protocol, conclude an ambitious pact for the environment in 2020, make the Convention on Biological Diversity's COP15 in 2020 a success, implement existing commitments, include environmental and social obligations in trade agreements, and mobilize sovereign funds to finance low-carbon policies. He also promised to improve the climate change commitments made at COP21 in Paris in 2015 and seek new coalitions and formats in order to do so.

When France's G7 presidency began, the second of the five priorities Macron announced was environmental equality through climate finance, a fair ecological transition, the oceans and biodiversity. Climate change thus had part of a pillar of its own, and links to other ones. Moreover, France wanted the Biarritz Summit to achieve an ISO 20121 certification and "Equality at a Major Event" label for a "green" event (G7 France 2019d).

Macron sought to overcome the perception of "six plus one versus seven" from the Taormina Summit in 2017 and the shadow of Trump's behaviour in 2018, which he thought had eclipsed Trudeau's effective summit. Macron assumed Trump would participate at Biarritz, and France would therefore have to reach the best agreement with the most ambitious agenda possible to bring everyone together.

Sherpa Process

The sherpas discussed the environment at their first meeting on February 4–5 (G7 France 2019a). They discussed it again at their April 16–18 meeting and visited CD2E, an eco-business hub in the former coal mining town of Loos-en-Gohelle (G7 France 2019b). On June 13–14, the guest sherpas were invited to speak on the climate emergency and biodiversity loss in a working session on biodiversity, oceans and climate change (G7 France 2019c).

Ministerial Meetings

At the seven ministerial meetings covering nine portfolios, there was more consensus on key environmental and climate issues than among the leaders.

On April 5–6, the foreign and interior ministers made one reference to climate change, thanking the Working Group on Climate Change and Fragility and noting its report.

Meeting in Metz on May 5–6, the environment ministers produced 131 commitments, almost double the previous high of 69 at Halifax in September 2018 (Kirton and Warren 2018b). They focused heavily on promoting biodiversity and issued a Declaration on Halting Deforestation Including through Sustainable Supply Chains for Agricultural Commodities. They made 14 commitments on international mobilization for climate change; eight applied to the United States alone and focused on the US withdrawal from the Paris Agreement and support for economic growth, energy security and isolationism. The "G6" ministers reiterated their determination to update their NDCs and to mobilize climate finance for developing countries.

Meeting with their education colleagues in Paris on July 4–5, development ministers made two commitments on climate adaptation and building resilience to climate-related disasters.

On July 17–18, finance ministers and central bank governors made one commitment on climate finance — the first since 2012.

On the Eve

The order of the summit agenda remained unclear, but the opening session would not include climate change.

Macron sought a consensus communiqué. There could be other announcements from different forums. There were rumours that Trump would not attend the summit, and France planned for this possibility. However, this would be unusual given that the United States would hold the G7 presidency next year.

France sought Biarritz to build larger alliances on the environment, climate and oceans. France judged that Macron had become a global hero at the Osaka Summit by refusing to agree to a communiqué that did not mention climate change. He intended to use Biarritz to boost the UN's Climate Action Summit a few weeks later and compel the United States to act. The session on the environment would consider the entire spectrum on climate change.

At Osaka, Macron had met all his G7 colleagues and bonded closely with European leaders and Trudeau on climate action (BBC Monitoring European 2019). He announced he would invite Modi to Biarritz to "work on a common agenda" on climate and "take very concrete action."

At the Summit

Just before the summit, Macron declared that the leaders would focus on the forest fires ravaging the Amazon in Brazil. G7 leaders responded to Colombia's request for help and offered affected countries technical, financial and other support to stop the damage. Following phone calls from Macron and Trudeau, Bolsonaro agreed to send his army to fight the fires. G7 members would provide specialized firefighting personnel, equipment and aircraft. But the leaders were criticized for not responding quickly enough and not committing enough money.

On August 26, the leaders started their session on climate change, biodiversity and oceans with the expanded group (Henden 2019). France sought to stress climate finance, specifically how to encourage financial institutions, including banks, to finance adaptation and the transition to a low-carbon economy. Macron had low expectations, but hoped Trump might recognize the huge market for green finance as opportunities for investors and insurance.

France remained committed to the climate finance goal of $100 billion a year by 2020. In 2018, the Agence française de développement, under the Ministry of Europe and Foreign Affairs, committed €4.8 billion of "financing with climate co-benefits" (Permanent Mission of France to the United Nations in New York n.d.-a).

France would also stress biodiversity and the preservation of the oceans.

Trump skipped the session on biodiversity and climate change on the final day. At his news conference, he avoided any mention of the existence of climate change and emphasized the value of economic growth based on exploiting natural resources and energy from hydrocarbons.

Results

The day after the summit, France published the Chair's Summary on Climate, Biodiversity and Oceans. It contained only four commitments, with none on climate change. Macron failed to have Trump or the full G7 adjust and act on climate change. Biarritz thus produced only a small performance, by keeping the issue discussed by the G6 leaders and their guests without the US leader.

Dimensions of Performance

Domestic Political Management

All G7 leaders attended all sessions of the Biarritz Summit, apart from Trump who skipped the climate session. There were no communiqué compliments on climate change (see Appendix C).

Deliberation

The private conversations among the leaders took place over three days, a 50% increase from the two-day summits of the previous seven-year cycle. As usual, some time was spent in bilateral conversations, most notably the opening pre-summit lunch between Macron and Trump.

Macron declared before the summit's start that he would dispense with any concluding communiqué. He produced a chair's summary of only 250 words. The summit issued 10 documents in all, with various configurations of leaders adhering to each. Some were released a day after the summit, after most of the media had left, and received less attention than they would typically attract.

The Biarritz Chair's Summary on Climate, Biodiversity and Oceans contained 1,217 words, but only 892 relating to climate change, although 12% of the total — 3% less than in 2018 but well above the 2% from 2017 (see Appendix C). It reported that the UN secretary general briefed the leaders on the Climate Action Summit scheduled for September 23. The leaders also discussed the Montreal Protocol. It indirectly recognized climate as a crisis.

Direction Setting

The Chair's Summary on Climate contained no affirmations of democracy or human rights (see Appendix C). There was an implicit causal link on coastal resilience. The summary noted "the interlinked challenges of global warming, land management, food security and diets, and the subsequent mitigation and adaptation policy responses to be considered" as emphasized by the 2019 IPCC Special Report on Climate Change and Land, but did not indicate by whom.

Decision Making

For the first time since 1998, the G7 made no commitments on climate change (see Appendix C). The Chair's Summary on Climate made three commitments on marine litter and one endorsing the Metz Charter on Biodiversity. A few of the Metz Charter's 30 commitments linked biodiversity losses with climate change.

Delivery

Two assessed environment commitments, both on biodiversity, averaged compliance of 91%. All members except the United States fully complied with both; the United States had only 25% compliance.

Development of Global Governance

There was one reference linked to climate change to international institutions inside the G7. There were 14 references to 12 outside bodies, including the Paris Agreement, COPs, the Green Climate Fund, multilateral development banks, the IPCC and the UN Climate Action Summit. They offered no specific instructions or support.

Causes of Performance

Biarritz's small performance was caused primarily by the shocking images of the Amazonian forest fires, with other climate action deterred by Trump and the yellow vest protests in France.

Shock-Activated Vulnerability

G7 leaders recognized no climate or environmental shocks or vulnerabilities in their communiqué (see Appendix E). There were six references to other subjects.

The Chair's Summary on Climate noted that civil society groups had mobilized and committed "to face systemic environmental challenges and a climate crisis, which poses existential risks" — thus only indirectly recognizing the urgency.

In the media, no new climate stories appeared on the front page of the elite daily newspapers. On US television network newscasts and Sunday morning shows on ABC, CBS and NBC, climate change stories totalled 238 minutes in 2019, with solutions capturing 37% of them (Macdonald 2021). This was a sharp rise from 142 minutes in 2018 and 50 minutes in 2016, but just below the 260 minutes in 2017.

Yet, physical ecological shocks were big and broad. The highly visible forest fires in the Amazon galvanized G7 action at Biarritz. Massive wildfires in Siberia in July sent smoke to the West Coast of Canada and the United States. Global temperatures broke records in the United Kingdom, France, Germany and elsewhere in Europe in July (Hook 2019c). Deaths from extreme weather events rose to 1,921 in 2019 from 857 in 2018 (see Appendix G).

The Biarritz documents noted the alarming science from the IPBES and IPCC reports, which reinforced their focus on biodiversity rather than climate change.

Multilateral Organizational Failure

Multilateral organizational failure continued. The ministerial-level UNFCCC process had not completed the Paris rulebook. Moreover, Macron held his One Planet Summit as a regional African meeting in the spring instead of alongside Biarritz. The Climate Action Summit at UNGA in September provided a rationale to do nothing at Biarritz a few weeks before.

Predominant Equalizing Capabilities

G7 members at Biarritz possessed a substantial share of the overall and specialized capabilities needed to control climate change. All members had positive economic growth, although slower than in 2018 (see Appendix H). US growth was the strongest at 2.2%, reinforcing Macron's deference to Trump.

The G7's emissions declined by an average of 3.04% from 2018, while only Canada's rose, for the third year in a row (see Appendix I).

Converging Characteristics

The G7's democratic characteristics rose and converged slightly (see Appendix J). As a group, it rose one spot to 22nd, although only France's ranking increased — from 29th to 20th.

On climate performance, the G7 remained 37th of 61 countries, but the United States dropped three ranks to 59th (see Appendix K).

Domestic Political Cohesion

Domestic political cohesion in almost all G7 countries was low. In France, Macron was still reeling from the yellow vest protests and his low popularity.

Trump, whose Republicans had lost control of Congress, continued to have negative net approval ratings, as the 2020 presidential elections approached. As of August 21, 2019, 53.7% of Americans disapproved of him and only 41.9% approved (FiveThirtyEight 2021). A Pew Research Poll on major threats to the United States, taken on July 10–15, ranked global climate change and Iran's nuclear regime tied for second at 57%, after cyberattacks (Pew Research Center 2019).

Germany's coalition government faced a surging Green Party, with Merkel having announced her departure in a few years. In Italy, Conte's coalition government was dissolving and could fall at any time. Johnson, with low approval ratings, was preoccupied with the October 31 Brexit deadline, which worried the G7's EU members. Trudeau was approaching an election on October 21, with his Liberals tied with the main opposition party. Only Japan's Abe, whose coalition government had just won the upper house, had strong domestic support.

Club at the Hub

The G7 at Biarritz was not a valued club for Trump, or for Macron this time. Macron began the summit with a long, televised bilateral lunch with the US president, and ended it by sharing his concluding news conference with Trump, telling the world, inaccurately, that this was a G7 tradition for the outgoing and incoming hosts.

Macron also produced a complex array of different sets of documents. He brought together leaders of major democracies from several global regions, but did not mobilize them on climate change. The African leaders prioritized development and conflict resolution, not climate change.

Special Climate Summits, March and September

Two special climate summits helped a little to fill the gap left by Macron's G7.

One Planet Summit, March

The third One Planet Summit was held in Nairobi, Kenya, on March 14, 2019, alongside the fourth session of the UN Environment Assembly. It produced a small performance. This first regional edition of the summit, which drew on the IPCC's (2018) report, was designed to highlight the unique role of Africa — responsible for 4% of emissions, with 65% of its people affected by climate change — as a global partner, especially regarding solutions for adaptation and resilience (France, Ministry for Europe and Foreign Affairs n.d.-a).

By convening government and private sector leaders, international organizations, donors, researchers, entrepreneurs, youth, civil society and other global stakeholders, the summit would help countries present their NDC plans at an early stage.

The summit mobilized over $47.5 billion for Africa (Onyango 2019). Pledges came from the Agence française de développement, the Global Environment

Facility, Sweden and others (One Planet Summit 2019). The major commitments came from the World Bank and the AfDB. As part of its 2025 targets announced at COP24 in Katowice in 2018, the World Bank committed $22.5 billion over five years (International Institute for Sustainable Development [IISD] 2019b; World Bank 2019). It also announced that, with Germany's support, it would intensively engage with Rwanda and Kenya to accelerate the implementation and ambition of their NDCs (World Bank 2019). The AfDB (2019) committed at least $25 billion over five years to align Africa's financial institutional flows with the Paris Agreement's goals.

However, not all this money was new, such as the commitment of the Subnational Climate Fund Africa to raise $350 million in four years to fund low-carbon infrastructure projects in Africa, announced first at the 2018 One Planet Summit (One Planet Summit 2019).

UN Climate Action Summit, September

For the first time, there were two special climate summits in a year. The Climate Action Summit on September 23 was the first embedded in the multilateral UN as UNGA opened. UN secretary general António Guterres (2018) had promised this summit would "leverage unprecedented ambition, transformation, and mobilization" to accelerate the implementation of the Paris Agreement. In his opening remarks, he called on governments, cities, businesses, financial actors, civil society and youth to cut emissions by 45% by 2030, reach carbon neutrality by 2050 and "limit temperature rise to 1.5°C by the end of the century" (Guterres 2019b). He later called the summit "a springboard to set us on the right path ahead of crucial 2020 deadlines" (Guterres 2019a).

The summit was attended by over 60 world leaders, with 66 countries pledging to reach net zero emissions by mid century (Hook and Hodgson 2019). It launched initiatives in 12 areas for reducing emissions and strengthening adaptation and resilience (UN n.d.). Seventy-five countries committed to deliver their strategies for 2050 net zero emissions by 2020. The European Union announced that at least 25% of its next budget would be devoted to climate-related activities (Tusk 2019). Russia pledged to ratify the Paris Agreement, bringing the total number of countries to 187 (Sengupta and Friedman 2019). China committed to fulfilling its obligations and announced it would cut its emissions by more than 12 billion tons annually (Environment News Service 2019). Other countries promised to promote renewable energy and transition to clean power. Trump showed up briefly and unexpectedly during the opening session, but the leaders of Japan, Australia, Saudi Arabia and Brazil did not attend at all (Hook and Hodgson 2019).

The private sector also promised to work toward carbon neutrality: 87 global companies, with a combined market capitalization exceeding $2.3 trillion, and 130 banks representing a third of the global banking sector, committed to set climate targets aligned with the Paris Agreement (Environment News Service 2019). None, however, specified how they would achieve these goals.

The summit acknowledged the need for accountability, by providing "a base-line for monitoring the progress of the initiatives to ensure that promised results are achieved" (UN n.d.).

UN Madrid, December

Introduction

Taking place in Madrid on December 2–13, COP25 was the first since the United States had filed its intent to withdraw from the Paris Agreement. Trump's decision thrust the United States at the centre of Madrid's negotiations, as its obligation to maintain its commitment and report its emissions to the UN remained in effect until November 4, 2020 — one day after the US presidential election. The US withdrawal meant a 20% reduction in financial assistance to developing countries through the Green Climate Fund.

Debate

In assessing COP25, the first school of thought saw a key opportunity for the 27,000 delegates to finalize the Paris Agreement rulebook "by settling on rules for carbon markets and other forms of international cooperation under Article 6" (Evans and Gabbatiss 2019).

A second school saw renewed momentum, due to the growing global climate movement influenced largely by Greta Thunburg. In September, she sailed across the Atlantic for nearly three weeks to reach the Climate Action Summit, where she admonished the world's leaders for not being "mature enough to tell it like it is" (Thunberg 2019). After sailing back to Europe, in Madrid she declared that "the change we need is not going to come from people in power. The change is going to come from the people, the masses, demanding change" (BBC News 2019b).

A third school saw the "corrosive" Trump effect effectively cripple the US–China–EU coalition that had delivered the Paris Agreement, thereby "ensuring any calls for ambition, action and environmental integrity [would be] rolled back" (Galey and Hood 2019).

Argument

COP25 produced a small performance. It failed to provide a focal point for subject-specific coalitions to counter climate change. Article 6 of the Paris Agreement pit several G7 and G20 members against each other. Australia and the BASIC group of Brazil, South Africa, India and China demanded that key elements of the Kyoto Protocol on international carbon markets be subsumed under the Paris framework. The European Union and many small island and vulnerable states sought to prevent old Kyoto credits from being adopted. Funding adaptation, another key component of Article 6, remained unresolved. So did using the proceeds from selling carbon offsets to fund adaptation in the most vulnerable countries (Evans and Gabbatiss 2019; Timperley 2019c). The final Madrid text compromised by using voluntary language on funding adaptation (Timperley 2019c). At the end of the meeting, Guterres (2019d) himself said he was disappointed with the results.

On the Way to the Meeting

COP25 suffered setbacks from the start. In November 2018, Brazil's newly elected president Bolsonaro withdrew from hosting the event (Londoño and Friedman 2018; Viscidi and Graham 2019). During his election campaign, he had repeatedly expressed his desire to withdraw from the Paris Agreement, a position he retreated from only after an intense backlash from the international climate community (Viscidi and Graham 2019).

On March 7, 2019, Chile said it would assume the COP25 presidency, with a robust agenda focused on Article 6 commitments and added attention to Antarctica, oceans, renewable energy, the circular economy and electromobility (Darby 2019; Sauer 2019; UN Climate Change 2019b).

The annual intersessional meeting took place in Bonn on June 17–27. Participants attempted to generate text for ministers to agree on to replace Kyoto's Clean Development Mechanism (CDM) with a centralized system to trade carbon emissions anywhere in the world (Gabbatiss 2019). Article 6 would also allow countries to strike bilateral and voluntary agreements for trading carbon units so a country could transfer its emission reductions to another country but still have them count toward its own NDCs (Farand 2019; Germany, Federal Ministry for the Environment n.d.).

Brazil wanted inherited CDM credits to be allowed in the Paris carbon market scheme, but the European Union and other Latin American countries argued the CDMs were effectively "old, worthless credits" (Sauer and Darby 2019). Australia, responsible for 1.2% of global emissions in 2018, insisted at Bonn on the use of Kyoto carry-over credits (BP 2019, 57; Evans and Gabbatiss 2019; Hannam 2019). Double counting also raised concerns (Gabbatiss 2019; Kizzier et al. 2019).

The United Kingdom faced another year of Brexit negotiations, coupled with a snap election. However, on June 11, the United Kingdom became the first G7 country to commit to reaching net zero carbon emissions by 2050 (Twidale and Green 2019). With this commitment becoming law on June 27, the United Kingdom strengthened its position at COP25 and as host of COP26 the following year (UK, Department for Business, Energy and Industrial Strategy 2019).

Members sought to resolve these issues in San José, Costa Rica, on October 8–10. The BASIC bloc still supported trading old CDMs for Sustainable Development Mechanisms (SDMs). The issue of double counting, however, remained unresolved (Timperley 2019b).

On October 30, one month prior to COP25, Chile officially withdrew from hosting the conference, following anti-government protests in Santiago (BBC

News 2019a; Mathiesen and Farand 2019). The following day, Spain offered to host COP25 in Madrid (UN Climate Change 2019a).

Less than a week later, on November 4, the United States officially informed the UN of its intent to withdraw from the Paris Agreement, meeting the requisite three-year notification period (McGrath 2020). At the time, the United States accounted for 15.2% of the world's emissions — the highest absolute portion in the G7 and the second highest after China in the G20 and the world (BP 2019, 57, 2020, 13).

Prospects for finalizing the Paris rulebook became even more bleak. The US delegation at COP25, led by Nancy Pelosi, the Democratic Speaker of the House of Representatives, had no direct White House representatives. This signalled Trump's unwillingness to negotiate, despite the legal requirement that the United States remain in the Paris Agreement until its withdrawal formally took effect in 2020.

Germany, with the highest contribution of global emissions among the European G7 and G20 members in 2018 at 2.1%, sought to advance progress on Article 6. At the preparatory conference in Berlin on November 14, 2019, environment minister Svenja Schulze suggested an agreement was possible on the international carbon credit system "without loopholes and without double-counting" (Schulz 2019). She called on the European Union to act decisively. With Europe responsible for 10.3% of the world's emissions (the third highest among the G20 in 2018), the European Parliament declared a global climate emergency on November 28, stressing the need to commit at Madrid to net zero emissions by 2050 (BP 2019, 57; European Parliament 2019).

Brazil focused on creating a centralized governance structure for countries and the private sector to trade emissions credits anywhere in the world, with the new SDM replacing the CDM of the Kyoto Protocol (Farand 2019). It insisted that countries planning to reduce emissions under Article 6 would not be required to make corresponding adjustments when credits were sold, a process that would effectively enable double counting within the new SDM governance model (Evans and Gabbatiss 2019).

At the Meeting

COP25 opened with a stark address by Guterres (2019c) who said the world could either aim for carbon neutrality by 2050 or head down a "path of surrender … while the planet burned." His call did not catalyze a resolution of the major sticking points on Article 6 (Fotheringham 2019). Nor did Thunberg's passionate pleas (Britton 2019).

The issue of carbon markets under Article 6 proved so difficult that the conference had to be extended by an additional two days (John et al. 2019; Sengupta 2019). What began as 672 passages on Article 6 in square brackets for negotiators to agree on became 31 square-bracketed passages on the eve of the scheduled close. The two most persistent conflicts remained the transition of carbon credit units under the Kyoto regime and double counting (Evans and Gabbatiss 2019).

The four BASIC countries, accounting for over 37% of global emissions in 2018, favoured transitioning their unspent CDM units to meet their Paris commitments (China 2019). Brazil, China and India held the majority of global certified emissions reductions (Evans and Gabbatiss 2019).

Australia also sought to carry over its carbon credits (Morton 2019). It had accrued them by overachieving its emission reduction targets under Kyoto (McManis 2020). Laurence Tubiana, former French environment minister and a key architect of the Paris Agreement, called the carry-over "cheating," and said Australia was willing "to destroy the whole [Paris Agreement] system" through this approach (Hook 2019a).

The European Union and the blocs of the least developed countries, landlocked developing countries and small island developing states argued that these certified emissions reductions would undercut further ambition by using emissions reductions that had already occurred (Evans and Gabbatiss 2019). On December 14, a group of 31 countries, which later grew to 32 and included Germany, France, Italy and the United Kingdom, agreed on the San José Principles for High Ambition and Integrity in International Carbon Markets (Costa Rica, Ministry of the Environment and Energy 2020). These countries remained strongly committed to robust carbon market rules and protocols (Evans and Gabbatiss 2019; Timperley 2019a; Gigounas et al. 2020).

By the final day, it became clear that the most contentious sticking points over Article 6 would not be settled (Dennis and Harlan 2019; Sengupta 2019). Delegates delayed their decisions rather than force consensus on weaker, more incomplete Article 6 agreements.

Results

Nonetheless, Madrid made some small progress. The Coalition of Finance Ministers for Climate Action (2019) launched the Santiago Action Plan to align policy actions with Paris, promote capacity building, mobilize private climate finance and develop a financial sector that recognized mitigation and adaptation. It was endorsed by more than 50 finance ministries including those of Canada, France, Germany, Indonesia, Italy, Mexico and the United Kingdom. Some saw this as a "game-changing moment," as finance ministries would need to invest to transition to net zero carbon economies (John et al. 2019).

Italy and Mexico both committed to increase climate and environmental education (IISD 2019a).

Canada's environment minister Jonathan Wilkinson pledged that Canada would introduce net zero legislation with five-year milestones intended to exceed its Paris Agreement targets. For some this was welcome, as Canada ranked 15 in the world for global emissions per capita in 2018 (BP 2019, 57). It ranked highest in the G7 for per capita emissions and was the only G7 member whose emissions rose in 2019.

Japan faced heavy criticism over its continuing use of coal-fired generation for over 33% of its total energy mix (Johnston 2019a; Kyodo News 2019b; Volcovici

2019). It was second only to Germany within the G7 (at 35%) for its coal consumption (BP 2019, 57). But Germany had the highest percentage of renewable energy as a percentage of total electricity generation in the G7 at 32%, compared to Japan with only 10% (BP 2019, 57). Japan's environment minister Shinjiro Koizumi announced that phasing out fossil fuels could not be done in the immediate future, but options would be considered (D. Abe 2019; Volcovici 2019).

Conclusion

The G7, G20 and UN each performed differently in 2019.

Abe's G20 Osaka Summit in June produced a significant performance, with 13 climate commitments. A veteran whose country valued the G7 club, Abe adjusted at the last minute to the demands of his French, Canadian and German colleagues for climate action and persuaded a resisting United States to go along on some environmental issues, including "G19" action on climate change.

Despite global temperatures reaching historic highs in July 2019 in France, Germany and the United Kingdom, at the G7's Biarritz Summit host Macron was largely deterred by Trump's continued anti-climate posture and did little to advance climate action. Biarritz produced a small performance, with no core climate change commitments at all. Macron's decision not to confront Trump meant that climate change was abandoned as a summit priority and the environmental agenda shifted to biodiversity and oceans. The only shock sufficient to inspire climate action, against a backdrop of significant public pressure, was the burning Amazon rainforest.

Two special climate summits provided some continuity in the intensifying reconfiguration of global climate governance. The third One Planet Summit produced only a small performance, as did the new UN Climate Action Summit, with Russia's ratification of the Paris Agreement. Both summits generated new financing and public-private collaboration, although there was not much diversity of partnerships beyond the private sector or enough climate finance to meet critical targets.

The UN's ministerial-level COP25 in Madrid in December produced a small performance too. It largely failed in solidifying the international carbon market, funding adaptation measures and defining reporting requirements for transparency and common time frames for climate pledges. These and other key issues were left for 2020.

7   COVID-19 Crowd-Out, 2020

Introduction

The year 2020 was truly exceptional, both in the natural world and in the global summit governance of climate change. It was dominated by the very deadly COVID-19 pandemic and ensuing economic and social crises, and the often slower, still deadly effects of chronic climate change. The impacts of COVID-19 completely crowded out any attention to climate change at the G7's emergency virtual summit on March 16. Aided by the increasing re-election preoccupation of the G7 host, US president Donald Trump, they resulted in the cancellation of the regular annual G7 summit itself. The COVID-19 crisis similarly crowded out any attention to climate change at the G20's extraordinary virtual summit on March 26. Then, at the G20 Riyadh Summit on November 21–22, with a now lame-duck Trump denying both climate change and the November 3 US presidential election results and with oil-dependent Saudi Arabia and King Salman bin Abdulaziz Al Saud as host, the COVID-19 crisis led to a small climate performance. At the United Nations, it caused the cancellation of the long-scheduled and long-awaited climate summit in Glasgow in November, for the 26th Conference of the Parties (COP) of the United Nations Framework Convention on Climate Change (UNFCCC). Yet, progress was kept alive, if to a small degree, by special high-level climate summits at the UN in September in New York and with the better performing UN Climate Ambition Summit in December in Paris to celebrate the fifth anniversary of the Paris Agreement in 2015.

Thus, in 2020, the G7, G20 and UN all failed to address the mounting scientific evidence and reality of climate change as its extreme weather events grew. None of these established global institutions filled the gap the others left. Their failing or small performance was caused by the diversionary shock-activated vulnerability of the COVID-19 pandemic and its ensuing crises, with strong assistance from Trump, King Salman and the virtual diplomacy that eroded the essential character of their summits as valued personal clubs. Despite these serious shortcomings, the reconfiguration of global summit governance strengthened, with the addition of the special climate summits in September and December, and with Trump now due to disappear.

DOI: 10.4324/9780429055485-7

G7 Governance

Introduction

On January 1, 2020, Trump and the United States assumed the presidency of the G7 for the year. There were few hopes that he would use it to address climate change directly, but some possibility that he would act on clean air, water, land and energy, and on growing trees and other things that could contribute to countering it nonetheless. These hopes were quickly dashed when, at the virtual summit he called on March 16 to confront the new COVID-19 crisis and the subsequent call among the leaders on April 16, the environment and climate change were absent. The regular summit scheduled for June was delayed and then disappeared. With the COVID-19 crowd-out and Trump's disbelief in climate change, the G7 abandoned and thus fully failed in the global governance of climate change.

Debate

In the debate about the G7's 2020 climate performance, the first school saw nothing lost. Barber (2021b) concluded: "One looks in vain for evidence that either Trump or the other six leaders regretted the missed occasion."

The second school saw a fuller failure. Warren (2021d) said the G7's attention to climate change was overwhelmed by "continued political inertia combined with the chaos caused by the COVID-19 pandemic," although nothing stopped the rest of the G7 members from acting together on climate without Trump. Kirton (2020a) noted that the crisis was all-consuming at the March 16 meeting and delayed the UN summit in Glasgow by a full year.

For the third school, the G7 institution was an outdated "throwback to an era of European dominance that doesn't reflect today's power dynamics," exacerbated by Trump's antagonism toward US allies and friendliness toward its foes (McTague and Nicholas 2020).

Argument

For the first time since it began continuously governing climate change, the G7 failed in 2020 (see Appendix C) (Warren 2019b; Kirton 2020b).

This failure was primarily due to the determination of a climate change–denying Trump as host. From the start, he designed his G7 summit to exclude climate change and to feature the free market economic growth, government deregulation and hydrocarbon-based energy security that directly cause climate change. He easily dismissed climate shocks from California wildfires and southeastern hurricanes, believing that the declining US emissions and cleaner air under his presidency made his country invulnerable. He rejected the COVID-19 shock arriving early in the year as scientific fake news or portrayed it as a Chinese plot, with no link to climate change. His decision to withdraw the United States from the Paris Agreement and his antipathy toward multilateral organizations compounded the failure of the two G7 meetings he chaired, on March 16 and April 16. Even as COVID-19 drove the US economy into a steep decline, he spread misinformation about how well the United States was performing relative to its G7 peers, while he became increasingly preoccupied with his re-election as US president before and after the election on November 3. The COVID-19-created shift to virtual diplomacy for the G7 emergency response in the spring left no time for his relatively climate-committed G7 colleagues to get Trump to adjust to add climate action, using even Trump's prior unilateral commitment to help plant 1 trillion trees.

Plans and Preparations

In October 2019, Trump's four priorities for his G7 summit were announced: strengthening growth and prosperity, reducing regulations, eliminating trade barriers and opening energy markets (White House 2019). Climate change was absent. All four of these entirely economic priorities would increase climate change. His was to be an anti-environmental summit from the start, long before the COVID-19 shock appeared.

When COVID-19 did appear, the proliferating pandemic and its economic damage kept the economy first and added health. These health and economic shocks inspired G7 leaders to mount the first emergency summit in history, which took place by videoconference on March 16. This made the G7 the fast, first responder among global summit institutions to the escalating COVID-19 crisis. The summit produced 26 commitments, all addressing the impacts of COVID-19. The G7 called "upon the G20 to support and amplify these efforts." The G7 leaders would hold follow-up monthly calls until their summit on June 10–12. However, they held only one more conference call on April 16, which produced no communiqué.

The unprecedented COVID-19 crisis delayed the regular G7 summit into the summer and then September, with Trump finally saying he might hold it after the US election (White House 2020). In the end, no regular summit took place.

Causes of Performance

By far the most salient cause of the G7's full failure on climate change was Trump's personal commitment from the start to dismiss the scientific facts about climate change. He was supported by his core domestic political base and Republican Party, whose influence grew as the November 3 election approached. The COVID-19 and ensuing shocks and vulnerabilities in the United States and the G7 crowded out attention to the issue at the virtual summit and ministerial meetings, despite the increasing climate shocks and growing greenhouse gas concentrations and global warming as the year advanced. The delayed UN climate summit and the strong US position at the core of the G7 club at the hub reinforced these forces. The COVID-19 recession reduced G7 and global emissions for the first half of the year, allowing all G7 leaders to conclude that they could put off climate action to next year.

Shock-Activated Vulnerability

The March 16 summit did not recognize any climate shocks or vulnerabilities (see Appendix C). There were many references to COVID-19 and other shocks but none were linked to the natural environment.

Nor did climate change appear in the elite media coverage of the summit. More broadly, throughout 2020, coverage of climate change on US television network nightly newscasts and Sunday shows, including ABC, CBS, NBC and Fox, plummeted 53% from 2019 to only 112 minutes, the lowest level since 2016 (Macdonald 2021). There was no climate coverage in March, April or June. The nightly newscasts linked the coronavirus to climate change only 3 times and the morning broadcasts did so only 10 times. Climate coverage was dominated by the wildfires in the United States and Australia, other extreme weather events and the US election. PBS coverage declined 58%. Since monitoring began in 2002, 2020 was the fifth worst year for climate coverage — "almost certainly due to the deadly coronavirus pandemic that dominated news cycles" (Macdonald 2021).

In the physical world, the unprecedented COVID-19 shocks and vulnerabilities quickly overwhelmed the far less visible, but still deadly climate shocks, led by the wildfires in California and Australia. On January 15, Japan identified the first COVID-19 case in a G7 member (World Health Organization [WHO] 2020a). On January 23, 2020, the WHO (2020b) reported the first case in the United States. By January 31, the day after the WHO declared the outbreak to be a public health emergency of international concern, there were cases in all G7 members except the United Kingdom. On March 13, Europe became the epicentre of the pandemic with more reported cases and deaths than the rest of the world outside China, with 12,462 cases and 827 deaths in Italy alone (WHO 2020c). The G7's European members thus prompted Trump's United States to call the emergency G7 summit on March 16 to address the COVID-19 shock, and were too preoccupied with it to include climate change.

Yet, the World Economic Forum's 15th Global Risks Report, published on January 15, put climate change first. Its press release, titled "Burning Planet: Climate Fires and Political Flame Wars Rage," reported that the top five global risks, for the first time in 10 years, were all environmental, and included extreme weather events, government and business failure to mitigate and adapt to climate change, human-made environmental damage and disasters such oil spills, and major biodiversity loss and ecosystem collapse with "irreversible consequences" (World Economic Forum 2020).

Additional Causes

Multilateral organizational failure was complete, but did not spur G7 leaders to act on climate change. On April 1, UN Climate Change (2020b) postponed COP26, originally scheduled for November. Trump blamed the WHO for the COVID-19 crisis and in May began moving to withdraw the United States from the WHO (Ehley and Ollstein 2020).

By autumn, declining US relative capability provided an incentive for the United States to adjust to its G7 partners and their priorities, but Trump publicly disregarded this evidence. Declining emissions due to COVID-19 reduced the desire of all to advance climate control. Japan's highly experienced Shinzo Abe stepped down as prime minister for health reasons and was replaced by the domestically focused Yoshihide Suga. Trump's denial of COVID-19 and climate change increased as the November election approached. COVID-19 led to the cancellation of all in-person events, where leaders could try to convince Trump directly to act on climate change.

G20 Riyadh, November

Introduction

On November 21–22, the G20 leaders met virtually, hosted from Riyadh by Saudi Arabia's King Salman, assisted by Crown Prince Mohammad bin Salman. By November 20, the global pandemic had infected 57.2 million people and killed at least 1.3 million (WHO 2020d). The shock of COVID-19 and its ensuing crises crowded out almost all G20 attention to climate change. Combined with Saudi Arabia's leadership, this led to Riyadh's small performance on this escalating ecological catastrophe.

Debate

In the debate about Riyadh's performance, the first school stressed its launch of the circular carbon economy (CCE) initiative, a controversial and expensive approach that focused on managing rather than reducing emissions, including recycling carbon back into oil and gas production. al-Sudairi (2020) identified global warming as "the most critical future issue," and said the summit set

> four principles to deal with carbon dioxide more realistically without hindering modern renewable technologies, and curb carbon emissions without impeding the resources of countries that use fossil fuels, especially oil, which can be used to produce environmentally-friendly materials such as blue ammonia.

This performance was due to the economic, environmental and energy challenges directly affecting Saudi Arabia, and to increasing desertification, logging and emissions.

The second school saw a positive, principled step forward, due to Saudi Arabia's presidency and G20 members' globally predominant economic and ecological capabilities. Al-Khattaf (2020) argued that the summit promoted the CCE concept and acknowledged the need to tackle climate change. This was despite US withdrawal from the Paris Agreement and China's emergence as the world's top polluter, and due to Saudi Arabia's presidency, the G20's 85% of global gross domestic product (GDP) and 80% of global emissions, and Saudi Arabia's global leadership in exporting blue ammonia.

The third school saw a strong multilaterally supportive success. Aluwaisheg (2020) noted strong support for protecting the environment and fighting climate change and for the CCE Platform's framework of reduce, reuse, recycle and remove. The key causes were several breakthroughs on vaccines and Saudi Arabia's hosting of two G20 summits in one year.

The fourth school highlighted serious shortcomings. Kirton (2020c) emphasized that the climate change texts were "watered down." The central, long-unfulfilled commitment to phase out fossil fuel subsidies was simply repeated in a qualified way. G20 leaders could not even commit to doing their fair share to meet the long-agreed global goal of planting 1 trillion trees, even though Trump repeated his promise to do so in his video address prepared for the Safeguarding the Planet side event just a few hours before the communiqué appeared. Climate change was crowded out by the intense summit attention to and advances on COVID-19 and vaccine development, while virtual summitry left much less space to do more.

The fifth school similarly saw failure, but due more to Trump. Kami and Rappeport (2020) agreed the G20 leaders made little progress on climate change, due to deep divides between Trump and other G20 leaders. Yet, despite Trump's attacks, the G20 at least reiterated that climate change was among "the most pressing challenges of our time," and noted that the G20's Financial Stability Board (FSB) was "continuing to examine the financial stability implications of climate change." Fahim (2020) more harshly saw low expectations, unmet challenges and mixed results, with climate change left out, due to Trump's attack on the Paris Agreement.

The sixth school saw full failure, due to limited cooperation among the United States, Europe and China. Brown (2020) concluded that "in 2009 environmental investment comprised one-sixth of the stimulus — far below the near-50 per cent share that we now should target as the starting point for progress at the COP26." He recommended an emergency G20 summit in February 2021, immediately following the US presidential inauguration. Larry Summers, Mark Sobel and Matthew Goodman agreed (Morris 2020; Shalal 2020).

The seventh school saw failure caused by the shock of COVID-19, divisions and lack of trust among G20 members and Trump (Bhatia 2020). Messenger (2020) argued the G20 failed on intergenerational issues such as climate change.

Argument

Riyadh produced a small performance, making only three commitments on climate change. Although it continued the rising compliance trend since 2015, it did not keep pace with the growing severity and urgency of the climate crisis.

This small performance was caused primarily by the sudden, severe, shared shock-activated vulnerability of COVID-19, which crowded out the climate crisis's more chronic, less visible and less directly comprehensible effects. The complete multilateral organizational failure of the delayed COP26 summit did not spur Riyadh to fill the gap. The G20's predominant equalizing capability did grow, led by China's increasing rise and US decline, but not enough to offset COVID-19's diversionary shock. Moreover, convergence on democratic principles and environmental performance decreased, with Saudi Arabia in the chair and as complacency grew due to the temporarily declining emissions caused by COVID-19 lockdowns. Climate change–denying or –reluctant King Salman, Vladimir Putin, Jair Bolsonaro, Xi Jinping and even an electorally defeated Trump had sufficient domestic support to prevent much climate action abroad. Finally, the shift to virtual summitry severely damaged the G20's ability to act as a valued in-person club of leaders (Naylor 2020). In addition, its Saudi host at the hub belonged to few plurilateral summit institutions of global relevance and reach.

Plans and Preparations

The 2020 summit was scheduled to be held in Riyadh on November 21–22. After March, all preparatory and ministerial meetings were conducted virtually.

On assuming the presidency on December 1, 2019, Saudi Arabia launched an ambitious agenda, with the ecological preservation of the planet first, in a broad, well-targeted way (Kirton 2019b). King Salman identified the "environmental changes" transforming the world and said the G20 should "strive for sustainable economic policies to safeguard planet earth" (G20 Saudi Arabia 2019). The presidency referred to "climate and natural disaster threats" and proposed to "deliver concrete actions."

The first of Saudi Arabia's three aims, "Empowering People," promised to "scale up efforts for sustainable development, and foster inclusive and sustainable tourism" (G20 Saudi Arabia 2019). Its nine priorities included two explicitly environmental or ecologically sustainable development ones.

The second, fully ecological aim of "Safeguarding the Planet," started with climate change, including "advancing synergies between adaptation and mitigation" (G20 Saudi Arabia 2019). Four of its six priorities explicitly referred to climate change. One of the seven priorities of the third, security-focused aim of "Shaping New Frontiers" was on space and addressed the environment and climate change. Together, 9 of the 22 Saudi priorities were explicitly ecological ones.

"Safeguarding the Planet" began with "Managing Emissions for Sustainable Development," and noted the "urgency" of this "imperative" (G20 Saudi Arabia 2019). It sought a new economic model to reduce emissions with comprehensive measures in all sectors. It embraced sinks, selecting "nature-based solutions such as reforestation" and "restoring marine resources" as instruments of choice.

The second priority, "Combating Land Degradation and Habitat Loss," also prioritized forests, noting that deforestation and other land uses are responsible for 24% of emissions.

The third priority, "Preserving the Oceans," noted how climate change was endangering coral reefs.

The fourth priority, "Fostering Sustainable and Resilient Water Systems Globally," linked directly to the UN's Sustainable Development Goals (SDGs). It noted that improving financing was critical.

The fifth priority, "Promoting Food Security," noted that climate change harmed food security and "changing dietary habits" was an important response (G20 Saudi Arabia 2019).

The sixth priority, "Cleaner Energy Systems for a New Era," promised "to further advance cleaner energy transitions" through using all energy sources and innovative technologies. It committed to discuss the CCE.

These priorities built on those the crown prince had announced to the G20 leaders at the end of the G20's Osaka Summit in 2019. They were backed by the domestic priorities presented in Saudi Arabia's Vision 2030. Saudi Arabia knew that its economic and physical survival depended on climate change control, as the oil and gas on which its economy depended were becoming stranded assets and outdoor temperatures relentlessly rose toward levels the human body could not withstand.

Sherpa Process

The first meeting of sherpas and of finance deputies was held on December 3–7, 2019, in Riyadh. The meeting covered all of Riyadh's agenda, including the environment, climate change and water, food and energy security (Saudi Press Agency 2019). There was no discussion of climate change at the next two sherpa meetings.

Extraordinary Virtual Summit, March

The COVID-19 crowd-out was also complete on March 26, when G20 leaders held their first emergency summit ever and their first in virtual form. It followed a call from the G7 leaders, who had similarly met 10 days before.

G20 leaders produced the Statement on COVID-19. The environment and climate change were entirely absent from this 1,494-word, 30-paragraph statement (Kirton and Warren 2020c).

Ministerial Meetings

There were many meetings of G20 ministers during the Saudi presidency, some in addition to their originally scheduled meetings, to respond to the COVID-19 pandemic. But climate change was rarely discussed.

G20 finance ministers and central bank governors met on July 18, and issued a communiqué for the fourth time that year (Kirton 2020d). It contained four references to the natural environment but none to climate change.

At the end of the G20 environment ministers meeting on September 17, the Saudi presidency issued a brief press release with only one reference to climate change: "The COVID-19 pandemic along with climate challenges demand a commitment to build a more sustainable and prosperous future for all" (G20 Saudi Arabia 2020). The communiqué was quietly published on November 22, after the leaders' meeting had ended.

On the Eve

On the eve of the Riyadh Summit, it was left to the finance deputies, finance ministers and sherpas to make any advances they could. On November 19, negotiations went reasonably well. There was optimism about climate finance and reporting standards. However, major divisions remained on the language on climate change. The finance ministers also met on the evening of November 20.

The United States and others demanded "balance." They insisted that if the communiqué said anything negative about carbon and fossil fuels, it also needed to say something positive. There was no reference to evidence. These issues were stalled at the sherpa level and left to the leaders themselves to agree on the language.

At the Summit

During the summit, several leaders presented their views on climate change and the environment in pre-recorded, publicly broadcast messages and in the two virtual side events on Safeguarding the Planet: The CCE Approach and on Pandemic Preparedness and Response. For Safeguarding the Planet there were video addresses by King Salman, Trump, China's Xi, Australia's Scott Morrison, India's Narendra Modi, France's Emmanuel Macron, Japan's Suga and Italy's Giuseppe Conte.

Saudi Arabia's climate change agenda centred on the CCE concept. King Salman highlighted the Kingdom's goal to make 50% of its electricity mix from wind and solar by 2035. The carbon captured from this CCE project would be filtered back into the oil and gas sector to produce more fossil fuels. This climate plan was embedded within Saudi Arabia's Vision 2030, which included commitments to increase reliance on renewable sources and phase out fossil fuel subsidies.

Trump, in his video, bragged about his administration's efforts to reduce lead in water, plant 1 trillion trees, reduce marine litter and restore access to federal lands, including the Bears Ears national monument. He declared that he was protecting American workers by pulling the United States out of the Paris Agreement and claimed that US emissions had been reduced more than any other country. He declared that America's air was cleaner than when he had entered office and that renewable energy production had risen, and praised American leadership in horizontal drilling and hydraulic fracturing.

Morrison said Australia was committed to meeting its emissions reduction targets and expressed support for "unlocking" low-emissions technology, including hydrogen, the CCE, green steel and aluminum. He said Australia had overachieved its Kyoto Protocol targets by a significant margin and was on track to meet its Paris Agreement target, to which it was firmly committed.

Xi said the G20 "could strengthen our response to climate change" in three ways. The first was to "continue to take the lead" by pushing for the full, effective implementation of the Paris Agreement. He highlighted China's goals of reaching peak emissions by 2050 and carbon neutrality by 2060. The second way was to "deepen the transition toward clean energy." Xi applauded the CCE proposal, referred to the SDG goal of Sustainable Energy for All, touted China's domestic efforts to advance energy efficiency, new energy, green industries, and greener economic and social systems, and said China had the world's largest clean energy system. The third way was to "protect the ecosystem with a respect for Nature."

Modi opened by stating that focusing on fighting climate change was as important as "saving our citizens and economy from the effects of the global pandemic." He said that "climate must not be fought in siloes, but in an integrated, comprehensive and holistic way." Progress would be faster with greater support of technology and finance for developing countries. He highlighted India's success in implementing low-carbon and climate-resilient development practices, including renewable energy, smoke-free kitchens, reduced single-use plastics, biodiversity efforts that increased the lion and tiger population, land restoration, the circular economy and improving the resilience of critical infrastructure. Modi also stated that India was "exceeding" its climate targets.

Suga reiterated Japan's commitment to reach carbon neutrality by 2050, create a "green society" at home, make the environment a pillar of Japan's economic growth strategy, support the CCE alongside "various options," and conserve the marine environment and implement the 2019 Osaka Blue Ocean Vision. He called for international cooperation on the Paris Agreement and for a paradigm shift on economic growth, because "addressing climate change is not a constraint on economic growth." He emphasized the need to transform industry as well as the economy and society.

Conte stated that climate change would be a top priority for his 2021 G20 summit and described climate change and biodiversity loss as an "existential threat." He identified several priorities, which included the climate–energy nexus and a swift transition to a carbon neutral future, without relying on the CCE alone. Special attention would be paid to smart cities and green jobs. Conte reiterated Italy's commitment to reach carbon neutrality by 2050. He announced that Italy would host a youth conference on climate change at the end of 2021, as part of its co-hosting role for COP26 with the United Kingdom. He said, "current social and economic systems fail to take climate change into account," thus acknowledging the need for systemic change.

In his video for the Pandemic Preparedness and Response event, Macron called for concerted efforts "for the common good of our planet … towards climate and biodiversity."

Boris Johnson, in his Leader's Message for the summit, highlighted the United Kingdom's new Ten Point Plan for a Green Industrial Revolution, praised the Saudi city of Neom's leadership on advancing green hydrogen and solar power, and called on the G20 "to make bold pledges and harness [its] collective ingenuity and resources to defeat the pandemic and protect our planet and our future for generations" ahead of the Climate Ambition Summit in December.

No other leader mentioned climate change in their public addresses, not even Angela Merkel. Without her and Justin Trudeau, and with Macron emphasising only COVID-19, the G7 coalition that had produced advances on climate change at Osaka had shrunk. The COVID-19 crowd-out remained.

Dimensions of Performance

Riyadh's small performance on climate change was led by its three general commitments on climate change (see Appendix D).

Domestic Political Management

Attendance was complete among G20 leaders, although virtual. But none of Riyadh's four communiqué compliments referred to climate change. Subsequent media approval was low. The *Financial Times* (2020) lead editorial on November 24 ignored climate change.

Deliberations

Riyadh's private deliberations were very low. Virtual summitry eliminated in-person conversations of any sort among the leaders and enabled some key leaders to leave early (Follain et al. 2020). Trump left during Johnson's speech and was replaced first by treasury secretary Steven Mnuchin and then by White House economic advisor Larry Kudlow. Virtual bilaterals were scarce.

In its public deliberations, Riyadh gave the fifth lowest number of words to climate change of any summit since the first two summits. Its 679 words were significantly lower than Osaka's 2,034 words and the G20's overall average of 1,308 words (see Appendix D). Riyadh's 12% was much lower than Osaka's 24% the year before.

These small climate deliberations did cover many subjects: the CCE, nature-based solutions, implementation of the Paris Agreement and submission of updated nationally determined contributions (NDCs), climate finance for developing countries' adaptation and mitigation efforts, natural disasters and extreme weather events, the financial stability implications of climate change, and environmental degradation including biodiversity, oceans, and clean air and water.

Direction Setting

Riyadh's direction setting was solid (see Appendix D). The climate change passages made two affirmations of promoting global financial stability. One was to advance environmental stewardship for future generations while maintaining healthy economies. The other noted the FSB's continued study of the implications of climate change. On the second mission of making globalization work for all, there was one reference to the environment, on "building a more environmentally sustainable and inclusive future for all people." The climate change passages had no priority placement, or affirmations of the principles of democracy or human rights.

Decisions

Leaders made only 3 climate commitments, a sharp drop from the 13 at Osaka (see Appendix D). Two reiterated past climate commitments: to continue to support climate action and to "implement the Paris Agreement … as we promote economic growth." The third endorsed the CCE. There were also reiterated commitments on biodiversity loss and phasing out inefficient fossil fuel subsidies.

Riyadh mobilized $11 trillion for COVID-19, but very little for climate change.

Delivery

By July 29, 2021, G20 compliance with the two assessed climate commitments was 80%. The Paris Agreement one had 95% and the CCE one had 65%. Compliance was only 55% with both the environment marine plastic litter and energy fossil fuel subsidies commitments.

Development of Global Governance

Riyadh made six climate-related references to institutions both inside and outside the G20 (see Appendix D). This was lower than the average for both inside and outside institutions.

Two references to inside institutions went to the FSB and the Climate Stewardship Working Group. The former noted that the FSB was continuing to examine the financial stability implications of climate change. The latter recognized a "toolbox" developed by the Climate Stewardship Working Group to address sustainability and climate change.

Four highly focused references went to two outside institutions, with two references to the UNFCCC broadly, and two to COP21 and COP26 specifically.

Causes of Performance

Shock-Activated Vulnerability

Riyadh's small performance was driven primarily by the sudden diversionary shock of COVID-19, which crowded out the more chronic effects of the climate crisis.

On March 26, the G20 leaders' statement on COVID-19 did not mention climate change or the natural environment. "COVID-19" and "pandemic" dominated their November declaration, which made 38 references to these shocks, exceeding the references to the economic crisis (see Appendix F). Those terms appeared in 27 (71%) of Riyadh's 38 paragraphs. The leaders linked the pandemic to many subjects, but only once to energy and never to climate change. They twice recognized the pandemic's impacts on the environment, and on water, food and energy security, and referred in general to natural disasters and extreme weather events.

In the elite media, COVID-19 similarly crowded out attention to climate change and the environment. Every day between November 7 and November 17, COVID-19 and the economy appeared on the *Financial Times* front page. In sharp contrast, climate change appeared only on four days, including November 9 and November 17. In the three days immediately before November 19, health shocks dominated, health-economy links remained strong and climate change disappeared.

The *New York Times* front page was dominated by the November 3 US election. On November 8, 50% of the front-page stories were on the COVID-19 crisis, 25% on the economic crisis and none on climate change.

In the physical world, however, there were many extreme weather events. Between November 3 and November 19, Hurricanes Eta and Iota hit Central America, including Mexico, and killed more than 200 people, leaving at least 200,000 homeless and increasingly vulnerable to poverty-related illnesses (BBC News 2020b; Palencia and Lopez 2020). A "gigafire" in California tore through 1 million acres and other wildfires burned 4 million acres there (Milmand and Ho 2020). They resulted in 33 direct deaths and between 1,200 and 3,000 deaths from smoke and air pollution (Bay Area News Group 2020; Vainshtein 2020). Australia had 240 days of bushfires from 2019 into 2020 (BBC News 2020a). By February 2020, they had directly killed 33 people, over 1 billion animals and hundreds of billions of insects (Richards and Brew 2020). Another 400 people died from smoke pollution (Pickrell 2020). "Zombie fires" in the Arctic covered Canada, Russia's Siberia, Alaska and Greenland in smoke (Witz 2020). They released more emissions in a single month than in the previous 18 years (Sengupta 2020).

By the end of 2020, the number of deaths from extreme weather events in G20 countries rose from 5,696 in 2019 to 8,454 (see Appendix G).

Yet COVID-19's shock was far more compelling. Its first wave surged just before the March emergency summit. The subsequent shocks of the second and third waves came just before the Riyadh Summit. G20 leaders failed to connect the health and climate crises, nor had any G20 summits before (Kirton and Warren 2020a).

The G20 also ignored the stark science showing a relentless global average temperature rise to a new critical threshold. The years from 2015 to 2020 were the warmest since records had begun (*Financial Times* 2021a). Recorded global temperature rise averaged 0.2°C per decade since 1970, peaking in 2016 and again in 2020, at 0.6°C above average (Hook 2021a). The year 2020 was 1.25°C warmer than the 1850–1900 period, despite emissions declining about 7% in 2020, due to decreased economic activity.

Many G20 members reached new heat highs, with Europe at 1.6°C higher in 2020 than the 1981–2010 average (*Financial Times* 2021a). In 2020, it was the world's fastest warming continent (Hook 2021a). In the Arctic areas of the United States, Canada and Russia, temperatures were 6°C above the historical average, with a record of 38°C in northern Russia in June.

Yet, with the global economy stalled and GDP declining in almost all G20 members due to COVID-19, global emissions and the accompanying air and noise pollution also temporarily declined. G20 leaders thus assumed that climate change control could wait.

Multilateral Organization Failure

Multilateral organizational failure was very high, led by the cancellation of COP26 in 2020 and its postponement for a full year. But this, surprisingly, did little to spur other multilateral bodies or the G20 summit to fill the gap.

The 15th COP to the UN Convention on Biological Diversity, originally scheduled for October 2020, was similarly postponed to May 2021. The UN did not transform either COP to a virtual format.

Predominant Equalizing Capabilities

The G20's predominant equalizing capability was not enough to overcome the COVID-19 shock. Only China and Turkey had positive economic growth in 2020, and neither was a leader on climate change control (see Appendix H).

Nonetheless, G20 members still produced almost all the world's valuable and used currencies, and about 80% of the global economy and its emissions. The US dollar's surge against other major G20 and global currencies at the start of the pandemic was consistent with the absence of G20 action on climate change at the March summit. So was its decline in the autumn and the reappearance of climate attention and action, however small, in three commitments at the Riyadh Summit.

In underlying capabilities and vulnerabilities, at the top of the G20, China's GDP and emissions both rose in a tightly coupled tandem, while those of the United States and the other G20 members continued their simultaneous decline.

Yet, as overall G20 economic capability declined, its climate vulnerability did too, when measured by the standard component of emissions (see Appendix I). By April 2020, global emissions were 17% lower than in 2019 (Le Quéré et al. 2020). By the end of 2020, the G20's emissions were expected to be 7.5% lower than in 2019. But these declines were due to the economic recession caused by COVID-19 and not to government policy.

Converging Characteristics

Divergence among G20 members on the core characteristics of open democracy constrained climate performance. Democracy remained complete in most G7 members, but eroded in the United States, Russia and Turkey, and in the European Union's Hungary and Poland.

From 2019 to 2020, the G20 dropped nine spots down the democratic ranking of 167 countries to 56th place, with five countries declining, nine rising and the rest remaining constant (see Appendix J).

On climate performance, the G20 as a group rose one spot on a 61-country index, but still ranked low at an average of 36th (see Appendix K).

Domestic Political Cohesion

Domestic political cohesion was a further constraint. Leaders with the greatest domestic support were the climate change–denying or –reluctant King Salman, Xi, Putin and Bolsonaro. The highly climate-committed Merkel, Macron and Trudeau were less secure at home.

King Salman, who assumed the Saudi throne in 2015 and attended the 2015 Antalya Summit, effectively co-hosted the virtual Riyadh Summit with Mohammed bin Salman, with both giving speeches. The king appeared frail, and his delivery and leadership were unenergetic as he read his script. But he and the crown prince had full control of their government.

Among the veterans, Putin, Recep Tayyıp Erdoğan and Joko Widodo were less committed to climate action, even if Xi and Modi had now started to adjust. Trump, despite his defeat in the November 3 election, still represented the United States. Bolsonaro and Morrison joined the climate skeptics.

Merkel, the G20's greatest veteran, was now a lame duck, having announced her intention to retire in 2021. Macron was still afflicted by the yellow vest protests in France. Trudeau had a minority government.

Club at the Hub

The COVID-19–created shift to virtual summitry crippled the G20's dynamic as its leaders' valued in-person club. The ailing King Salman and Saudi Arabia at the hub belonged to few of the plurilateral summit institutions of global relevance and reach. It linked only to the Organisation of Islamic Cooperation, to which only Turkey and Indonesia also belonged, and the Organization of Petroleum Exporting Countries (see Appendix L).

Nor did the invited guests bring much climate change commitment or global institutional reinforcement. They were Spain, a permanent guest of the G20, under the leadership of Pedro Sánchez, King Abdullah II of Jordan, Paul Kagame of Rwanda and chair of the New Partnership for Africa's Development, Lee Hsien Loong of Singapore, Simonette Sommaruga of Switzerland, Mohammed bin Rashid Al Maktoum of the United Arab Emirates and chair of the Gulf Cooperation Council, and Nguyen Xuân Phúc of Vietnam and chair of the Association of Southeast Asian Nations.

The heads of the invited international organizations all came from economic or health bodies: the International Monetary Fund, the UN, the Islamic Development Bank, the Organisation for Economic Co-operation and Development, the Arab Monetary Fund, the WHO, the World Bank, the Financial Action Task Force, the

FSB, the Food and Agriculture Organization, the World Trade Organization and the International Labour Organization. None were invited from any core environmental or climate bodies.

The Missing COP26

In 2020, the UN's climate summit, originally scheduled to start in Glasgow on November 9, was postponed. UN Climate Change determined that holding the universal climate negotiations virtually, due to COVID-19, was not feasible or inclusive enough.

After the WHO declared COVID-19 a global pandemic, UN Climate Change (2020b) announced on April 1 that COP26 would be delayed for a full year. This was considered an essential precautionary measure that would allow parties and the expected 30,000 delegates to concentrate on combating COVID-19 (Nature 2020). There was also doubt that such a large delegation could produce the needed deliverables by meeting virtually (Vetter 2020). The COP's preparatory meetings soon disappeared too.

The Consequences of COP26's Postponement

The COP26 delay left many governance gaps. There were still no definitive action plans or NDCs adequate to limit temperature increases below 2°C, let alone 1.5°C. There would be no action-enforcing deadline to advance energy efficiency and renewable energy, reduce fossil fuel subsidies or increase climate finance and other support for least developed countries (Tubiana et al. 2019).

Some argued that postponing the 2020 COP would have some benefits, as new ideas emerged for more ecologically sustainable economic growth as countries planned to reopen their post-pandemic economies (Harvey 2020c). COP26 would now not be overshadowed by the US election on November 3, or by Trump's decision to withdraw the United States from the Paris Agreement, which would take legal effect on November 4. With the new president assuming office only in January 2021, the Trump administration would have represented the United States at COP26 regardless of the election result. This may have discouraged other participants from announcing more ambitious national goals (Harvey 2020a; Nature 2020). But with Joe Biden as president-elect, participants could expect the United States to re-enter the Paris Agreement and re-establish its global climate leadership (Harvey 2020c).

Yet, several climate diplomats, officials, experts and non-governmental organizations urged COP26 to continue, insisting on the dire need for a stringent climate mitigation regime. Delaying the conference would break the momentum of the hard-won progress already achieved, particularly on carbon markets, plans for cutting emissions to net zero, mobilizing finance for less wealthy countries and stricter accountability measures (Harvey 2020a; Nature 2020). Postponement signalled that global priorities had shifted away from climate change. Indeed, the financial and economic strain from the pandemic made even the wealthiest and most powerful countries relax emissions limits and ease climate mitigation standards (Harvey 2020a).

The delay was ecologically costly. In July, the World Meteorological Organization (2020) released scenarios predicting that the world was warming at an alarming rate. If emissions patterns continued at current rates, the end of the 21st century could see temperature increases of 4°C (UCAR Center for Science Education n.d.). With global rainfall growing by 3%–5% and with rising sea levels from melting ice sheets and thermal expansion, large-scale flooding would put coastal regions and small island states at increased risk. Warming ocean waters would heavily influence ocean currents, increasing the frequency of cyclones, hurricanes and typhoons. Higher levels of atmospheric carbon dioxide would reduce pH levels, increasing acidity and further threatening aquatic life.

These projections did not account for tipping points that could render climate recovery practically impossible, such as the breakdown of major Arctic ice sheets, unexpected releases of methane from the Arctic's permafrost and the collapse of the ocean's thermohaline circulation (UCAR Center for Science Education n.d.).

However, within a few weeks of the mid-March lockdowns, about 2.6 billion metric tons of emissions were saved (Millan Lombrana and Warren 2020). Decreased pollution cleared skies in smog-choked places around the globe, and the Himalayas could once again be seen (Millan Lombrana and Warren 2020). In its October World Energy Outlook, the International Energy Association (2020) estimated global energy demand would drop by 5% for 2020 and energy-related emissions by 7%, with annual carbon dioxide emissions returning to where they were a decade earlier. Declining industrial activity and air pollution led to striking reductions in atmospheric levels of nitrogen dioxide (Watts and Kommenda 2020). With approximately 16,000 aircraft grounded worldwide and flight activity in Europe plunging by more than 80%, over 10 million metric tons of emissions were saved (Millan Lombrana and Warren 2020).

Unilateral Action

Without a COP meeting or summit to spur them on, several key countries took unilateral action. As COP26 co-host, the United Kingdom invested in harnessing offshore wind energy to end its dependence on coal by 2025 in an effort to reach its national target of net zero emissions by 2050. It invested in financing hydrogen technology, passed clean air legislation to reduce emissions from road transport and endorsed several trade agreements to ramp up climate-related progress.

In December, the European Union (n.d.) adopted the Next Generation EU recovery plan, with a budget of €806.9 billion, with 30% of the funds aimed at fighting climate change. The plan would also support members' "fair climate and digital transitions" and biodiversity protection. It would help fund the EU goal of a net zero carbon economy by 2050 (Boffey and Rankin 2020).

China announced its shift in focus to "sustainable growth," committing to investments in current technology aimed at reducing fossil fuel consumption, greening infrastructure and providing cleaner energy sources (Nature 2020). It encouraged its decision makers to draft policies that promoted environmental well-being while building more resilient economic systems.

Special Climate Summits

With action at the UN and G7 summits absent, and with the G20's small performance, the centre of global climate governance in 2020 shifted to the UN's High-Level Roundtable on Climate Action and the Summit on Biodiversity in September, and the Climate Ambition Summit in Paris in December. Although these were all held in a virtual format, they took the reconfiguration of global summit governance of climate change to a new, stronger level.

High-Level Roundtable on Climate Action, September

UN secretary general António Guterres convened a 90-minute high-level roundtable on September 24, during the UN General Assembly (UNGA), intended to raise ambition to achieve the goals of the Paris Agreement and encourage creating recovery packages that lowered emissions and improved sustainability (UN 2020b). The event included an address by Johnson and a moderated discussion on finance among Ursula von der Leyen, Trudeau, Conte and several other leaders (UN 2020c). Leaders noted China's recent pledge to become carbon neutral by 2060, and von der Leyen referred to the European Union's proposal to reduce emissions by 55% by 2030 and to issue €200 billion in green bonds. Conte stressed divestment from fossil fuels and, along with Trudeau and Johnson, stressed the need to design recovery plans that relied on climate action and green jobs. Chile's Sebastián Piñera announced Chile would phase out all coal-powered generation and electrify public transport by 2040. On climate finance, especially regarding the urgent needs for developing and vulnerable countries, Trudeau called for a reorientation of capital flows. Participants also included representatives from business and civil society.

Guterres (2020c) and Johnson used the opportunity to announce the Climate Ambition Summit on December 12, to maintain momentum for the Paris Agreement goals in the absence of COP26 (UK, Prime Minister's Office 2020).

UN Summit on Biodiversity, September

With climate and biodiversity inextricably linked, climate issues also became a focal point for discussion at the UN Biodiversity Summit on September 30, also held during UNGA and streamed live from New York. Yet, it was also only a small success. Two days earlier, Guterres (2020b) called on leaders to "make decisive commitments to protect our planet" to deal with "dual threats of climate change and biodiversity collapse."

The Biodiversity Summit inspired several clean energy, carbon neutral initiatives as well as nature-based solutions for achieving the Paris Agreement targets. Inspired by the European Union's recently proposed goal of reducing emissions by 55% by 2030, Xi announced that China planned to achieve peak emissions by 2030 and carbon neutrality by 2060, respectively (UNGA 2020).

Climate Ambition Summit, December

The 2020 reconfiguration culminated at the Climate Ambition Summit on December 12, which produced a significant performance.

Convened by the UN, the United Kingdom and France, with Chile and Italy, to mark the fifth anniversary of the Paris Agreement, it was streamed from Paris. This virtual "sprint to Glasgow" served as a small-scale session for state and non-state actors to negotiate new climate commitments on adaptation plans, NDCs, long-term strategies for net zero emissions and climate finance pledges for the most vulnerable countries (IISD 2020).

Guterres (2020a) opened the conference by stating that the trillions of dollars needed to recover from COVID-19 cannot be used to "burden future generations with a mountain of debt on a broken planet." He said the international community could not rest until net zero emissions targets were met and called on "all leaders worldwide to declare a State of Climate Emergency in their countries until carbon neutrality is reached." This would require that all COVID-19 recovery endeavours promote a green and resilient future with little to no fossil fuel dependence (Doig 2020).

At the summit, 75 government, business and civil society leaders announced new commitments (Doig 2020; UN Climate Change 2020a). The number of strengthened NDCs grew to 71 and included those from Canada, Argentina, Barbados, Colombia, Iceland and Peru. The momentum encouraged other countries to adopt or adhere to similar strategies (UN Climate Change 2020a). On long-term strategies for net zero emissions, 24 countries set new deadlines and approaches, including the European Union, which raised its goal of reducing emissions from 40% to 55% by 2030 from 1990 levels (Doig 2020).

G20 members China, Japan, Korea and Argentina made new pledges aimed at net zero emissions by 2050 (UN Climate Change 2020a). China also committed to shift 25% of its primary energy consumption to renewable sources by 2030. India promised to increase its capacity for renewable energy by 450 gigawatts by 2030. The United Kingdom, France and Sweden committed to cutting international financial support for fossil fuel consumption. Canada announced it would raise its carbon tax to CA$170 per ton of greenhouse gas emissions by 2030. The United Kingdom, Portugal and Spain indicated they would consolidate adaptation measures (UN Climate Change 2020a). Twelve countries pledged to extend climate finance to the developing world, with Germany committing an additional €500 million and France €1 billion per year.

The summit shone a harsh spotlight on Australia, Russia and Brazil, which offered no new climate commitments (Doig 2020). Although the Trump administration did not attend, President-elect Biden's administration released a statement promising to rejoin the Paris Agreement and committed to arrange a meeting with leading economies to discuss climate mitigation shortly after assuming office on January 20, 2021 (Doig 2020).

The Climate Ambition Summit helped fill the gap left by the postponement of COP26. It added 45 NDCs, 24 net zero commitments and 20 adaptation and resilience plans to the existing pledges (Doig 2020). It ensured that the COVID-19 pandemic would not completely crowd out action on climate change. However, COP26 president Alok Sharma (2020b) said four major goals remained: increasing mitigation strategies, strengthening adaptation measures, providing uninterrupted finance to developing countries and enhancing international collaboration. These four pillars would form the basis of negotiations at COP26 in November 2021.

Conclusion

The global summit governance of climate change largely failed in 2020. The G7 did nothing at its emergency summit on March 16, and failed to hold a regular summit later in the year. Likewise, the G20 did nothing at its emergency summit on March 26 or at its scheduled Riyadh Summit on November 20–21. And the UN delayed its long-awaited climate summit for a full year.

The overwhelming cause of this great failure was the diversionary shock of the COVID-19-created health and economic crises, and the continuing inability of G7, G20 and most UN leaders to make the synergistic link between health and climate change. These causes were compounded by the postponement of COP26 in 2020, the continuing absence of multilateral environmental organizations at G7 and G20 summits, the rising capability of a still climate-reluctant China and the decline of democracy in Russia, Brazil and Turkey. A key cause was Trump's continuing climate denial as G7 host in 2020. Moreover, the shift to virtual summitry damaged the G7 and G20 dynamic as leaders' personally valued, in-person clubs. And fossil fuel–dependent, non-democratic Saudi Arabia as G20 host lay outside the core of the supportive network of global summit institutions.

The reconfiguration of global summit governance of climate change reached a new stage with two special summits and a round table, all hosted under the UN umbrella. These events kept climate change on the global governance agenda in the face of an unprecedented global health scare. One — the Climate Ambition Summit — performed as well as the 2017 One Planet Summit, the first ad hoc climate summit that had launched the reconfiguration of summit governance. But the other two produced more limited results. Eyes now turned to the year ahead.

8 Combined Leadership, 2021

Introduction

As 2021 began, there was hope that the global summit governance of climate change would revive in an ambitious way, with the expanding special climate summits reinforcing the restored, reinvigorated G7, G20 and United Nations ones. The year started with two special virtual summits, the substantially performing One Planet Summit on Biodiversity from Paris on January 11 and the Climate Adaptation Summit from the Netherlands with its small performance on January 25. They were followed by US president Joe Biden's strongly performing Leaders Summit on Climate from Washington on April 22–23. The United Kingdom's Boris Johnson hosted a special virtual G7 summit on February 19 and then produced a strong performance on climate change at the G7 Cornwall Summit on June 11–13. Italy's Mario Draghi skilfully prepared the G20's Rome Summit he would host on October 30–31, with "People, Planet and Prosperity" as its theme. Johnson and Draghi would co-chair the Glasgow Summit of the 26th Conference of the Parties (COP) to the United Nations Framework Convention on Climate Change (UNFCCC) on November 1–12. The G7, G20 and UN were leading together, robustly reinforced by an expanded array of successful special climate summits, as never before.

This unprecedented progress was propelled by several strong forces. Soaring climate shocks took centre stage, as the rollout of safe, effective vaccines reduced the COVID-19 crowd-out of climate governance. The major multilateral organizations still struggled to produce an adequate response by themselves. A growing China and reviving United States increasingly realized they must cooperate abroad and converge at home to confront the shared, existential threat of climate change. The inauguration of a climate-committed Biden as US president on January 20 provided new power and leadership for climate action. Two climate-committed G7 leaders hosted the G7 and G20 and co-hosted the COP for the first time. Above all, the return of in-person summitry and the addition of several special climate summits helped expand both the G7 and G20 as valued clubs at the hub of a network of global summit governance to an unprecedented extent.

To be sure, these summits faced several constraints: new COVID-19 waves, rising oil prices that emboldened the world's fossil fuel economy and powers,

DOI: 10.4324/9780429055485-8

Chinese–US geopolitical competition, Biden's narrow legislative control and the departure of a politically powerful Angela Merkel as German chancellor in early October. Moreover, this strong surge of global summit governance in the first half of 2021 was still nowhere near the level needed to solve the soaring climate and ecological crises now at hand. It remained to be seen if the deadly extreme weather events fuelled by climate change and the ominous scientific evidence escalating into the autumn could force the combined leadership from the G7, G20, UN and special climate summits to produce a genuine physical and human success.

Starting Special Summits

The year opened for the first time with two special climate summits that promised to push global climate governance in 2021 to new heights.

One Planet Summit for Biodiversity, January

The One Planet Summit for Biodiversity from Paris on January 11 produced a solid performance. Attended by 11 country leaders, including host France, Germany, Italy, the United Kingdom, Canada and the European Union, and the UN secretary general, it sought for the first time to preserve the climate and natural ecosystems in an integrated way. It made important advances in its four priorities of protecting terrestrial and marine ecosystems, promoting agro-ecology, mobilizing funding for biodiversity and protecting forests, species and human health. It raised new climate finance for several initiatives and launched the High Ambition Coalition for Nature and People among 52 states and the Task Force on Nature-related Financial Disclosure.

Climate Adaptation Summit, January

The Climate Adaptation Summit hosted by the Netherlands on January 25 produced a small performance. Here, the United Kingdom launched the Adaptation Action Coalition (UK, Foreign, Commonwealth and Development Office 2021). The coalition would mould international pledges to the UN Call for Action on Adaptation and Resilience into practical support for the most vulnerable communities. The United Kingdom also promised to host a climate and development ministerial meeting in March and push action at the spring meetings of the International Monetary Fund (IMF) and the World Bank. Johnson said he would make adaptation and resiliency focal points of the United Kingdom's G7 2021 presidency, setting the tone for COP26 in Glasgow.

G7 Virtual Summit, February

Significance

On February 19, the G7 held its first full-strength inter-sessional summit. It was the leaders' first meeting since April 2020, Johnson's first as host and Biden's first major multilateral meeting — scheduled for just after his inauguration — as well as Draghi's first. The United Kingdom also announced it would chair a meeting of the UN Security Council (UNSC) to address climate change on February 23.

Debate

In the lead-up to the February summit, the first school saw an opportunity for the G7 to work with the G20 and UN to advance climate action (Forsyth 2021).

The second school predicted a strong success: Kirton (2021a) declared Johnson was "off to a swift, strong start to bring G7 summitry back … [as] an integral part of a well-prepared plan for global summit governance as a whole," in which the G7, G20 and UN work together to solve the challenges facing the world.

At the summit's end, the third school saw significant success in bridging the transatlantic divisions left by former US president Donald Trump and in reaching out to the G20 (Jarvis 2021).

The fourth school saw successful US re-engagement in multilateral cooperation, through the G7 and the UN at COP26, due to Biden's desire to swiftly undo the deep damage done by Trump (Dodwell 2021).

The fifth school saw a solid start on climate change. Warren (2021c) said the leaders "laid a solid foundation" for their Cornwall Summit, "but, as usual, they need to do much more."

Argument

The February summit produced a substantial performance on climate change. It made new commitments to control climate change and boost COP26. This performance was driven by recent climate shocks affecting several G7 members, including the United States. Several multilateral environmental organizations still failed to control such shocks, but a supporting pull came from the subsequent G7, G20 and UN summits. The new climate-committed leaders from the United States, Italy and Japan had strong popular support. Above all, with no regular summit in 2020, all G7 leaders valued their club, with Johnson seeking to bring out the best of the G7's global leadership as Biden's America adjusted to its partners on climate change.

Plans and Preparations

Johnson's preparations publicly began in his video address to the UN General Assembly (UNGA) on September 26, 2020. He promised "to use the UK's G7 presidency next year to implement a five-point plan to prevent future pandemics and global health crises" and address climate change (Gross and Pickard 2020).

When G7 finance ministers and central bank governors met virtually on February 12, British chancellor of the exchequer Rishi Sunak said he would prioritize "nature considerations … paving the way to a truly green global economic recovery" (UK, Her Majesty's Treasury 2021). He urged his colleagues to match the UK goals before COP26 and transition smoothly to net zero. Janet Yellen, US treasury secretary, said the United States understood "the crucial role that the United States must play in the global climate effort" and promised "engagement on this issue to change dramatically relative to the last four years" (US Department of the Treasury 2021b).

Promises to revive climate action came when Biden and India's Narendra Modi spoke on February 8 (White House 2021b). That day, Johnson and Merkel also agreed to cooperate through the G7 on climate change (UK, PMO 2021). Speaking to Japan's Yoshihide Suga on February 16, Johnson "welcomed Japan's commitment to reach net zero carbon emissions by 2050 and looked forward to an ambitious 2030 emissions reduction target" by COP26 (UK, PMO 2021a).

At the Summit

Shortly after 2pm in London, Johnson opened the Virtual Summit, waving to his colleagues and saying it was "great to see all of you" (Giordana 2021). After addressing the global economy and COVID-19, he called climate change "the other great natural challenge about which we've been warned time and time again" and said, "the warnings have been even clearer than they were for Covid" (Brown 2021).

Results

Domestic Political Management

Attendance at the Virtual Summit was complete. The *Financial Times* headlined the US formal return to the Paris Agreement before COP26 (Manson et al. 2021).

Deliberation

G7 leaders met virtually for about two hours. The Joint Statement of G7 Leaders' 711 words devoted one paragraph to the natural environment.

Direction Setting

The conclusions on climate change made one affirmation of open democracy.

Decisions

Leaders made three commitments on climate change. One linked climate change to biodiversity and another to energy and to employment — reflecting Biden's central message that climate change control brought good jobs. The climate change and biodiversity commitments on net zero goals for 2050, however, looked to the distant future.

Delivery

The three commitments were constructed in several ways likely to increase compliance, as all were highly binding and two were synergistic. Compliance would also benefit from the G7 environment ministers' meeting on May 20–21 and the Leaders Summit on Climate and COP26.

Development of Global Governance

The statement referred once to the UNFCCC and to UN Biodiversity, for 11% of all references to institutions outside the G7.

Causes of Performance

Shock-Activated Vulnerability

This substantial performance was propelled primarily by the recent extreme weather events affecting several G7 members, especially the United States, just before and during the summit itself, although the statement recognized no climate shocks or vulnerabilities. Climate change appeared on the front page of the *Financial Times* on February 17, and then expanded the next day as a severe cold wave and winter weather paralyzed the United States and prevented the distribution of COVID-19 vaccines. Climate-related physical shocks in Europe had started on January 10, when Spain's worst snowstorm since the 1980s killed four.

On January 8, the UK Met Office forecast that in spring 2021, carbon dioxide concentrations would be 50% higher than before the Industrial Revolution (Abnett and Green 2021). The World Economic Forum's 16th Global Risks Report, released in January 2021, reported the highest likely risks of the next 10 years included "extreme weather, climate action failure and human-led environmental damage" and the highest impact risks were "infectious diseases … in the top spot, followed by climate action failure and other environmental risks" (World Economic Forum 2021, 7).

Multilateral Organizational Failure

Multilateral organizational failure was high with the 2020 postponement of COP26 and the inability of the multilateral environmental organizations to meet at the summit or ministerial level to help control these new shocks. But a supporting pull came from the G20 Rome Summit in October and the rescheduled COP26 in November.

Predominant Equalizing Capabilities

The G7's increasing internally equalizing capabilities also pushed performance. At the end of 2020 and the very start of 2021, the US dollar declined almost 7% against other major currencies (Alabi 2021). By January 7, driven by the shock of the violent attack on the US Congress, the value of the US dollar against a group of its peers had fallen from a peak of $1.03 in March 2020 to $0.98, its weakest since April 2018, while the euro rose to multiyear highs.

Converging Characteristics

G7 members' high political convergence on democratic values and climate performance rose, with Biden replacing Trump. Outside the G7, global democracy declined, strengthening this core common bond within the G7.

Domestic Political Cohesion

Domestic political cohesion was high. Johnson had a majority Conservative government, which faced re-election only in 2024. He had now attended two G7 and two G20 summits and had recently been foreign minister. He had a political commitment to and popular support for his climate change priorities, although his personal and party popularity waned as COVID-19 cases and deaths soared. An opinion poll reported 58% support for making international climate agreements legally enforceable, including 47% of Conservative, 73% of Labour and 68% of Liberal Democratic voters (Stone 2021).

In the United States, Biden was starting his four-year term with a convincing electoral mandate and his Democrats narrowly controlling both legislative chambers. He had strong and rising approval ratings and extensive international experience as a senator and as Barack Obama's vice president for eight years. He was fully committed to climate action. Yet, he had no direct G7 summit experience and had to cope with the polarizing aftermath of the coup attempt on the US Congress on January 6.

In Japan, Suga's coalition controlled both legislative houses, but faced possible elections soon. He had no in-person G7 summit experience, and no personal political mandate.

In Germany, Merkel, a former physical scientist and environment minister, was the longest serving veteran at G7, G20 and UN summits. She remained personally and politically committed to climate priorities. But her personal power as chancellor would end after Germany's federal election on September 26. As 2021 began, her coalition government was challenged by the second-placed Green Party in the polls. Merkel herself remained highly popular too.

In France, Emmanuel Macron had substantial summit experience and complete executive and legislative control. His approval rate had risen to 41% from a low of 23% during the yellow vest protests in 2019 (Ifop 2021b). He remained committed to climate change control.

In Italy, Draghi was sworn in as prime minister on February 13, with support from all major parties in parliament. He had no G7 summit experience in this capacity, but brought great experience and respect as head of the European Central Bank and the Bank of Italy, plus expertise in economics. He immediately appointed a cabinet with Roberto Cingolani, a physicist and IT expert, as minister for ecological transition, a new portfolio combining the old environment and energy ones, and to lead an interministerial committee on the green transition (Balmer and Jones 2021). This new ministry was critical to secure parliamentary support from the far right but ecologically committed Five Star Movement and consistent with the need for stronger climate polices to secure funds from the European Union's major recovery plans, with Italy the largest designated recipient and where 37% of the funds must go toward transitioning to a low-carbon economy (Jewkes 2021). At his first cabinet meeting on February 13, Draghi declared "ours will be an ecological government" (Jewkes 2021).

In Canada, Justin Trudeau had attended five G7 summits including one he hosted. His Liberals now had a minority government following its re-election in 2019. By early February, his party's political support had shrunk to tie with the Conservatives. A poll conducted between February 11 and 19 showed Canadians chose as the top issues in Canada COVID-19 at 47%, health care 36%, climate change 29% and the economy 27% (Angus Reid Institute 2021).

The European Union's Charles Michel and Ursula von der Leyen had no in-person G7 summit experience, as their terms started in December 2019. In December 2020, EU leaders agreed to cut emissions by at least 55% from 1990 levels by 2030, upgrading their previous 40% target.

Across all G7 and G20 members, public support for climate change control was strong. A poll by the United Nations Development Programme found that 64% of 1.2 million people surveyed in 50 countries from October to December 2020 thought climate change was a "global emergency" (Carrington 2021). Agreement was led by the 2021 summit hosts of the United Kingdom and Italy at 81% each, with Australia at 72%, the United States and Russia at 65% each and India at 59%. The strongest support came from small island developing states (74%), followed by high income countries (72%), middle income countries (62%) and least developed countries (58%).

Club at the Hub

Above all, leaders at the first G7 summit in 10 months valued their club at the hub of a network of global summit governance that would expand throughout the year. At the core were climate-committed leaders from the United Kingdom, the United States and Italy.

Leaders Summit on Climate, April

Introduction

A major advance came from Biden's Leaders Summit on Climate on April 22–23. It followed Biden's initiative on his inauguration day, to have the United States rejoin the Paris Agreement and to chair the first Quadrilateral Summit of the Indo-Pacific democratic powers of Japan, India and Australia, which made three commitments on climate change (Kirton 2021b). These moves were backed by his appointment of former secretary of state John Kerry as his special climate envoy.

Debate

In the debate about the Leaders Summit, the first school saw a welcome start that built on the February 19 and Quadrilateral summits to provide momentum for the G7's Cornwall Summit and the G20's Rome Summit (Kirton 2021b). Biden sought immediate action in a forum more inclusive than the G20 but more manageable than the universal UN, which had so frustrated Obama at Copenhagen in 2009 (2020, 503–516).

The second school saw US-centred summit theatre. Rachel Kyte thought it could be "extremely impactful if there is a big centerpiece … the U.S. plan" (Friedman 2021).

The third school saw a cooperative China creating summit success by issuing a stronger emissions cutback commitment (White and Hook 2021).

The fourth school saw at least an important climate finance opportunity (Sharma 2021e).

The fifth school saw potential failure, due to countries' lack of trust in Biden's ability to deliver his own improved cutback plans (Friedman 2021).

Argument

In the end, the Leaders Summit on Climate performed strongly. All G7 members there improved their targets and timetables with pledges to cut back emissions by about 50% by 2030. New commitments came from the majority of G20 members, including the leading polluters of China, Brazil, South Africa, Korea and Indonesia. Still, several key emitters offered nothing new and little climate finance came. The Climate Action Tracker (2021) estimated that if all the promises were implemented in full, along with those made since September 2020, by 2100 global heating would decline to only 2.4°C.

Propelling this strong performance was the inclusion of all G20 leaders, and strong support for Biden and climate action among the US public, for Germany's rising Green Party and from the United Kingdom and Italy as G7, G20 and COP26 hosts. The vast foreign policy experience of Biden, Kerry and their team helped make many key players credibly promise enough to make the summit succeed.

Plans and Preparations

On March 26, Biden announced the virtual Leaders Summit, which would be streamed live on April 22–23 (White House 2021a). It would "underscore the urgency — and the economic benefits — of stronger climate action" and be "a key milestone" on the road to Glasgow, seeking to "catalyze efforts" to keep within reach the goal of limiting warming to under 1.5°C.

Biden promised a new, ambitious US 2030 emissions target and urged his 40 invited leaders to do so too. Invited leaders included the 17 members of the old US-led Major Economies Forum on Energy and Climate (MEF) (Happaerts 2015; Kirton and Kokotsis 2015). Biden added leaders of countries showing strong climate leadership and those most vulnerable, as well as business and civil society (White House 2021a).

The summit's efforts to spur participants to improve their very inadequate nationally determined contributions (NDCs) well before COP26 were especially important as the UN preparatory process had stalled due to the absence of in-person meetings.

To prepare the summit, Kerry toured Europe and Asia. On April 17, the most important signal came: the US–China Joint Statement Addressing the Climate Crisis, which said the world's two leading economic powers and climate polluters agreed on the critical fact that climate change constituted a crisis and promised "concrete actions in the 2020s" (US Department of State 2021). Canada's budget on April 19 raised its emissions reduction target to 36% below 2005 levels by 2030 and added major money for domestic climate change control (Radwanski 2021). Two days later, Canada promised to announce an ambitious new target at the summit (Graney et al. 2021).

On April 20, Johnson announced that by June he would legislate "cutting emissions by 78% by 2035 compared to 1990 levels," improving on the United Kingdom's previous pledge to cut them to 68% by 2030 (UK, Department for Business, Energy and Strategy 2021). By including aviation and shipping emissions, the United Kingdom would be "more than three quarters of the way to net zero by 2050." Set to speak in the summit's opening session, Johnson would urge others to follow the United Kingdom.

Biden planned to announce cutbacks up to 50% by 2030, or about double the 26%–28% cuts the Obama administration had pledged for the Paris Agreement (Eilperin and Dennis 2021).

On April 21, the European Union agreed to legislate its 55% cut below 1990 levels by 2030. Japan promised a substantial improvement to its 2020 NDCs (Harvey 2021a). Brazil promised to end illegal deforestation in the Amazon rainforest by 2023 (Harris 2021). India, Russia, Korea, Indonesia, South Africa, Saudi Arabia and Australia remained reluctant, while Mexico had recently reduced its promised NDC (Harvey 2021a).

At the Summit

During the summit's first session, several G20 leaders made important new commitments, led by the United States, Japan and Canada.

Biden declared the United States would reduce its emissions by 50%–52% by 2030. Suga promised to cut 46% by 2030. Trudeau promised 40%–45% cuts by 2030. With the United Kingdom and the European Union's promised cuts by then, all G7 members except Italy thus identified 50% cuts by 2030 as their collective target and timetable.

Xi Jinping said China would "strictly control coal-fired power generation projects, strictly limit the increase in coal consumption over the 14th Five-Year Plan period," phase down coal use from 2025 for five years and have its emissions peak before 2030 (Hook and Hodgson 2021; Xi 2021).

Jair Bolsonaro said Brazil would reduce its emissions by up to 40% by 2030, and reach net zero by 2050, 10 years earlier than previously pledged. He also promised to end illegal deforestation by 2030 and to reduce greenhouse gas emissions from deforestation by 50%.

Cyril Ramaphosa promised that South Africa's emissions would decline by 2035, 10 years earlier than previously announced.

Korea's Moon Jae-in promised to end financing for overseas coal plants and quickly phase out 10 of its domestic ones. Mexico's Andrés Manuel López Obrador promised more reforestation.

Performance

The Leaders Summit produced a strong performance.

Domestic Political Management

Domestic political management was strong. All G20 leaders attended and spoke at the opening session. They were joined by 19 other leaders from a well-balanced array of countries from all regions and levels of development, with democratic polities dominating. The heads of the IMF, the World Bank, the UN and the African Development Bank also spoke.

Biden stayed for much of the summit's opening session, returned several times and closed the conference.

Most G20 leaders complimented Biden for calling the summit and for rejoining the Paris Agreement. A few noted his historic cutback commitment. Some also referred in a positive tone to other leaders such as Xi.

The April 23 edition of the *Financial Times* (2021d) headline read "Earth Day Climate Goals Stepped Up." It reported the new commitments of the United States, Japan, Canada, Korea and China, but it included Greta Thunberg's judgement that "the new targets were 'very insufficient' and full of loopholes."

Deliberation

Public deliberation was strong. No collective statement was issued, but each speech was live-streamed and most were released in textual form. Those from the 21 G20 leaders totalled 13,879 words. They were led by Biden (who spoke twice) with 13.2% of the total, followed by Xi with 10.8% and Bolsonaro with 6.2%. G7 members led with 45% of the total words, followed by the BRICS (Brazil, Russia, India, China and South Africa) members with 30%.

Private deliberation was slender, as the virtual format prevented the spontaneous and casual contact of in-person summits. Covering 12 time zones made personal interaction even more difficult, as interruptions in Macron's pre-recorded address showed.

Yet, some informality arose. Most came from Johnson, who addressed Biden as "Joe" and used colloquial language, with "bunny hugging" his most memorable phrase.

Direction Setting

There was little affirmation of key principles. Leaders did affirm the urgency and existential nature of the climate crisis, that controlling it was necessary for economic growth and human health, that they were each responsible for doing so and that they must take immediate action to meet the goals of the Paris Agreement. Yet, some still emphasized the principle of common but differentiated responsibilities (CDR), and their limited capabilities to act.

Decision Making

Decision making was strong, as G20 leaders made 125 individual climate commitments, including 26 new ones and 99 that reaffirmed existing ones. G7 members made 57 commitments, including 11 new ones. BRICS members made 29, including 5 new ones. Most G20 leaders offered at least one new commitment, led by China, the United States and Japan with five. Every G20 member made at least two new or old ones.

Many of the major polluters announced improved NDCs, with a medium-term deadline of 2030. As Macron (2021) said, "2030 is the new 2050."

Several leaders committed to curb coal financing, production, transportation or use, including Korea and China. Many promised nature-based solutions, including planting trees, stopping the loss of rainforests by Brazil and expanding marine-protected areas, although some described moves already made. There were few clear promises on climate finance, although on the sidelines the United States made a multibillion-dollar pledge.

Delivery

The delivery of these decisions remained in doubt. Kerry said the private sector would make the United States comply. There were no "G40" ministerial and official-level institutions or another Leaders Summit to follow up.

US secretary of state Antony Blinken (2021) said the United States would give a strong message in May to the G7 and at the Arctic Council ministerial meeting, with Russia, Canada and some EU members there.

Development of Global Governance

The development of global governance was small. No new institutions were created. Leaders referred often to COP26 and the Paris Agreement, but infrequently to other bodies.

The IMF, the World Bank and the UN did not commit to making climate action their central mission or to using their full array of instruments to meet participants' annual targets and Paris goals. The World Bank had only begun to take small, selective steps toward this (Rappeport 2021).

These established multilateral organizations were primarily the finance and economic ones from 1944, and excluded newer environmental ones, such as the World Meteorological Organization (WMO), the UN Environment Programme, UN Climate Change, UN Biodiversity and the World Organisation for Animal Health.

Causes of Performance

Several forces propelled the Leaders Summit's strong performance.

Shock-Activated Vulnerability

Shock-activated vulnerability was high. Several leaders recognized climate change as an "existential threat," and many referred to recent and rising extreme weather events.

During the first three months of 2021, the front pages of the G7's elite daily newspapers remained dominated by the new COVID-19 shocks in health and the economy. Yet, climate change appeared continually on the *Financial Times* front page during the Leaders Summit.

Physical climate shocks were led by fresh memories of the deadly deep freeze in February that did so much damage in Texas and other US states.

Recent scientific reports offered stark results. On March 2, the International Energy Agency ([IEA] 2021a) reported that world energy-related emissions were higher in December 2020 than in the same month in 2019. They surged during the second half of 2020 to rise 2% year on year in December. China was the only country to increase its annual emissions in 2020 from the year before. US emissions in December 2020 remained the same as in December 2019.

On April 19, the WMO's (2021a) State of the Global Climate Report confirmed that 2020 was one of the three hottest years ever. In 2020, greenhouse gas concentrations rose to a new high of 410 parts per million (ppm) and were on track to hit 414 ppm in 2021. There were new highs in ocean heat in 2019 and sea level rise in 2020. The report's summary of extreme weather events in 2020 identified those in the United States, China, India, Russia, Japan, Brazil, Argentina, South Africa, Korea, Australia, France, the United Kingdom and Canada.

On April 20, the IEA (2021b) said the world was heading for the second largest ever increase in energy-related carbon emissions in 2021, due to renewed coal use in Asia, especially in China. This was the highest increase since 2010, reversing the drop in 2020. Coal demand in 2021 would rise 4.5% to approach its 2014 peak, as it increased in China, the United States and the European Union.

On April 20, UN secretary general António Guterres urged no new coal plants be built and coal be completely phased out everywhere by 2040 (Hodgson 2021a).

That day, Petteri Taalas, WMO secretary general, gave a 20% chance that the world would warm temporarily by 1.5°C in the next five years (Hodgson 2021a).

Multilateral Organizational Failure

Multilateral organizational failure remained high. Preparations for COP26 were still delayed, as some members refused to engage in the virtual format. The UN reported that if all members fully implemented their existing NDCs, it would produce only 1% of the progress needed to reach the Paris Agreement goal.

Predominant Equalizing Capabilities

Summit participants produced more than 80% of the world's gross domestic product (GDP) and emissions. Both China and the United States led in the levels and increases in the first quarter of 2021. The IMF predicted growth in advanced economies would increase from 2019 to 2020 by just 1% less per capita than in January 2020, and emerging and developing countries would drop by 4.3% (Wolf 2021a). But short-term shifts in exchange rates increased the internal economic equalization among key participants. On April 19, the US dollar declined to a six-week low (Lockett and Parkin 2021).

In the specialized capabilities for controlling climate change, the summit's global predominance and internal equality were very high. China led in producing and exporting solar panels, wind turbines, electric vehicles and batteries, and held almost one third of the world's renewable energy patents (Manson and Hook 2021). The United Kingdom and Europe led in offshore wind power, where the United States lagged far behind.

Converging Characteristics

Upwardly converging democratic characteristics and climate performance were high. Among G20 members, democracies dominated at 85%. Retreats in China over Hong Kong and in Russia over Alexei Navalny were offset in the United States with Trump's departure as president and his followers' violent assault on the US Congress on January 6 fading from view.

Domestic Political Cohesion

Domestic political cohesion in key participants was high. In the United States, Biden had passed his first major initiative through Congress and his approval ratings remained strong.

In Germany, on March 14, Merkel's Christian Democratic Union did poorly in regional elections, as the Greens increased (Barber 2021a). A poll released on April 20 showed the Green Party in first place, 7% ahead of Merkel's centre-right coalition with the Christian Social Union (*Financial Times* 2021e). In France, an Ifop (2021a) poll published on April 19 said Macron's popularity, down two points at 34% and the lowest since he took office, was higher than his predecessors at the same stage in their mandate.

In Canada, a Nanos (2021) poll released on April 20 showed that Trudeau's Liberals, with a 7% lead over the Conservatives, would probably win a majority if an election were held.

China's Xi, with over half a decade of G20 and BRICS summit experience, would be in power for many more years, and had complete control of his government and much of China's economy as well.

Globally, public support for national governments tackling climate change was high (*Financial Times* 2021c). An Ipsos (2021) poll on April 22 reported that an average 65% of global respondents agreed their government would be failing them if it did not act now on climate change. China, Russia and Saudi Arabia had the lowest percentage of respondents who agreed.

Club at the Hub

The Leaders Summit was the first gathering of this large group of 40 countries. Both Biden and Xi valued such plurilateral summitry. The MEF-centred formula had proven its worth (Happaerts 2015; Kirton and Kokotsis 2015, 13–14; Kirton 2021c). The Leaders Summit was the first in which both Biden and Xi participated. It extended the network of global summit governance radiating outward, by doubling the number of participating countries from the G20's 20 plus guests to 40 equal participants.

G7 Cornwall, June

Introduction

The impact of the Leaders Summit depended in part on how the G7's Cornwall Summit seven weeks later followed up (Patrick 2020; Nardelli 2021; Parker 2021). It took place at Carbis Bay, a resort on the Atlantic coast of southwest England, as a three-day, in-person event on June 11–13 (Holder and Santora 2021). G7 leaders were joined by guests from four major democratic G20 members: India's Modi, Australia's Scott Morrison, Korea's Moon Jae-in and South Africa's Cyril Ramaphosa.

Debate

In the debate on Cornwall's prospects and performance, the first school saw strengthened democratic preparedness for climate action, due to Biden replacing Trump, the pull of the UN climate summit in Glasgow and the move toward a "Democratic 10" (D10) with Australia, India and Korea invited (Forsyth 2020; Patrick 2020).

The second school saw favourable conditions for success, due to Johnson's hosting, the supporting UN summit and the United Kingdom's partnership with the European Union and other like-minded countries to combat climate change (Cavendish 2021).

The third school saw a potential, broad, British-led success, due to Johnson and effective vaccination in the United Kingdom (2020). The *Economist* (2021b) cited as causes the United Kingdom's membership in the North Atlantic Treaty Organization (NATO), the G7, the G20, the Commonwealth and the UNSC. The *Economist*'s Briefing (2021a) added the United Kingdom's assets including its fifth-ranked economy, extensive overseas territories, its queen as head of the 54-member Commonwealth and Johnson's experience as a foreign minister.

The fourth school saw a US-led success on climate change, with Biden back (Walker 2020). Yet, the *Financial Times* (2021b) saw a domestic democratic distraction, following the January 6 coup attempt on Congress.

The fifth school saw a tough task and test for Johnson. Parker (2021) noted that Johnson ambitiously sought to produce a G7 stepping stone to Glasgow by investing in green technology and lead the thrust toward global net zero, but was constrained by the United Kingdom's withdrawal from the European Union and his strained relationship with Biden.

The sixth school foresaw failure, as the G7 lacked the power to shape the world and its D10 extension would bring more divisions than reinforcing capability, even with Biden there (Islam 2021).

After the summit, the seventh school saw failure too. Sachs (2021) concluded that "G7 leaders rightly embraced the goal of global decarbonization by 2050, and called on developing countries to do so," but, instead of proposing a plan, "reiterated a financial pledge first made in 2009 and never fulfilled" — possibly because G7 countries' share of global GDP had dropped to 31% from 51% in 1980.

The eighth school saw partial success, with Cornwall's climate commitments offset by failure to do anything new on climate finance, to recognize the interdependence of climate, biodiversity, pollution and sustainable development and to add "practical actions, financial commitments or follow-through mechanism" to ensure compliance with its commitments (Mohieldin 2021). Mirza (2021) saw progress on ending coal finance and coal itself, but not on climate finance, due to Biden's failure to secure his partners' full support. Casci (2021) added the diversionary row over Northern Ireland. Norton (2021) welcomed the "build back better world" initiative but noted the need to cooperate with China, invest in nature-based solutions such as mangroves and coral reefs, address inequality and respect Indigenous peoples' rights.

The ninth school saw more comprehensive success. Seetharaman (2021) highlighted Cornwall's support for conserving and protecting 30% or more of land and sea by 2030. Luce (2021) noted Biden promoting his climate change priorities in an America-alongside-its-partners approach. Wolf (2021c) concluded "the alliance of democracies is back, due to democratic decline," the power shift from the United States to China, and the need to protect "humanity against global threats, such as pandemics and environmental disasters." Mann (2021) judged that Cornwall produced "some real climate wins but was... still dwarfed by the scale of the problem." Its success was due to Biden replacing Trump and the rising climate shocks. Mann concluded the summit was a small success but that big failures could still come.

Argument

The Cornwall Summit produced a strong performance on climate change. It created the most comprehensive, ambitious and innovative performance of any of the previous 46 annual G7 summits, led by its historic focus on ending reliance on coal and on enhancing nature as the key carbon sink. Yet, it did not meet the extreme, urgent global demand to control the existing climate emergency.

Cornwall locked in the commitment of the G7 members' promises at the Leaders Summit on Climate to cut their emissions by 50% by 2030, and backed them with several specific plans and actions, led by ending unabated international coal financing and use by the end of 2021, rapidly reducing international financing for oil and gas for energy, restoring nature as a carbon sink, ending inefficient fossil fuel subsidies in the near term and promising to provide over $100 billion annually for developing countries by 2025. However, several of these came with qualifications.

This strong performance was propelled by several new, severe, climate-related extreme weather shocks from drought and heat, signalling unprecedented vulnerabilities in the United States, Canada and Europe. The UN still failed to get its members to meet their Paris targets to prepare a successful Glasgow Summit. The G7's overall relative capabilities grew, and its internal equality increased. Its key specialized climate strengths remained, above all in finance and technology, even as its share of global emissions declined to only 20%. Biden's arrival as the most powerful member restored the G7's political convergence on core democratic characteristics and climate policy, and US domestic political support for bold G7 climate action. Above all, the G7's status as a personally valued club at the hub of an expanding network of global summit governance was strengthened by its return to in-person summitry, the emergence of a G7-centred "D11" (now with South Africa), G7 members' hosting the subsequent G20 and UN summits and the many special summits that arose on the road to and from Cornwall.

Plans and Preparations

In announcing his priorities for his 2021 G7 summit at UNGA in September 2020, Johnson gave climate change a prominent place. The summit website launched in early 2021 listed "tackling climate change and preserving biodiversity" as the third priority, after COVID-19 and trade. More detail included a "move to zero emissions," "financial support for developing countries to do so," support for a "green recovery" and "sustainable development financing." There was no reference to the Paris Agreement, suggesting the United Kingdom's G7 would not be confined by the UN regime.

Ministerial Meetings

Momentum came from ministers at the March 2 meeting of the Powering Past Coal Alliance, founded by Canada and the United Kingdom in 2017 (O'Malley 2021). Here, Guterres asked the Organisation for Economic Co-operation and Development members to cancel all coal projects by 2030 and other countries to do so by 2040, and for private and central banks to stop financing coal plants. He called on G7 members to announce this coal phase-out at the Cornwall Summit.

G7 foreign and development ministers, in London on May 4–5, made nine climate change commitments (Kirton 2021c). This confirmed that climate change was seen as a security issue, as Johnson had argued before the UNSC several weeks before. Eight climate commitments focused on finance. The nine noted nature-based solutions, agriculture, water, the Paris Agreement, gender equality, Indigenous peoples, the private sector and COP26.

Major momentum came from the meeting of G7 climate and environment ministers on May 20–21 (Warren and Kirton 2021). Hosted by Alok Sharma, the minister responsible for COP26, and environment minister George Eustace, it showed that the G7 and UN were offering combined leadership. Its unusually long and broad communiqué contained 183 commitments, more than double any G7 environment ministerial before. Two thirds were highly binding. Commitments covered many subjects, giving robust attention to nature-based solutions, including 10 on forests and the first ever on peatlands. Ministers boldly promised to use a 1.5°C rise as their goal, end direct financing of thermal coal power plants in developing countries in 2021 and protect 30% of nature by 2030. They made three affirmations of the rights of Indigenous peoples, local communities, women and marginalized groups.

They recognized the existential threat to nature and two vulnerabilities — the threats of deforestation to the climate, biodiversity, food security and livelihoods, and the "significant" threat of human action on the ocean and the seas.

On the Eve

As the summit approached, leaders were poised to accept much of what their ministers had agreed.

They would also encourage China to make much stronger climate commitments, do more to reduce its emissions and make the outcomes of the UN Biodiversity conference it would host in October support serious climate action. Macron, expressing relief that a new US president was in power, planned to prioritize the climate crisis along with the pandemic, vaccine distribution and global inequality (Rose and Holland 2021). Von der Leyen stressed that the European Union would argue for a green and resilient transition and recovery from the pandemic (European Commission 2021).

At the Summit

At the summit, climate change was discussed on all three days. On June 11, in the first session, on "Building Back Better from COVID-19," Johnson's opening remarks put climate change in a priority place. At the session on health, Johnson said he wanted the G7 to be "building back better, building back greener" (Widakuswara 2021).

The next day, the economic session and the foreign policy session also addressed climate change. In the first 90-minute session, Biden presented his "Build Back Better World" (B3W) initiative to invest in clean, green infrastructure in developing countries. Draghi, with support from Merkel, Michel and von der Leyen, resisted presenting the B3W initiative as competing with China's Belt and Road Initiative, but Trudeau, Biden, Johnson and Macron sought to show the G7's positions and values through more action-oriented efforts and coordination.

The United Kingdom's similar efforts, focused more on climate change, received some attention. The United States judged that these fit within the larger B3W effort. Leaders agreed to enhance their collective infrastructure effort by identifying the details of the B3W initiative and including Johnson's initiative. His proposal to create a working group or task force to do this received a warm response. The infrastructure initiative thus appeared in the communiqué (*Taipei Times* 2021).

On June 13, the summit's final two-hour session, attended by the four guest leaders, focused on climate and biodiversity and included a speech by Sir David Attenborough. Most major advances were confirmed, led by the signature achievements on coal financing and the Nature Compact. Several G7 leaders pushed the United States and Japan to set a date to phase out coal at home, to no avail (Mann 2021). The United States and the United Kingdom unsuccessfully pressed Morrison to commit before the Glasgow Summit to net zero by 2050.

Johnson (2021a) listed climate change as the fourth of the summit's results at his concluding news conference, mentioning COP26 first and continuing: "G7 countries account for 20% of global emissions, and we were very clear this weekend that action has to start with us." He noted that Cornwall was the first net zero G7 summit, that all members had agreed to eliminate their emissions as fast as possible and that they would help developing countries at the same time. He noted that Australia, India, South Africa and Korea had joined the G7 in their resolve to tackle climate change. He would also push his G7 colleagues to support an infrastructure plan, in part to counter China's Belt and Road Initiative.

Merkel said she supported the new G7 infrastructure task force, and that the G7 was still struggling to agree on a joint date for ending the use of coal (Reuters 2021). Suga was successful in getting "green growth" under the G7's infrastructure initiative (Bartlett 2021).

After the summit, the host was proud of the summit's advances on coal finance, climate finance from all countries but Italy and increasing its quality, and land-based solutions. The summit was necessary to get the United States and Japan to agree to the "30 by 30" nature goals on land and sea.

Results

Among all 47 annual G7 summits, Cornwall stood first in the number of climate commitments made and first by far in the climate and nature ones combined. It stood second in the portion of conclusions, affirmation of democratic principles and references to international institutions both inside and outside the G7 made on climate change.

Dimensions of Performance

Domestic Political Management

Cornwall's domestic political management was substantial. Of the four communiqué compliments to G7 members, one went to the United Kingdom on climate change. But the *Financial Times* front page reported on stumbling preparations for the Glasgow Summit "after last week's G7 summit in Cornwall failed to produce specific plans for new climate funding" (Hook 2021b).

In parliament on June 16, Johnson (2021b) highlighted the summit's achievements on climate change in fifth place. He cited the G7 leaders' pledge to reach net zero by 2050, end support for coal power generation, improve climate finance by 2025 and protect 30% of members' land and seas. He added that the G7's B3W initiative would help developing countries "construct new and clean and green infrastructure in a way that is transparent and environmentally responsible."

As of July 13, polls showed that Johnson had 44% approval and 50% disapproval ratings (Morning Consult 2021). Biden fared better: 52% approval and 39% disapproval. Draghi had 65% approval and 28% disapproval; Merkel had 54% approval and 39% disapproval; but Macron had only 36% approval and 55% disapproval. Suga had the lowest approval rating of only 27%, with 61% disapproval. Trudeau had 48% approval and 45% disapproval.

Deliberation

Climate change took 3,612 (19%) words in the five Cornwall declarations (excluding the summary), the third highest number and second highest portion of any G7 summit, and the highest since 2009 (see Appendix C). They covered COP26, reaching net zero emissions, providing climate finance, protecting nature, promoting green growth and infrastructure, decarbonizing energy systems and transportation, phasing out coal, improving adaptation and carbon pricing.

Direction Setting

Cornwall's communiqués affirmed democracy three times in the context of climate change, the second highest ever (see Appendix C). Cornwall also gave climate change significant priority placement.

Decisions

Cornwall produced 54 commitments on climate change, as in 2008, and constituting 13% of its historic high of 429 commitments on all subjects (see Appendix C).

Climate commitments ranked third, after health with 89 (21%) and the environment with 55 (13%). Combining climate and the environment, nature came first by far, with far more commitments than at any G7 summit before.

In the comprehensive communiqué, the first four climate commitments came in the preamble. They promised a 1.5°C rise limit, net zero emissions no later than 2050, halving emissions over the two decades to 2030 and increasing and improving climate finance by 2025. The environment commitment promised "to conserve or protect at least 30 percent of our land and oceans by 2030." All were highly binding.

Of the 54 climate commitments, 75% were highly binding. Of the 55 environment or nature ones, 70% were highly binding.

Money Mobilized

Cornwall mobilized some new money for climate change. Canada, Germany, the United Kingdom and the United States promised $2 billion for the Climate Investment Funds to support developing countries to stop using unabated coal, a sum expected to mobilize up to $10 billion in co-financing (Edwardes-Evans 2021). The United Kingdom promised £500 million for its Blue Planet Fund. The United Kingdom offered £120 million and Germany €125 million for their new fund with the United States for climate protection for the most vulnerable communities in Africa, South East Asia, the Caribbean and the Pacific. Canada doubled its climate finance contribution to $5.63 billion from 2021 to 2025, its largest multiyear climate finance pledge ever. The infrastructure initiative, with climate change one of its four pillars, sought to raise trillions of dollars in public, international financial institutional, private sector and philanthropic finance. Nonetheless, these fell short of the annual $100 billion the G7 promised to help mobilize for over 10 years.

Delivery

Compliance with these commitments was likely to be high. Highly binding language is a good predictor of compliance with G7 climate change commitments. The highly productive performance of the environment ministerial meeting suggested that Cornwall would perform similarly on climate change in the number of commitments it made and in members' compliance (Warren 2021a).

Development of Global Governance

Cornwall's communiqués in a climate context made 2 references to institutions inside the G7 and 31 to those outside (see Appendix C). These were both the second highest ever.

Causes of Performance

Shock-Activated Vulnerability

The Cornwall Summit's strong performance was spurred primarily by the severe climate-related extreme weather shocks in G7 members, as the diversionary shock of COVID-19 became more manageable.

Cornwall's documents contained three references to climate shocks and six to related biodiversity shocks. This was the first time that G7 summit communiqués recognized climate shocks (see Appendix E). It was the second time since 1975 that they had recognized biodiversity shocks, tied at six with Charlevoix in 2018. Cornwall twice recognized climate change and its twin biodiversity shocks as "existential threats." There were no communiqué-recognized climate vulnerabilities.

Media-highlighted climate shocks and vulnerabilities were also high. Climate change stories appeared on the *Financial Times* front page on May 18 and 19, just before the environment ministers met. They reappeared on May 26 and 27, and again on June 3 and 4, and continuously from June 11 to June 16. In the Sunday *New York Times*, they appeared on May 23 and May 30, but none did before June 14.

In June, globally, media attention to climate change increased 26% compared to the previous month (Boykoff et al. 2021). It was also up 27% in the first six months of 2021 compared to 2020. There was a large spike in the United States, where print coverage of climate change increased 27% and television coverage increased 147% from May.

Physical shocks were very strong. In the two months before the summit, the US Midwest, California and Canada's prairies had extreme drought, heat, water shortages and wildfire risk. The Arctic regions had unusually high heat. Europe also experienced high heat, with wildfires erupting in Greece. The United Kingdom had one of its wettest Mays on record. By May 15, California had experienced severe drought everywhere, as April was the sixth warmest ever, and wildfires grew into June (Lowe 2021). By June, the severe drought and record-setting heat spread throughout the southwestern United States. On May 21, the year's first sub-tropical storm, Ana, formed near Bermuda, 10 days before the official start of hurricane season, which was forecast to produce between 6 and 10 named hurricanes in 2021.

In India, on May 18, an unusually intense and deadly typhoon struck Gujarat, Modi's political base, killing at least 33 people.

On May 18, the IEA declared that all new oil and gas exploration projects must stop in 2021, that spending on low-carbon technologies must rise from $2 trillion now to $5 trillion a year by 2030 and that fossil fuels must fall from 80% of energy supply today to 20% by 2050 (IEA 2021c; Hook and Raval 2021). On May 27, the WMO (2021b) reported there was a 40% chance of the annual average global temperature reaching 1.5°C above pre-industrial levels in one or more of the coming five years and a 90% chance it would be the hottest on record.

Multilateral Organizational Failure

The COP26 preparatory process stalled as COVID-19 continued to prevent in-person meetings and some members resisted the virtual format.

On February 26, UN Climate Change (2021) reported that only 75 of its members, representing only 30% of global emissions and 40% of the signatories to the Paris Agreement, had submitted their promised updated, improved NDCs by the end of 2020. The United Kingdom and the European Union had submitted improved commitments (Hodgson 2021b). But even if those 75 countries implemented their promises in full, emissions would fall by only 0.5% by 2030, rather than by the required 45%. Japan, Russia, Australia and Korea had submitted plans with no improvements. Brazil's lacked any goals for 2030. The world's biggest emitters of China, the United States and India did not submit any plans.

Predominant Equalizing Capability

Relative capability changes helped push Cornwall's strong success. The World Bank's base scenario had seen rich world growth in 2021 averaging 3.3%, with China's at 7.9%, other emerging and developing economies at 3.4% and global growth at 4% (Strauss and Wheatley 2021). However, as the spring advanced, the actual and projected GDP growth of the United States and the G7 surpassed that of China and the BRICS. In April, the value of the US dollar resumed its slide through to the end of May.

In the most climate-relevant specialized capabilities, the G7 had global predominance in several, starting with finance. The United States ranked high in producing electric vehicles and charging stations, with European producers joining quickly.

Converging Characteristics

Political convergence among G7 members at a higher level of democracy and environmental policy also grew. In the United States, Biden issued climate-supportive executive orders and bargained on a major green infrastructure bill with Congress. A ruling by Germany's constitutional court raised the deadline for its net zero pledge from 2050 to 2045. More broadly, the Leaders Summit on Climate had inspired all G7 members to increase their promised climate actions significantly. The G7 also led in countries and firms that had committed to net zero emissions by 2050.

Domestic Political Cohesion

Domestic political cohesion was high in the UK host, and substantial in Germany, the United States, Japan and Canada. In Germany, by May, some polls showed the Green Party in first place over Merkel's coalition government for the federal election in October. In Canada, in May, some polls showed Trudeau's lead narrowing. But in France, in May, a few polls showed Marie Le Pen's far right party narrowly ahead of Macron's for the presidential election in April 2022.

Club at the Hub

The G7's status as the leaders' valued club at the hub of a network of global summit governance was very high. The return to in-person summitry enhanced the interpersonal bonding just before, during and after the summit, following a year and a half's absence. It continued for most G7 members at the NATO summit on June 14.

The UK host was at the core of a G7 hub of an expanding network of global summit governance, as a member of the summit clubs of NATO (along with all G7 countries except Japan) and the Commonwealth (along with Canada and guests India, Australia and South Africa), which would hold its biennial summit in Rwanda in 2021 (see Appendix L). While no longer in the regional European Union, the United Kingdom would host and co-chair COP26 in November. Expansion in the summit network in 2021 had already come from the several new summits and would continue with the UN Food Systems Summit in September, the UN Biodiversity COP15 in October and the Summit of Democracies that Biden had promised to create and host, with a virtual one in 2021 and an in-person one in 2022.

G20 Rome, October

Introduction

Global summit governance of climate change was set to strengthen at the G20's Rome Summit on October 30–31. The first regular G20 summit hosted by Italy, it would bring G20 hosting back to a major democratic power and one with a long Mediterranean coastline at the crossroads of Europe, Africa, the Middle East and the Atlantic worlds, making oceans and the natural environment a core concern. It would take place immediately after the UN's long-delayed biodiversity conference in Kunming and a day before the UN's similarly postponed climate summit in Glasgow on November 1–12, which Italy would co-chair with the United Kingdom.

Debate

In the debate about the Rome Summit's prospective performance, the first school saw US leadership. Sobel and Goodman (2020) urged Biden to have Italy call a special G20 summit early in 2021 to address critical issues including climate change.

The second school saw a promising push for climate finance. Goldar and Jain (2021) said that while Italy was "leveraging global financial flows toward the

Paris targets," the G20 should "pursue formal standardisation of green finance; explore innovative financing instruments; assist developers in project preparation phase; and support developing countries in developing green finance roadmaps." Others gave credit to the G20's central bankers, such as Mark Carney and Lael Brainard (Singhai 2021).

The third school saw the G20 as a very weak institution that paid only lip service to the SDGs, caused the UN to decline and had low expectations for Rome, due to its poor Riyadh performance, a Germany without Merkel and Italy's very fragile government (Savio 2021).

The fourth school emphasized the G20's failure to meet the climate challenge, due to US–China tensions and tribalism everywhere. In July, Wolf (2021b) called the US promise of $5.7 billion in annual climate finance by 2024 trivial, and concluded that G20 leaders would not meet the climate challenge in Rome.

Argument

By August 2021, the G20's Rome Summit was poised to produce a significant performance on climate change. It would advance climate finance for developed and developing countries, and do more on nature-based solutions than previous G20 summits. It would shift energy from hydrocarbon to renewable sources and provide momentum for the UN Glasgow Summit starting the following day. But it would not do much to limit the global temperature increase to 1.5°C.

Rome's significant performance would be driven by the soaring climate shocks in G20 members, the failure of COP26 preparations to make the needed progress, and growing internal equalization among G20 members' economic and ecological capabilities and vulnerabilities, led by the United States and China. Yet, the continuing reluctance of China, Saudi Arabia, Russia, Brazil and India to take bold climate-controlling steps, congressional constraints on Biden's climate leadership and the absence of politically powerful Merkel all constrained success. Still, the G20's virtual Rome Health Summit on May 21 introduced Biden to G20 summitry and intensified the G20's position at the hub of an expanding network of global summitry. Above all, with the highly respected Draghi co-hosting the Glasgow Summit with G7 host Johnson, the Rome Summit would help the G7, G20 and UN work together to provide combined leadership to an unprecedented degree.

Plans and Preparations

Italy's summit priorities of "People, Planet, Prosperity" put climate change in second place in the top tier. First presented by Giuseppe Conte in his UNGA address in September 2020, they highlighted the new priorities of health and climate change. For the first time at G20 summits, health and the environment came first.

On December 1, 2020, Italy's G20 presidency highlighted, as the third of its five more detailed priorities, climate change, including green growth, renewable energy and the environment. It emphasized that the economic recovery must be more inclusive, greener and smarter, with a cross-cutting green agenda. The G20's central finance track similarly sought a transformative economic recovery that was healthier, greener, digitized, inclusive and backed by reformed multilateralism.

Among its six initiatives, Italy included four on climate change. The first was a G20 early warning mechanism that included economic, health and environmental risks to enhance pandemic preparedness. The second was green, shock-resilient, sustainable traditional and digital infrastructure, including in transportation and urban areas, nature-based solutions and finance. The third was integrating sustainable finance into financial stability, including by having central banks promote climate change control. The fourth was sustainable, progressive, steadily increasing environmental taxation. This involved eliminating fossil fuel subsidies and increasing carbon taxation to rapidly decarbonize the global economy.

Further momentum came when Draghi became prime minister on February 12 and appointed a cabinet with Roberto Cingolani, a physicist and technology expert, as minister for ecological transition (Balmer and Jones 2021).

Sherpa Meetings

The sherpas discussed climate change at their first meeting on January 25. At their second meeting on July 12–13, Italy opened with an overview of the progress on energy, climate change and the environment. Korea stressed that the G20 should promote renewable energy, low-carbon technology and energy efficiency, and hydrogen, and highlighted Korea's commitment to biodiversity and the importance of a G20 agreement on liberalizing trade in environmental goods and services.

Ministerial Meetings

In a letter to the G20 finance ministers and central bank governors published the day before their February 26 meeting, US treasury secretary Janet Yellen prioritized "supporting a green recovery" and "the existential threat of our time: climate change" (US Department of the Treasury 2021a). She pledged to help the G20 prepare for COP26, co-chair the re-launched G20 sustainable finance group and proposed elevating it to a working group and to "support low-income countries as they pursued their climate goals." The following day, China's central bank governor Yi Gang said his bank was pleased to co-chair the study group (Hua 2021).

At the G20 meeting, the IMF's Kristalina Georgieva expressed strong support for the G20's proposal on climate risks and environmental taxation and committed the IMF to help with " integrating climate in public revenue and spending policies, climate-related financial stability risks and data" (IMF 2021).

World Bank president David Malpass (2021) said that as the biggest source of climate finance to developing countries, the World Bank would provide more to implement the targets he had announced at the 2020 Climate Ambition Summit to support developing countries update their NDCs, and launched "Country Climate and Development Reviews to integrate climate change into all our country diagnosis and country strategies." The World Bank would focus on a "just transition" from coal to sustainable energy and a framework for sustainable fiscal policy and growth including carbon taxation.

The meeting's two-page press release issued on February 26 signalled that the updated G20 Action Plan in April would include "climate and environment-related risks," and that the G20 would hold a climate conference in Venice in July. It endorsed "initiatives to fill data gaps for assessing climate-related financial risks and to promote more consistent climate-related disclosure." It re-established the Sustainable Finance Study Group.

On May 4, tourism ministers in Rome committed to act on a "green transformation, managing tourism to sustain global and local environments."

On June 29, ministers for foreign affairs and development issued the Matera Declaration on Food Security, Nutrition and Food Systems. It included a promise to accelerate the adaptation of agriculture and food systems to climate change, and pledged support through COP26 and COP15.

On July 10, G20 finance ministers in Venice made several climate change commitments, including two that synergistically integrated climate and health risks and shocks in a forward-looking, preventative way. A highly binding one promised to implement disclosure requirements building on the Task Force on Climate-related Financial Disclosures framework.

On July 22, environment ministers in Naples issued a communiqué with 7 commitments on climate change and 60 on the natural environment. Six climate commitments were highly binding although very general. They met with energy ministers the following day, but that communiqué appeared only on July 25, due to disagreements among members. It contained 10 highly binding and 6 low binding climate change commitments and 25 energy ones. Ministers promised to devote an ambitious portion of their COVID-19 economic recovery plans to climate mitigation and adaptation, according to national circumstances and priorities. The energy commitments on fossil fuel subsidies promised only to increase implementation efforts, again according to national circumstances, and included no dates. Most members sought more ambitious action on coal and fossil fuel subsidies, but strong opposition came from China, India and Russia, with only Russia relenting by the end (Jones 2021). Cingolani said Italy would take these issues up at the Rome Summit.

Further progress could come at the ministerial meetings scheduled for health on September 5–6, agriculture on September 17–18, trade on October 12 and finance on October 12–13.

Support also came from G20 engagement groups. The Urban 20 (2021) communiqué issued on June 17, co-chaired by the mayors of Rome and Milan, called on the G20 leaders to work with cities to achieve "carbon-neutral, climate-safe, inclusive and prosperous societies," including by divesting from fossil fuels, investing in public transit and building cities with "nature for climate resilience." The Think 20's Task Force 2 on the Climate Change, Clean Energy and the Environment prepared over 15 policy briefs for G20 leaders to act on at Rome.

Small support came from the G20's Rome Health Summit, held virtually on May 21. It brought new leaders led by Biden directly into G20 summit governance, although its declaration did not mention climate change.

Causes of Performance

By August, the proven propellers of G20 summit performance suggested a significant performance on climate change at Rome.

Shock-Activated Vulnerability

The first propeller was the soaring climate-associated shocks in G20 members in the months before Rome. Communiqué-recognized climate shocks and vulnerabilities increasingly arose at G20 finance ministerial meetings.

Elite media attention to climate change grew. On July 3–9 and 15–22, climate stories appeared on the *Financial Times* front pages almost daily, equalling or exceeding those on health and exceeding those on the economy. They featured the launch of the European Union's climate plan and deadly floods in Germany and its European neighbours. On the Sunday *New York Times* front pages on July 11 and 18, coverage surged due to the deadly heatwave in the United States and floods in Germany.

Physical shocks proliferated. Summer in the northern hemisphere brought unprecedented peaks in extreme heat, droughts, wildfires, rainstorms, hurricanes and typhoons. Emissions, concentrations and temperatures rose to new highs and critical thresholds, as most G20 economies, led by China and the United States, opened up and expanded. The 2020 COVID-19 shock declined in many G20 members, even as the more infectious Delta variant caused spikes in India, Indonesia, Europe and, by August, the United States, the United Kingdom, Canada and elsewhere. Potential financial and food crises intensified, which G20 leaders and ministers could easily link to climate change. By May, a severe drought in Brazil led Bolsonaro to declare it "the worst hydrological crisis in history" and in June it was the worst in 91 years, due to record deforestation in the Amazon (Andreoni and Londoño 2021; Harris and Ribeirao Preto 2021). Heatwaves throughout the United States and Canada hit record highs: Seattle and Portland set all-time heat records; British Columbia broke 44 heat records by June 29, including 49.6°C on June 29 in Lytton, destroyed the next day by fire. An early, exceptionally hot summer came to much of the northern hemisphere, including North Africa, the Arabian peninsula, Eastern Europe, Iran, northwestern India, western Russia, Turkey, Greece and the Caspian Sea. By July 16, extreme floods had killed hundreds in Germany, Belgium, the United Kingdom and beyond. On July 20, a major flood struck a city in China, killing at least 25 people. Major fires devastated Turkey and Greece. At the end of August, Hurricane Ida left a trail of destruction in Louisiana and the southeastern United States.

The latest assessment released by the Intergovernmental Panel on Climate Change ([IPCC] 2021) on August 9 starkly warned that the climate crisis had already reached alarming and some irreversible levels.

Multilateral Organizational Failure

Multilateral organizational failure to control these soaring climate shocks was large. The COP26 preparatory process struggled to prepare the Glasgow Summit. Some suggested that developing countries might not even attend, due to the Cornwall Summit's failure to provide them with enough climate finance. However, some support might come from the UN Food Systems Summit in September and COP15 in October.

Predominant Equalizing Capability

Predominant equalizing capability and ecological vulnerability strengthened. Economic recovery in the G20 economies bolstered their 80% global predominance in overall economic and specialized ecological capabilities, starting with green technology and clean energy. The United States and China continued to grow faster and stronger than a recovering Europe and other BRICS members, even as rising oil, gas and mineral prices helped growth in Saudi Arabia and Russia.

In the United States, economic growth grew due to Biden's $1.9 trillion fiscal stimulus package. The European Union started disbursing its multitrillion euro, multiyear spending package too.

Converging Characteristics

Changes in converging characteristics, however, would constrain Rome's performance. The degree of democracy declined in the G20 and the internal divide deepened between a slowly democratizing United States and G7, and a faster de-democratizing China, Russia, Turkey and Saudi Arabia. Moreover, all G20 members, including the G7 ones, supported fossil fuels in some form.

Domestic Political Cohesion

Domestic political cohesion would be constrained by congressional resistance to Biden's climate leadership, and politically powerful Merkel's absence at Rome. However, Germany's election could bring to power a coalition government containing the Green Party in a prominent role.

In Italy, Draghi's position was strong. But in Japan, Suga's popularity suffered from rising COVID-19 cases, and uncertainty about its general election in the fall. China's Xi had complete control as his country celebrated the 100th anniversary of the Communist Party's creation. Much depended on how he would implement his proclaimed support for multilateral cooperation for climate change control.

Club at the Hub

In-person summitry promised to return to both the G20's Rome Summit and the sideline BRICS one traditionally held there. As G20 host and Glasgow co-host,

Italy connected the G20 to the G7 (with India, Australia, Korea and South Africa at Cornwall) and the UN Glasgow Summit. The G7's Virtual Summit in February, Biden's Leaders Summit on Climate in April and the G20's Rome Health Summit in May added more spokes to the hub. So did the Asia-Pacific Economic Cooperation Informal Leaders' Retreat on COVID-19, hosted by New Zealand's Jacinda Ardern on July 16, which promised to promote a green and sustainable economic recovery for all. These summits would help G7, G20 and UN summitry work together to provide combined, unprecedented leadership.

UN Glasgow, November

Introduction

The culmination of the reconfiguration of global climate summit governance, based on the combined leadership of the G7, G20 and UN and support from the special climate summits, would come at the long-awaited COP26 Glasgow Summit on November 1–12.

Debate

In the debate about its prospective performance, the first school of thought had high hopes, expecting it "to set a path for clean resilient recovery; ensure greater finance commitments and cooperation; and establish adaptation plans and policies" and revise the NDC targets (Goldar and Jain 2021).

The second school argued that the United Kingdom's decision to cut its commitment to providing 0.7% of its gross national income in official development assistance significantly threatened a good outcome (Norton 2021).

The third school saw a significant blow from Greta Thunberg's decision not to attend and her call for the conference to be postponed once again (Rowlatt 2021).

In the fourth school, McGrath (2021) noted "sharp divisions" among the major global emitters. Although several major carbon-producing countries committed to achieving net zero by 2050, Raj Kumar Singh, India's minister for power, saw this target as a "pie in the sky."

The fifth school argued that with a virtual COP26, developing countries would be less able to effectively participate, thereby rendering the negotiations unfair (Harvey 2021b).

Argument

The Glasgow Summit promised to produce a substantial performance. It would see its most carbon-intensive members commit to major reductions, targets and timetables. It would agree that a 1.5°C rise was the only acceptable target that science now supported and that efforts must be made faster than before. It would mobilize enough new climate finance and action on loss and damage so developing countries would agree to the deal. But it would make only limited progress on Article 6 and carbon pricing, and no progress on carbon border adjustment mechanisms, ending fossil fuel subsidies, eliminating coal by a specific medium-term date or agreeing to hold COP meetings more often to meet both the old and new goals.

This substantial performance would be driven by the climate shocks soaring to unprecedented, now critical heights around the world. The major multilateral organizations would still fail to mount an adequate response, while the UN Food Systems Summit and COP15 would advance the key nature and agricultural components needed for Glasgow to succeed. The IMF's August decision, generated by the G7 and G20, to create $650 billion in special drawing rights, would help produce adequate climate finance. The success of the US economic recovery, climate policies and green infrastructure initiatives, followed by China's in a global competition, would provide the upward convergence that spurs summit climate performance, even if no upward democratic convergence appeared. Sufficient domestic political cohesion supporting climate action would exist in almost all G7 members as well as in the BRICS leaders of China, India and Russia, and several G20 members beyond. Above all, the unprecedented combined, coordinated leadership of the G7 and G20 summits and systems on climate change and the expansion of special climate summits would propel the in-person Glasgow Summit, co-chaired by Johnson and Draghi, to greater success than Paris in 2015 had produced, even if the compounding climate crisis now demanded much more.

Plans and Preparations

Much depended on COP26. The United Kingdom's Sharma (2020a) argued that the conference "will make or break the zero-carbon economy." At the centre would be the resolution of the Paris Agreement's Article 6 on international carbon markets.

When the postponement of COP26 was announced, Guterres said that despite the global pandemic, "the science on climate has not changed, with emissions at a record high while global warming impacts compound the socio-economic challenges that this crisis will intensify" (UN 2020a). Patricia Espinosa, UN Climate Change's executive secretary, noted that despite the urgency of COVID-19 "we cannot forget that climate change is the biggest threat facing humanity" (UN Climate Change 2020b).

Delaying COP26 gave countries time to work out their individual responses to the COVID-19 emergency and re-evaluate commitments to shift to a low-carbon economy. It also allowed time, as Nicholas Stern suggested, to develop a new approach to "a sustainable and resilient economy in closer harmony with the natural world," in addition to giving Biden a chance to re-establish US climate leadership (Harvey 2020c). There was pressure to ensure that COVID-19 recovery plans did not entrench investment in high emission sectors and that countries did not use the pandemic to roll back environmental commitments and protections (Harvey 2020b).

The United Kingdom remained committed to hosting the summit in Glasgow, and continued its logistics planning alongside the Scottish government. It initially anticipated that an in-person meeting could take place in the spring of 2021. However, on May 29, 2020, UN Climate Change, the United Kingdom and Italy announced it would take place on November 1–12, 2021 (UK Presidency of the UN Climate Conference 2020). The new date would allow the United Kingdom as well as Italy to use their "G7 and G20 presidencies in driving climate ambition." Johnson invited Guterres to the G7 Cornwall Summit in June (UK, PMO 2021b).

With over 30,000 registered global attendees, Glasgow would be the largest summit the United Kingdom had ever hosted (ICAEW 2021; Walker 2021).

In early 2021, Johnson handed the COP26 reins to UK business secretary Alok Sharma to focus full time on the crucial summit preparations and appointed Anne-Marie Trevelyan, minister of state for business, energy and clean growth and the United Kingdom's International Champion on Adaptation and Resilience for COP26, to oversee implementing his 10-point plan for a green industrial revolution (ICAEW 2021).

In a UN address in February, Sharma (2021b) articulated his key objectives of emissions reductions and adaptation. He emphasized unhindered public and private finance flows, notably by developed countries providing $100 billion in climate finance. The UK COP presidency had published its finance priorities and would discuss climate finance access at a climate and development ministerial in March. Sharma noted the establishment of several new forums including the Energy Transition Council, Zero Emission Vehicle Transition Council and the Commodity Trade Dialogues through which all COP delegates could contribute.

Sharma (2021b) urged leading economies to demonstrate collective leadership, promising the United Kingdom would use the G7 to encourage more active roles in the climate emergency. He visited several developing countries and initiated monthly meetings with heads of country delegations to assist the preparatory process.

At the World Sustainable Development Summit on February 14, Sharma (2021a) again emphasized making robust adaptation plans, investing in green recoveries and committing to net zero emissions. He encouraged businesses, investors, cities and regions to enter the Race to Zero Campaign to reach net zero emissions by 2050. He also acknowledged the important contributions of civil society, women, young people, Indigenous peoples, and marginalized communities and groups, and noted the COP26 presidency's Civil Society and Youth Advisory Council. He also met with India's Modi, acknowledging his leadership and commitment to reduce emissions by delivering 450 gigawatts of renewable energy by 2030 (UK, Cabinet Office 2021a). He emphasized that the India–UK partnership was a key example of climate collaboration, encouraging other countries to follow suit. He met with several other G20 leaders as well, including Indonesia, Turkey, Brazil and Saudi Arabia.

Deciding on an In-Person Summit

By April, a virtual platform remained the preferred option for the Glasgow meeting given potential new COVID-19 waves. As well, a virtual summit would prevent releasing tons of emissions into the atmosphere from thousands of delegates flying to Scotland. In-person delegations would need to be scaled back, and access for non-state actors and social events would be limited.

But in-person events also facilitated the "continuous process of bilateral and multilateral discussions ... where hard negotiations take place and tough decisions have to be made ... [and] physical presence matters" (Vetter 2020). With numerous meetings during the typical two-week period, words and body language count, especially as complex discussions are translated in real time into the six UN languages. Social interactions are equally important for building trust and understanding.

A virtual summit would further limit developing countries' representation, as many had limited or unreliable internet access (Energy and Climate Intelligence Unit [ECIU] n.d.). Without their full participation, robust and meaningful negotiations would be extremely difficult.

A virtual setting would further harm negotiating blocks such as the Alliance of Small Island States and the Most Vulnerable Group, which depend heavily on in-person meetings to reach agreement among themselves. With many key negotiations occurring behind the scenes in smaller group settings, an online gathering could exclude several small states (ECIU n.d.). It could further prevent numerous non-governmental organizations, academics and business groups from making meaningful contributions. Some suggested that limiting the number of participants might improve productivity (edie 2021).

Just ahead of the Cornwall Summit, the United Kingdom announced it would provide vaccines to COP26 participants otherwise unable to get them, so "countries most affected by climate change are better able to participate fully in discussions about creating a greener future for the planet" (UK, PMO 2021c).

Post-G7 Summit Preparations

At the Cornwall Summit, G7 leaders reiterated their goal of keeping 1.5°C within reach, facilitating the transition to net zero, bolstering resilience and adaptation, closing the funding gap, encouraging innovation and presenting aligned NDCs at the earliest opportunity ahead of Glasgow. They encouraged others to join them.

On June 17, Sharma (2021c) stressed the criticality of responding to the latest science, including the IPCC report that would be published in August and would inform COP26 negotiations.

The following day, the COP26 presidency announced the broad "Visions for a Net Zero Future" collaboration among academics, business, civil society, youth, government, Indigenous and rural communities. It would produce a study of the impacts of electricity generation, agriculture, waste, water management, building design, reforestation and ocean conservation in transitioning to a net zero economy to be published in September, also to inform COP26 negotiations.

On June 20, at a Powering Past Coal Alliance round table in London, Sharma (2021d) stressed the need to phase out coal and create green jobs for workers whose livelihoods would be affected through the decarbonization of power systems across the globe.

On July 7, the United Kingdom unveiled the presidency's agenda for COP26 (UK, Cabinet Office 2021b). It detailed the World Leaders Summit on November 1–2, plus deliberations on clean energy, zero-emission transport, adaptation, loss and damage, gender, finance, science and innovation, nature and youth. Cross-cutting elements of science, innovation and inclusivity would underpin the entire two-week program. Noting that finance, energy, gender and adaptation were vital issues, Sharma stressed that "Glasgow is our best hope of safeguarding the planet for everyone, building a brighter future and keeping the 1.5C target alive" (UK, Cabinet Office 2021b).

Prospects for Performance

By August, some tentative judgements could be made on the likely performance of the Glasgow Summit.

All the leaders of COP members are likely to attend, although some may participate virtually. Preparations would remain flexible in case a new COVID-19 wave required Glasgow to lock down.

Private deliberations will likely be high, unless COVID-19 strikes with severity. Public deliberations in the Glasgow declaration will likely exceed the 7,353 words in the Paris Agreement and the 8,943 in the 2009 Copenhagen declaration, and cover a broader range of subjects, including health, in more detail.

Its principles will likely affirm that a 1.5°C rise is the primary or only target. It could slightly relax the differentiation between rich and poor countries. However, it will likely include some of the inherited references to CDR and national circumstances and capacity.

Glasgow will likely make more commitments, with a more highly binding balance and more ambitious content, than Paris did. Members' compliance with these commitments, however, will likely be low, as were those of the Paris Agreement.

The institutional development of UNFCCC governance will improve, possibly even with the next COP climate summits scheduled sooner than half a decade hence.

Causes of Performance

The propellers of this performance could be predicted with more confidence, except for the severity and impact of a new COVID-19 wave.

Climate shocks are already much stronger and widespread than those in 2015 and will likely remain so.

Multilateral organizational failure will likely continue to be high in the official-level preparatory process and among the ministers in Glasgow, prompting the leaders themselves to do much more than the previous COPs and more than the Paris Summit did.

Internal equalization will rise, as China, India, Russia and Brazil and, secondarily, the G7 countries, become more vulnerable, as the developing countries have long been.

Upward convergence on democracy will likely remain stable, but environmental performance will likely rise.

Domestic political cohesion will likely remain high in key countries, led by the United States and China.

The Glasgow Summit will be its key leaders' valued hub in an expanding network of global summit governance due to the unprecedented co-chairing by the G7 and G20 hosts, the coordinated support from their respective summits and the recent expansion of successful special climate summits.

Together these propellers suggested that Glasgow would be a substantial success, and could conceivably be the strongest UN climate summit ever, since the first at Rio in 1992. But its performance depends on the decisions made by its most powerful leaders, above all China's Xi.

Conclusion

By August, the three successful special climate summits, the strongly successful G7 summit and the potentially productive G20 and UN summits in 2021 gave some hope that the global governance of climate change would work as never before. This hope began with the solid performance of the One Planet Summit on Biodiversity and continued with the additional advances at the Climate Adaptation Summit in January, and the strong performance of Biden's Leaders Summit on Climate in April, with 40 major leaders. This hope and real progress was reinforced by Johnson's virtual G7 summit in February and his strongly performing G7 Cornwall Summit in June. Prospects looked promising for Draghi's G20 Rome Summit in October to produce a substantial performance, and thus for Johnson and Draghi as co-chairs at the COP26 Glasgow Summit in November to create the culminating fully global success.

These major advances and hopes for more progress were propelled first by the unprecedented climate shocks that took centre stage, as the vaccine rollout diminished COVID-19 as an overwhelming diversionary threat. The major multilateral organizations, led by UN Climate Change, still struggled to produce an adequate response. An economically growing China and reviving United States with world-leading emissions increasingly understood that they must cooperate abroad and converge at home to confront the shared, existential threat of climate change. Biden's arrival added climate-committed American leadership and power. With

Johnson secure as UK prime minister and Draghi's appointment as Italy's prime minister in February, two new G7 and G20 leaders would host their own summits and co-host the COP for the first time. Above all, the return of in-person summitry and the addition of several special climate summits, including new ones, helped make the G7 and G20 valued clubs at the hub of a network of global summit governance expanding to an unprecedented extent.

To be sure, formidable constraints came from the start of COVID-19's fourth wave, rising oil prices that emboldened the world's embedded fossil fuel economy and powers, Chinese–US geopolitical competition, Biden's narrow legislative control and the departure of Germany's politically powerful Merkel. Above all, given the soaring climate and ecological crises, the strongest ever start of global summit governance in the first half of 2021 produced results still nowhere near the level needed to solve the problem now at hand. It remained to be seen if deadly extreme weather events fuelled by climate change and ominous scientific evidence escalating through the autumn could force the combined leadership from the G7, G20, UN and special climate summits to produce a genuine physical and human success.

9 Conclusion

From 2015 to 2021, the global summit governance of climate change passed through four phases. In the first phase, in 2015, the G7 performed strongly at the Elmau Summit hosted by Germany's Angela Merkel to help create the United Nations' landmark Paris Agreement in December. That agreement, with its appropriate but ambitious target of limiting the global temperature rise to no more than 2°C, and ideally 1.5°C, above pre-industrial levels, provided a solid foundation.

In the second phase, from 2016 to 2019, the summits of the UN Framework Convention on Climate Change (UNFCCC) disappeared, as planned, for half a decade, leaving the annual Conference of the Parties (COP) ministerial meetings to do the work. The annual G7 and G20 summits took turns in leading climate action, through the substantial performance of Shinzo Abe's G7 Ise-Shima Summit in 2016, the strong performance of Merkel's G20 Hamburg Summit in 2017, the significant performance of Justin Trudeau's G7 Charlevoix Summit in 2018 and the significant performance of Abe's G20 Osaka Summit in 2019. Moreover, a reconfiguration of global climate summit governance developed, with the arrival of ad hoc, plurilateral, increasingly institutionalized, climate-dedicated summitry. It began with the significant performance of the One Planet Summit in 2017, added the substantial performance of the second One Planet Summit in 2018, the small performance of the third, now regional One Planet Summit and the significant performance of the UN Climate Action Summit in 2019.

During the third phase, in 2020, G7, G20 and UN leadership all disappeared. The deadly COVID-19 shock crowded out their action on climate change, and substituted virtual summitry for the more intimate in-person kind. Climate change–denying Donald Trump hosted an emergency G7 summit in March but did not host a regular summit that year. Climate change–reluctant Saudi Arabia hosted G20 virtual summits in March and November, which together produced only a small performance on climate change. The reconfiguration of global climate governance from the special summits continued this small performance at both the UN High Level Roundtable on Climate Action and Summit on Biodiversity in September, and contributed more with the significant performance of the Climate Ambition Summit in December.

In the fourth phase in 2021, global summit governance returned in full force to an unprecedented degree. G7/G20 members the United Kingdom and Italy would

DOI: 10.4324/9780429055485-9

host three regular, scheduled summits that highlighted climate change — starting with the strong performance at the United Kingdom's G7 Cornwall Summit in June, followed by Italy's G20 Rome Summit in October and the co-chaired UN climate summit in Glasgow in November. The United Kingdom added, for a fast start, an ad hoc G7 virtual summit in February. Moreover, the special climate summitry continued, with the solid success of the fourth One Planet Summit on Biodiversity in January, the small success of the Climate Adaptation Summit and the strong performance of the Leaders Summit on Climate, hosted by US president Joe Biden in April.

Together, this new configuration of old institutionalized and ad hoc, plurilateral summitry worked together on climate action. If sustained with proliferating results to the end of 2021, it could contribute to meeting and improving the Paris goals to help contain the climate emergency at hand. However, this is by no means certain.

Global Climate Governance, 2015–2021

These four phases were defined initially by how the particular configuration of G7, G20 and UN meetings each year acted on, or avoided, climate change and, subsequently, by how they were reinforced by the new generation of special climate summitry since 2017. Together this reconfiguration of the global summit governance of climate change rose to new heights by the end of 2021.

Producing Paris, 2015

The first phase of producing Paris began in 2015. The year began with high hopes and intense preparations for the culminating landmark UN climate summit in Paris at year's end — the first since the one that failed at Copenhagen in 2009. The G7 Elmau Summit in June, hosted by Germany's Merkel, produced a strong performance that prepared the path to Paris. But the G20's Antalya Summit, hosted by Turkey's Recep Tayyıp Erdoğan in November, with its small performance, did little to bridge the traditional divides between the major emerging and established economies. Instead, it deferred climate action to the UN, fully backing the upcoming COP21 and the broader UNFCCC.

At the fully multilateral COP21 Paris Summit at the end of November and early December, 196 members produced a solid performance, with their new, bottom-up, voluntary, all-in approach. All parties agreed that, given the scientific consensus, the world could withstand no more than a 2°C temperature rise beyond pre-industrial levels, and would be better off with a stricter 1.5°C limit. They slightly relaxed the principle of common but differentiated responsibilities (CDR) that had dominated since 1997 and replaced the old, strictly dichotomous categories with the language of "developed" and "developing countries," if with the qualifying addition of "in light of national circumstances." This encouraged more countries, including the rapidly rising China, to assume greater responsibility for their current and future emissions as their economies grew (Pauw et al. 2019).

This sequence of G7 success, G20 failure and UN progress, relative to their own past summits, was caused less by the climate shocks and scientific evidence of 2015, and more by the hope that the fully inclusive UN, operating again at the summit level after six years, could succeed where it had failed at Copenhagen. In 2015, the rising capabilities, democratic convergence and domestic political support of G7 and G20 leaders helped. But the G7 leaders' climate change convictions were not shared by the G20's climate change–denying or –resistant, less democratic leaders of Turkey, Saudi Arabia and Russia, and even China's Xi Jinping (Kirton 2016a). Above all, when all the leaders met in Paris at the UN climate summit for the first time in six years, they were confined by the UN's procedures and struggled to transcend the fixed positions, interests and identities that their officials, ministers and they themselves had brought. They ended with an agreement that still reflected CDR and voluntary cutbacks, in an agreement not designed to be substantially improved for another five years. Even if its commitments were fully complied with, they would not achieve its agreed goal. The Paris Agreement had a flawed design from the start. But it was the best the world had.

Relying on Paris, 2016

In 2016, the second phase of the rotating G7–G20 leadership began. Attention turned to ratifying the Paris Agreement, raising climate finance and taking the necessary steps to make it work. As world leaders were not scheduled to meet for another five years at the UNFCCC, these tasks were left to their ministers, as well as to the G7 leaders at Ise-Shima in May and the G20 leaders in Hangzhou in September. Both plurilateral summit institutions brought compact participation in a club at the hub of an expanding network of global governance and — especially at the G7 — the flexibility for leaders to decide on the spot, which UN summitry lacked. Both faced rising climate shocks and scientific evidence, as the average global temperature reached new heights. Moreover, the G7 summit was hosted by Japan, a powerful island country vulnerable to typhoons and sea level rise, with a leader who politically cared about the G7 and the environment. The G20 summit was hosted by China, a rapidly rising, soon second-ranked power that had become the world's largest absolute climate polluter, and whose leader was a professional engineer, had acquired more personal control of his government and still positioned his country as a disadvantaged developing one.

The G7 Ise-Shima Summit produced a substantial performance, with half as many climate commitments with less compliance than Elmau's strongly performing G7 summit the year before. The G20 Hangzhou Summit, in its small performance, generated fewer commitments, with less compliance than even the small performance at Antalya in 2015. And the UN had a small performance at the ministerial-level COP22, to try to implement and improve the newly minted, inadequate Paris regime.

In a repeat of the 2015 cadence, the G7's advances came from its democratic commonality and compact club cohesion, now reinforced by host Japan's geographic vulnerability and ability as the G7's second most powerful member to lead on climate change, even when a lame-duck US president Barack Obama, without legislative control, could not. In contrast, the G20 lacked these propellers, especially with Xi still largely wedded to his country's developing country identity and only beginning to redefine himself and his country as green.

Tackling Trump, 2017

In 2017, the global summit governance of climate change was dominated by the need to protect the UN's Paris Agreement against Donald Trump. He became US president on January 20, with a promise to pull the United States out of the Paris Agreement, which Obama had ratified the year before. Trump launched the most aggressive anti-environmental convictions and crusade of any G7 and most G20 leaders. At the G7's Taormina Summit in June, hosted by Italy's Paolo Gentiloni, G7 leaders passionately tried to convince Trump to reverse his position. They failed. Taormina's small performance was followed a few weeks later, by Merkel, hosting her first G20 summit in Hamburg, trying to tame Trump another way. It worked. Hamburg with its strong performance was the most successful G20 summit on climate change, even with Trump there. This showed that the G20, when led by a powerful G7 member with a highly committed and experienced host, could overcome the adamant resistance of its most powerful member, even when it was backed by Russia and Saudi Arabia. However, Hamburg's success prompted only a small performance at the UN's ministerial COP23 at Marrakesh in November.

Then, the reconfiguration of global summit governance of climate change began. The first ad hoc, plurilateral, climate-dedicated summit was the One Planet Summit in December, under the leadership of G7 member France's Emmanuel Macron, working with World Bank president Jim Yong Kim and UN secretary general António Guterres and without Trump present. It produced a significant performance.

G7 Leadership, 2018

In 2018, the G7 returned to lead global climate summit governance. Canada, the G7's least powerful member but America's highly integrated next-door neighbour, hosted the Charlevoix Summit in June. Trudeau was as politically committed to controlling climate change as Trump was to denying it. While standing by the core climate change priorities that his citizens and his other G7 colleagues said they valued, Trudeau focused on protecting the global oceans, an appropriate choice for a country located on three major oceans, with the longest coastline in the world. His strategy largely worked. Charlevoix performed significantly on climate change, despite Trump's impulsive, tweeted post-summit repudiation of the Charlevoix communiqué he had just accepted. Charlevoix did not secure US agreement on climate change, but did at least mobilize the "G6" to do more on climate change and oceans than the year before.

The G20 summit hosted by Argentina's Mauricio Macri in Buenos Aires on November 30–December 1, produced only a small performance. He had little commitment to countering climate change, and depended on a US administration that controlled the International Monetary Fund (IMF) and thus the financial support Argentina needed to survive its new financial crisis. Macri did not even try to use his summit to advance climate action. With no push from the G20 summit, COP24 — in coal-laden Katowice beginning two days later — produced only a small performance too.

However, the reconfiguration of global climate summitry became institutionalized. The second One Planet Summit, in New York in September, produced a substantial performance.

G20 Leadership, 2019

In 2019, G20 leadership returned. An experienced Abe hosted the Osaka Summit in June, before the G7's Biarritz Summit in August. Osaka produced a significant performance, making selective, if largely defensive progress on the looming climate crisis. All members but the United States reaffirmed their support for the Paris Agreement. All backed COP25, and innovatively introduced into the communiqué nature-based solutions and Indigenous peoples, who constitute 5% of the global population but protect 80% of the world's biodiversity and irreplaceable carbon sinks. However, Osaka did little to phase out fossil fuel subsidies, reduce emissions or reinforce sinks.

Osaka's performance was constrained by trade tensions and financial fragilities throughout the G20, while the UNFCCC remained ineffective. G20 members were the globally predominant source of emissions and the capabilities to reduce them. But its big, emerging members of China and India were growing very rapidly, and all G20 members invested more in fossil fuels than renewables. G20 members converged upwardly in the ecological domain, but not the democratic one. Domestic political cohesion was high in host Japan and also India and China, and still solid in Trump's United States. The G20 summit hub added to its network China's Belt and Road Forum, where climate concerns were weak. A consensus-oriented Abe was reluctant to push climate change control against an adamantly opposed Trump. But at the summit itself, Abe accepted a limited "G19" deal when a determined Macron, Merkel and Trudeau threatened to ruin his summit if he did not.

In contrast, the G7 Biarritz Summit in August produced a small performance, despite its climate-committed host. The key causes were Trump's antagonism to climate action as his re-election year in 2020 approached and Macron's desire to avoid Trump's post-summit tantrum at Charlevoix. Other causes included diversionary economic, financial and security shocks, and the pronounced domestic political weakness of most G7 leaders including Macron, who faced a severe backlash from the yellow vest protests for his attempts to impose a carbon tax. These forces hindered the ministerial-level COP25 in Madrid in December, which produced a small performance too.

However, the reconfiguration of global climate summit governance expanded. The third One Planet Summit was held in Nairobi in March. Here, the developing countries in Africa, along with the industrialized world, produced a small performance. And, with a more global form and focus, the UN Climate Action Summit in New York in September produced a significant performance.

The COVID-19 Crowd-Out, 2020

The third phase of 2020 was truly exceptional, with COVID-19 crowding out any attention to climate change at the G7's emergency virtual summit on March 16. Aided by the increasing re-election preoccupations of G7 host Trump, it helped delay and then cancel the regular G7 summit itself. Thus, Trump's G7 completely failed.

The COVID-19 crowd-out continued at the G20's virtual summit on March 26. The G20's regular Riyadh Summit hosted by King Salman bin Abdulaziz al Saud in November produced a small performance, due to COVID-19, a climate-reluctant Saudi host and a still present Trump. The UN fully failed, as the COVID-19 health shock caused the postponement of the long-scheduled COP26 climate summit in Glasgow in November.

In 2020, none of the G7, G20 or UN summits rushed to fill the void left by the others. Their largely failing performance was caused above all by the diversionary shock of COVID-19 and the ensuing economic and social crises, with assistance from Trump, King Salman and the damage done by virtual diplomacy to the summits' dynamic as in-person clubs.

Yet, in 2020, three new special climate-focused summits replaced the virtual disappearance of climate action at the G7, G20 and postponed COP26. The UN's High-Level Roundtable on Climate Action and Summit on Biodiversity in September both produced a small performance. Then, a significant performance came from the Climate Ambition Summit in December on the fifth anniversary of the Paris Agreement.

Combined Leadership, 2021

In 2021, the fourth phase of comprehensive, combined centrality arrived. The G7, G20 and UN each worked together toward an ambitious, adequate cumulative result. The arrival of safe, effective COVID-19 vaccines in December 2020 meant that the COVID-19 crowd-out of climate change could decline and in-person summitry return. A climate-committed Biden becoming president on January 20 brought US leadership back in the G7, G20, UN and beyond. On his first day in office, Biden had the United States rejoin the Paris Agreement.

The United Kingdom's Boris Johnson had already promised to highlight climate change at the G7's Cornwall Summit in June and spur ambitious action at the UN Glasgow Summit the United Kingdom would co-host with Italy in November. Italy's Mario Draghi declared "People, Planet and Prosperity" as priorities for the Rome Summit in October.

Johnson started quickly by hosting a virtual G7 summit in February. His scheduled Cornwall Summit in June produced the strongest performance of any G7 summit on climate change. Preparations for Draghi's G20 summit featured ministerial meetings that included climate action, even if his special Rome Health Summit in May did not. The G7, G20 and COP26 preparatory process together suggested the Glasgow Summit would be a substantial success.

Some progress came from the expanding special climate summits, with three held in the first four months of the year. They were led again by France, with the One Planet Summit, this time focused on biodiversity, on January 11 and its solid performance. The UN's Climate Adaptation Summit came from EU member the Netherlands on January 25, with a small performance. Then, in a major innovation, Biden's Leaders Summit on Climate, including all G20 leaders plus 20 others, took place in April and produced a strong performance. In all, 2021 started a year saturated in climate summits, with many more ambitious climate targets and action announced, but still much left to do.

This unprecedented climate summit sequence came as greenhouse gas emissions and concentrations, ominous scientific evidence and deadly extreme weather events soared to new heights, as the impending climate catastrophe closed in. Rising Chinese–US competition and the world's embedded fossil fuel economy remained major constraints. Yet, with climate-committed or -sensitive leaders from G7 members hosting or chairing the five climate consequential summits in the first half of 2021, they could finally work together to help contain the climate threat.

However, doubt remained, as G20 members committed almost $300 billion of public money to the fossil fuel industry in 2020 alone, while claiming since 2009 to be unable to raise $100 billion a year for climate action or end fossil fuel subsidies. Global climate strikes, and grassroots and Indigenous-led movements would continue to put sustained pressure on the G7, G20 and UN. Along with the proliferating extreme weather events and scientific evidence, this sustained pressure on the world's most powerful people could not be ignored or waited out.

Improving the Systemic Hub Model

These conclusions highlight several ways to improve the inherited systemic hub model of G7, G20 and global climate governance to account for the dynamics of the rapidly changing world since 2015 (Kirton 2013; Kirton and Kokotsis 2015).

The application of the systemic hub model in the *Global Governance of Climate Change: G7, G20 and UN Leadership* from 1975 to the approach of the Paris Summit in 2015 ended with a key, if careful, contingent prediction. Kirton and Kokotsis (2015, 306) concluded:

> The G20 is now becoming the leader in effective climate governance. It could even return to its cadence in 2009 and 2010 of having two summits a year to address the compelling global crisis … It would do so when the US leadership is constrained by the presidential election in November 2016, but when

the polls are already showing a shift to a US public more supportive of climate change control. Moreover, it would help Japan and China, as leading Asian rivals, supported by the US and a democratic India, work together on this most compelling global cause.

Six years later, these predictions proved partly true. The G20 did lead at Hamburg in 2017 and again at Osaka in 2019. But the G7 also led at Elmau in 2015, Ise-Shima in 2018, Charlevoix in 2018 and, especially, Cornwall in 2021. Each of these G20 and G7 summits led, with important caveats and with much depending on who hosted. And no G7 or G20 summit took the ambitious action needed for the transformational systemic change that the science and the changing climate demanded.

US domestic politics had a complex impact. Trump's 2016 election heavily constrained G7 climate performance in 2017 and 2019 and ended it in 2020, but did not prevent G6 action in 2016 and 2018. It similarly constrained G20 climate performance in 2016 and 2018, but less so in 2017 and 2019. For 2020, Saudi Arabia's initial priority on climate change almost disappeared at its summit in November, due to COVID-19, Riyadh's own climate hesitancy and Trump as a climate change–denying ally. Broad US public opinion supporting climate action had little effect, until Trump was defeated by a climate-committed Biden and Democratic Party on November 3. Public opinion also supported many US actors beyond the federal government to engage directly in the global governance of climate change, in its multi-stakeholder forms arising from the special summits since 2017.

Japan and China, supported by the United States and India, increasingly came together to combat climate change. Japan hosted relatively successful summits on climate change in 2016 at the G7 with the United States and in 2019 at the G20 with China and India and despite Trump's United States. And all the Indo-Pacific powers came together in April 2021 for Biden's Leaders Summit on Climate.

Yet, the 2015 book did not anticipate the central story of the subsequent six years — the reconfiguration of global climate governance from the addition of the special, increasingly institutionalized climate summits since 2017, with one each in 2017 and 2018, two in 2019, three in 2020 and three in the first half of 2021. The G7, G20 and UN summits also came back in full force in 2021 with G7 members chairing them all. Johnson, with his two G7 summits, worked closely with Draghi as host of the G20 and as co-host of the UN's Glasgow Summit. The G7's Macron hosted the One Planet Summit, the European Union's Mark Rutte of the Netherlands followed with the Climate Adaptation Summit, and Biden's Leaders Summit on Climate were all designed to spur success at Glasgow. Yet, this combined, coordinated leadership of the old and new summits would not be enough to control the compounding climate emergency. That depended on what a mutating SARS-CoV-2 virus and a de-democratizing China and its US competitor would do, and whether the world could end subsidizing and burning fossil fuels, preserve biodiversity, ensure climate justice and more.

These dramatic developments since 2015 point to 10 important improvements to the systemic hub model in the 2015 analysis that have been introduced here.

1. The Shifting Performance Referent

The first improvement is a shift in the central referent by which G7, G20, UN and special summit performance should be judged. Ultimately, only real-world results count. It remains appropriate to judge success by how well each summit did on each dimension of performance, relative to its predecessors, the contemporaneous summits each year and their successors. This referent of the past and present summit supply competes with the alternative referent of the global demand for global governance, defined by the 2015 Paris Agreement target of keeping the rise in post-industrial global temperature below 2°C and especially 1.5°C. Together these two referents yield the criteria of how much the supply of global summit governance each year reduced the gap to meet the growing demand, before the critical target threshold and irreversible tipping points were surpassed. In 2015, when emissions and concentrations were far lower than in 2021, Kirton and Kokotsis (2015) deemed it reasonable to focus on the supply of G7, G20 and UN governance, in the hope that incremental increases would produce enough cumulative results in time. However, by 2021, with the world's carbon budget shrinking swiftly to a low level, it was reasonable to focus more on the unmet demand and the growing gap. By this referent, all G7, G20 and UN summit leaders and ministerial meetings, alone or together, have failed.

2. Beyond the Old Big Three to the New Reconfiguring Summits

The second improvement in the systemic hub model is looking beyond the established G7, G20 and UN to how the new generation of ad hoc, plurilateral, climate-focused summits since 2017 substituted when those big three failed and stimulated and supplemented them when they succeeded. This book begins that work. Further research requires a more detailed, systematic examination of the performance and propellers of these reconfiguring special summits since 2017 and their relationship with the G7, G20 and UN summits and their hosts. One relevant question here is whether such summits dedicated to a specific subject such as climate change, as the COP ones are, produce siloed climate governance that erodes the synergistic whole-of-global-governance approach needed to solve the climate crisis. However, no one summit institution, of a siloed or synergistic sort, can do the job alone, as a deep paradigm shift across all systems is required for the sustainability and survival of humankind.

3. Substituting, Supporting, Synergistic Summit Sets

The third improvement concerns the relationship among the G7, G20 and UN summits. The most recent seven years show that both the G7 and G20 can work together to build successful UN climate summitry, as they have been doing in

2021. In the many years without COP summits, either the G7 or the G20 can lead in a given year, doing so when the other one does not. They are replacements and reinforcements, rather than rivals or redundant (Kirton and Larionova 2018b). Global climate governance needs both, especially as they add their weight at different times during the year. Climate change takes place every day, but global climate summits do not.

Indeed, in overall performance and the number of climate commitments, leadership alternated between the G7 in 2015 and 2016, the G20 in 2017, the G7 again in 2018 and the G20 again in 2019 (see Appendix B). However, on the key dimension of compliance, the G20 led only in 2015, while the G7 did in 2016, 2017 and 2018 (see Appendices C and D). Between 2015 and 2018, the smaller, richer, all-democratic G7 averaged 4% higher compliance annually than its G20 colleague did.

And as the first half of 2021 shows, the G7 and G20 summits can work together, along with the UN summits, and with the special climate summits that have bloomed. Moreover, as the special summits suggest, climate leadership is not limited to the leaders of nation-states, but now includes non-state and sub-state actors as networked leaders in their own right, spurred by the connecting force of an increasingly digitized world.

4. Performance Dimensions Varying across a Wider Range

The fourth improvement is charting the dimensions of performance across a much wider range of variation. On domestic political management, the disappearance of a regular G7 summit in 2020 and the arrival of a public emergency virtual summit in March, followed by a private one in a conference call in April, stretched the range of variation at both ends. Private deliberation was similarly stretched to the point of disappearance with the arrival of virtual summitry for the G7 and G20 in 2020 and during the first five months of 2021. It was similarly stretched for the special climate summits in 2020 and 2021. This brought new forms of public deliberation, as the Leaders Summit on Climate had only individual, public leaders' speeches and no collective communiqué or commitments. In delivery, summit compliance reached a new high of 94% with the priority commitments made by the G7 on March 16, 2020, although none was on climate change. Overall, the G7 failed on climate change in 2020, but had the strongest performance ever in 2021.

5. Complex Shock-Activated Vulnerability

The fifth improvement comes in the cause of shock-activated vulnerability. It clearly has more components, interacting in more complex ways, than the old systemic hub model contained. Shock-activated vulnerability remains a primary, powerful force propelling G7 and G20 performance, as 2020–2021 clearly show in the field of health. A succession of same-subject shocks produces particular strengthening of G7 and G20 performance on this subject, as 2021 also shows on climate change. However, this book highlights the great puzzle of why a rising succession of climate shocks, manifested as extreme weather events and supported by ever more certain climate science, did not spur all G7 and G20 summits since 2015 to stronger performance on climate change. This puzzle points to the importance of several additional components of a more complex configuration of shocks.

The first component of this complexity is the strength of diversionary shocks. Diversionary shocks drive down performance on a subject when they come, and are seen to come, from a different substantive source and where no synergistic link is made or exists. In the past, while relevant shocks steadily rose, and such shock-activated vulnerability spurred G7 and G20 success on climate change, shocks on several other subjects recurrently diverted and crowded out more climate performance from 2015 to 2021.

These diversionary shocks started with the Ebola health and terrorist shocks at G20 Antalya in 2015. The increasing collection of extreme weather events, coming at different times in different places, outside and inside G20 and G7 members, did not have the same force then, even when they arose on similar subjects such as hurricanes assaulting the Caribbean, Mexico and the United States in most years.

In particular, a major diversionary shock arose from finance for Argentina's G20 in 2018, soon before the summit in the G20's host country, for the first time since 2012. It came on the core subject of the G20's first distinctive foundational mission of promoting financial stability. Financial shocks were the cause of the G20's creation, and its successful response to the financial crises from Asia in 1997, the United States in 2008 and Europe in 2010–2012. This latest Argentinian financial crisis also created worries about international contagion, even into the United States itself when its market suddenly plunged in early 2018.

The greatest diversionary shock came from COVID-19, erupting in China in December 2019. It crowded out G7 and G20 climate governance almost completely in 2020. It began as a global health shock, then quickly became an economic and potentially financial one, and spurred the first G7 and G20 emergency summits in March. By the time of the scheduled G20 Riyadh Summit in November, the economic shock had receded somewhat, but the fear of a financial shock remained. Few G20 leaders thought that climate shocks caused or could come from financial and economic shocks, in the fast, full way that COVID-19 had just connected finance and the economy to human health.

A further component of this complex shock-activated vulnerability is how familiar, similar shocks from the real physical world acquire political force in G7 and G20 governance, even when they transcend the competitive diversionary shocks to become recognized in the summit communiqués. They do so when highlighted by the elite media that helps put them there. Such media-highlighted shocks matter. They bring the subject to the attention of summit leaders and can also make the connection between a diversionary shock on a different subject and the one of concern. On US television network coverage of climate change, ABC, CBS and NBC aired 50 minutes in 2016, 260 minutes in 2017, 142 in 2018, 236 in 2019 and 112 in 2020 (Macdonald 2021, 8). This coincides well with the performance surge to a historic high at G20 Hamburg in 2017, the significant performance at G7 Charlevoix in 2018 and at G20 Osaka in 2019, and the virtual disappearance at G7 and G20 summits in 2020, with Trump present at all. PBS Newshour aired only 46 segments in 2016, 69 in 2017, 72 in 2018, 121 in 2019 and 58 in 2020 (Macdonald 2021). This coincided with the performance spike at Osaka in 2019 and the plunge to almost nothing in the G7 and G20 in 2020. Moreover, in 2020, US television network news almost never connected coronaviruses, such as COVID-19, to climate change or ecological destruction, contributing to the COVID-19 crowd-out. But this over-time pattern of news coverage does not explain why only one summit — of the G7 or G20 — took the lead in every year from 2016 to 2019, and a different one from each group in every successive year.

Moreover, shocks from new scientific reports produced by authoritative mainstream sources must compete with and pass through these layers of physical, media-highlighted and communiqué-recognized shocks before they propel summit performance. Science does not speak for itself. The same sequence from sound science through physical shocks, media-highlighted shocks and communiqué-recognized shocks shows the diminishing force of the initial science as it passes through each stage. In all 12 G7 and G20 summits from 2015 to 2020, the initial science on climate change and even business leaders' risk perception of it were stronger than the communiqué-recognized shock-activated vulnerability on climate change. This suggests that although science matters, the way its results are communicated to, understood and valued by leaders matters even more. The consensus of the medical science behind COVID-19 and the economic science behind combating the resulting recession through immediate Keynesian fiscal and monetary policy stimulus was much stronger than the much less easily comprehended, multidisciplinary, scientifically complex consensus behind the complex adaptive system of climate change and how its anthropogenic components caused or amplified specific extreme weather events or chronic, cumulative dangers such as heat each year. It also matters who is relaying the scientific information. The 2021 report released by the International Energy Agency, long supported by the G7, influenced the G7's pledge to end unabated coal financing by the end of 2021, whereas science communicated by activists such as Greta Thunberg or groups such as the Union of Concerned Scientists was not heeded.

Further components are psychological and path dependence factors. Humans naturally fear and resist change. The world has depended on fossil fuels such as coal for a very long time. Today's global leaders and their citizens do not know a world powered mostly by renewable energy, or one that allows forests and nature to regenerate before humans extract more of the resources they want to support a high-quality and comfortable lifestyle. Fear of the unknown future coupled with fear of lost status or power as a result of such change is a powerful psychological obstacle that keeps the world on a familiar, even if ultimately self-destructive, path. This helps explain the high salience of individual G7 and G20 host's leadership and climate championship in producing the success of a given summit, starting with a far-sighted Helmut Schmidt at the G7's Tokyo Summit in 1979 (Kirton and Kokotsis 2015).

6. Multilateral Organizational Failure as a Constant

The sixth improvement to the systemic hub model is focusing on the embedded constitutional core and organizational culture of the major multilateral organizations from the 1940s, notably the UN, the IMF and the World Bank, compared to the more recent ones more relevant to climate change. With the exception of the World Meteorological Organization, the founding generation in the 1940s did not recognize the existence, let alone the value, of the natural environment, and produced constitutions and institutions with procedures, rules and organizational cultures that were similarly focused on economic growth and blind to the natural world. The multilateral environmental organizations from 1972 (UN Environment Programme) and 1992 (UN Biodiversity and UN Climate Change) were much weaker in their legal powers, coverage, and financial and organizational resources, and, above all, were fragmented into geographically separated components for climate, biodiversity, desertification, chemicals and wildlife, alongside the inherited World Organisation for Animal Health. No G7 or G20 government was organized in such a functionally and geographically fragmented way. At the global level, the heads of those fragmented multilateral organizational parts were never invited to G7 or G20 summits the way the heads of the IMF, the World Bank, the International Labour Organization and, later, the World Health Organization were. Multilateral organizational failure here is thus profoundly constant on climate change. To be sure, by 2021, the executive heads and governing boards of the most powerful, all economic multilateral organizations focused far more on climate change, but their constitutions, culture and professional competence remain largely unchanged. Moreover, the founding white men of the major multilateral organizations of the 1940s and Bretton Woods era largely excluded women and people of colour, continued to hold overseas territories and implemented destructive anti-Indigenous policies there and at home.

7. Predominant Equalizing Capability Components

The seventh improvement relates to predominant equalizing capability. It still counts, as realist scholars have long claimed. But the G7's declining overall global predominance from 2015 to 2021 does not account precisely for its strong climate performance in 2015, substantial performance in 2016, significant performance in 2018 and strong performance in 2021. Nor does the G20's stable overall global predominance of about 80% of global gross domestic product coincide with its spikes to strong performances in 2017 and significant performance in 2019, or its small performance in other years. Although the internal equalization of capability does better as a cause, neither it nor overall predominance coincides in any way with the collapse of G7 climate performance in 2019 and in the G7 and G20 in 2020. Nor does it explain why resources are easily found for the things that cause rather than counter climate change. More attention to the most relevant special-ized capabilities, and their concentration and distribution within the G7 and G20, is an important task for future research.

8. Converging Characteristics

The eighth improvement concerns converging characteristics in G7 and G20 members' degree of democratic openness and domestic climate change perfor-mance. Both the level and the degree of convergence have been consistently higher among G7 members and summits than among G20 ones. This coincides with the finding that the G7 has produced two strongly performing summits, in 2015 and 2021, a significantly performing one in 2018 and a substantially per-forming one in 2016, while the G20 has produced only one strongly performing one, in 2017, and one significantly performing one, in 2019. The pattern for the summit host is much clearer, as all of these successful summits have been hosted by G7 and thus democratic members. This matters for 2021, when all five of the old summits of the G7, G20 and UN and the three new special summits have been or will be hosted by G7 members.

9. Political Cohesion behind the Personal Summit Host

The ninth improvement emphasizes, among the many components of domestic political cohesion, the presence of an experienced, expert, committed leader host-ing the summit in a given year. These endogenous causes can be more salient than the largely exogenous ones of external shock-activated vulnerability, mul-tilateral organizational failure, predominant equalizing capability and converg-ing domestic characteristics and performance. Here, the greatest propeller and ensuing performance came from Merkel, the only leader to have hosted both a G7 and a G20 summit, as she did in 2015 and 2017, that produced strong success on climate change at both. Japan's Abe had a similar if less strong impact at the G7 in 2016 and G20 in 2019. So did Canada's Trudeau when he hosted in 2018, despite Trump's tantrum, which shows that the relative power of the hosting pol-ity is less salient.

Merkel and Abe hosted twice and succeeded twice, in both cases at the G7 first and the G20 next. They presumably learned from the G7 first to succeed later at the bigger, broader more diverse G20, even with Trump there. Abe acted on cli-mate change reluctantly at Osaka in 2019, but adjusted to most of his G7 partners when they told him he must.

Indeed, the United States alone does not preclude G7 or G20 action on cli-mate change, even with its leader in deep denial. Trump was outmanoeuvred on climate change by Merkel at the G20 at Hamburg in 2017, by Trudeau at the G7 at Charlevoix in 2018 and by Abe at the G20 at Osaka in 2019 — three years in a row when he was not the summit host. Conversely, a United States led by a climate-committed Obama did not guarantee climate success, as the G20 Antalya in 2015 and G20 Hangzhou in 2016 show. Moreover, the arrival at the G20 of fellow climate change–denying Jair Bolsonaro of Brazil gave Trump the backing he needed to prevent serious advances, along with the continuing resistance from Russia and Saudi Arabia.

10. The Compact, Common Valued Club at the Hub

The 10th improvement is to restore compactness and add commonality as components that make the club at the hub of the growing global summit network an interpersonally valued one. The smaller, all-democratic G7 summits did better than the bigger more diverse G20 ones. The G20 summits hosted by G7 members did better than the others. Together these two features account for all the successful summits, but not the G7's poor performance in 2017 and 2019 and its failure in the United States under Trump in 2020. Moreover, the institutionalizing, special climate summits, pioneered by France and hosted mostly by G7 members, added more radiating spokes to the network with G7 members at the common core.

Summary

These 10 improvements, guiding a refined analysis, point to promising prospects for combined success in 2021. By August 2021, the diversionary shock of COVID-19 had finally begun to retreat, the COP summit had finally reappeared and all old and new summits were hosted by G7 leaders who did not deny climate change. Yet, many risks remain, led by the continuing eruptions of COVID-19 variants, the absence of politically powerful Merkel after September, the temptation of the Rome G20 to leave climate change to the UN's Glasgow Summit immediately after, policy loopholes embedded into the G7 and G20 summit commitments and their individual policies at home, and more.

Conclusion

With the initial empirical application of these 10 analytic improvements to the systemic hub model, this book offers a new account of global summit governance of climate change from 2015 to 2021. It is centrally a story of the growing gap between the proliferating physical demand for climate change control, and the increasingly inadequate, if periodically improving, political supply of such governance from the G7, G20 and UN summits, reinforced by the reconfiguring special climate summits since 2017.

Midway through 2021, four things are clear. One is that the climate crisis and thus demand for global climate summit governance have soared to new heights. The second is that the supply from the G7, G20, UN and reconfiguring summits is also soaring to new heights. The third is that a great gap between the two is growing as the physical demand increasingly outstrips the political supply. The fourth is that the complex of old and new global summitry still has a small chance to close that gap enough now to avert catastrophe in the coming years.

Whether it will do so remains unclear and is in much doubt. But global leaders can no longer ignore the sustained demand from global citizens for real climate action. If they turn away or offer only half solutions, they risk irrelevance. If the current leaders will not lead, new leaders will emerge, including from the bottom up, and the balance of power will shift. Thus, should the leaders of countries fail to meet this unprecedented political challenge, there is still hope for humanity and the planet.

Appendix A

The Systemic Hub Model of G7 and G20 Governance

Performance Dimensions (Dependent Variable)

1. Domestic political management (credit)
   Attendance, communiqué complements, public and political approval
2. Deliberation (public communiqué conclusions, private conversations)
3. Direction setting (consensus affirmations of principles and norms)
   Distinctive foundational missions, priority placement, causal links
4. Decision making (commitments)
   Binding level, components, synergies with other subjects, compliance catalysts
5. Delivery (compliance with commitments)
   Average by year, subject, member, causes
6. Development of global governance (institutional constitutionalization)
   Inside, outside, by number, breadth, rank, inside–outside balance

Causes (Independent Variable)

1. Shock-activated vulnerability
   Communiqué recognized, media emphasized, physical, scientific reports, risks
2. Multilateral organizational failure
   Dedicated organization, fragmentations, weaknesses
3. Predominant equalizing capability
   Overall and specialized capabilities and vulnerabilities
4. Converging political characteristics
   Democracy, environmental performance
5. Domestic political cohesion
   Government control, electoral proximity, popularity, experience, expertise, conviction
6. Club at the network hub
   Host, hub size and intensity, members' membership in plurilateral summit institutions, guests' membership in plurilateral summit institutions

Appendix B
Summit Performance on Climate Change, 2015–2021

G7 and G20 Summits, 2015–2021

Date	Summit	Location	Host	Performance
2015 June 7–8	G7 Elmau	Germany	Merkel	Strong
2015 November 15–16	G20 Antalya	Turkey	Erdoğan	Small
2016 May 26–27	G7 Ise-Shima	Japan	Abe	Substantial
2016 September 4–5	G20 Hangzhou	China	Xi	Small
2017 May 26–27	G7 Taormina	Italy	Gentiloni	Small
2017 July 7–8	G20 Hamburg	Germany	Merkel	Strong
2018 June 10–12	G7 Charlevoix	Canada	Trudeau	Significant
2018 November 30–December 1	G20 Buenos Aires	Argentina	Macri	Small
2019 June 28–29	G20 Osaka	Japan	Abe	Significant
2019 August 24–26	G7 Biarritz	France	Macron	Small
2020 March 16	G7 Virtual	United States	Trump	Failure
2020 November 21–22	G20 Riyadh	Saudi Arabia	Salman	Small
2021 June 11–13	G7 Cornwall	United Kingdom	Johnson	Strong
2021 October 30–31	G20 Rome	Italy	Draghi	Substantial (projected)

UN COP Summits and Ministerial Meetings, 2015–2021

Date	Meeting	Location/type	Presidency	Performance
2015 November 30–December 13	COP21	Paris Summit	France	Solid
2016 November 7–18	COP22	Marrakech Ministerial	Morocco	Small
2017 November 6–17	COP23	Bonn Ministerial	Fiji	Solid
2018 December 2–15	COP24	Katowice Ministerial	Poland	Small

2019 December 2–13	COP25	Madrid Ministerial	Chile and Spain	Small
2020	COP26	Glasgow Summit	United Kingdom and Italy	Failure
2021 November 1–13	COP26	Glasgow Summit	United Kingdom and Italy	Substantial (projected)

Special Climate Summits, 2017–2021

Date	*Summit*	*Location*	*Host*	*Performance*
2017 December 12	One Planet Summit	Paris	Macron	Significant
2018 September 26	One Planet Summit	New York	Macron	Substantial
2019 March 14	One Planet Summit	Nairobi	Macron	Small
2019 September 21–23	UN Climate Action Summit	New York	Guterres	Significant
2020 September 24	High-Level Roundtable on Climate Action	New York (Virtual)	Guterres	Small
2020 September 30	UN Biodiversity Summit	New York (Virtual)	Guterres	Small
2020 December 12	Climate Ambition Summit	Paris (Virtual)	France/UK/UN	Significant
2021 January 11	One Planet Summit on Biodiversity	Paris (Virtual)	Macron	Solid
2021 January 25	Climate Adaptation Summit	Paris (Netherlands)	Rutte	Small
2021 April 22–23	Leaders Summit on Climate	Washington (Virtual)	Biden	Strong

Appendix C
G7 Climate Change Performance, 1975–2021

Summit	Domestic political management		Deliberation			Direction setting			Decision making			Delivery	Development of global governance		
	Compliments		Words		Documents	Priority placement	Democracy	Human rights	# made	Assessed		Score (%)	Inside	Outside	# bodies
	#	%	#	%						#	%		# references	# references	
1975	0	0	0	0	0	0	0	0	0	–	–	–	0	0	0
1976	0	0	0	0	0	0	0	0	0	–	–	–	0	0	0
1977	0	0	0	0	0	0	0	0	0	–	–	–	0	0	0
1978	0	0	0	0	1	0	0	0	0	–	–	–	0	0	0
1979	0	0	28	1.3	0	0	0	0	0	–	–	–	0	0	0
1980	0	0	0	0	0	0	0	0	0	–	–	–	0	0	0
1981	0	0	0	0	0	0	0	0	0	–	–	–	0	0	0
1982	0	0	0	0	0	0	0	0	0	–	–	–	0	0	0
1983	0	0	0	0	0	0	0	0	0	–	–	–	0	0	0
1984	0	0	0	0	0	0	0	0	0	–	–	–	0	0	0
1985	0	0	88	2.9	1	0	0	0	1	1	100	76	0	0	0
1986	0	0	0	0	0	0	0	0	0	–	–	–	0	0	0
1987	0	0	85	1.5	1	0	0	0	1	1	100	65	0	0	0
1988	0	0	140	2.7	1	0	0	0	0	–	–	–	0	0	2
1989	0	0	422	6	1	0	0	0	4	4	100	47	0	3	2
1990	0	0	491	5.9	1	0	0	0	7	4	57	45	0	3	2
1991	0	0	236	2.4	1	0	0	0	5	2	40	69	0	2	2
1992	0	0	137	1.8	1	0	0	0	7	3	43	86	2	1	1
1993	0	0	154	3.1	1	0	0	0	4	2	50	79	0	2	1
1994	0	0	107	2.6	1	0	0	0	4	2	50	86	1	2	2
1995	0	0	87	0.7	1	0	0	0	7	1	14	65	1	0	0

Summit	Domestic political management — Compliments #	Compliments %	Deliberation — Words #	Words %	Documents	Direction setting — Priority placement	Democracy	Human rights	Decision making — # made	Assessed #	Assessed %	Delivery — Score (%)	Development of global governance — Inside (# references)	Outside (# references)	# bodies
1996	0	0	167	0.8	1	0	0	0	3	1	33	79	1	2	2
1997	0	0	305	1.6	1	0	0	0	9	4	44	66	1	0	0
1998	0	0	323	5.3	1	0	0	0	10	3	30	100	1	0	0
1999	0	0	198	1.3	1	0	0	0	4	1	25	39	1	1	1
2000	0	0	213	1.6	1	0	0	0	4	1	25	72	1	1	1
2001	1	11	324	5.2	1	0	0	0	4	4	100	50	2	2	2
2002	0	0	53	0.2	1	0	0	0	1	1	100	95	1	0	0
2003	0	0	62	0.3	1	3	0	0	4	2	50	94	1	0	0
2004	0	0	98	0.3	1	5	0	0	3	3	100	85	0	20	6
2005	0	0	2,667	9.3	3	10	0	0	29	5	17	90	3	10	5
2006	0	0	1,533	3.1	3	2	0	0	20	9	45	68	1	16	7
2007	4	44	4,154	12.0	5	10	0	0	44	4	9	78	1	22	11
2008	0	0	2,568	17.5	3	8	0	0	54	5	9	77	2	19	10
2009	0	0	5,559	33.3	7	17	5	1	42	5	12	82	1	5	3
2010	1	11	1,282	12	1	1	2	0	10	5	50	67	0	7	6
2011	0	0	1,086	5.9	1	1	1	0	7	1	14	84	0	4	3
2012	0	0	789	7.1	2	0	1	0	5	1	20	56	0	5	4
2013	1	11	525	3.9	1	0	0	0	12	2	17	61	0	7	6
2014	0	0	747	14.6	1	0	1	0	16	3	18	84	0	44	3
2015	1	13	2,379	18.8	2	1	0	0	23	5	22	80	0	17	10
2016	0	0	3,802	16.5	2	1	3	2	12	3	25	73	0	0	0
2017	0	0	201	2.3	1	0	2	0	1	1	100	86	0	0	0
2018	0	0	1,696	15.1	3	1	2	0	12	5	42	84	1	5	8

Summit	Domestic political management		Deliberation			Direction setting			Decision making			Delivery	Development of global governance		
	Compliments		Documents	Words		Priority placement	Democracy	Human rights	# made	Assessed		Score (%)	Inside	Outside	
	#	%		#	%					#	%		#	# references	# bodies
2019	0	0	1	892	12.4	0	0	0	0	0	0	–	1	14	12
2020	0	0	0	0	0	0	0	0	0	0	0	N/A	0	0	0
2021	0	0	5	3,843	19	2	3	0	54	N/A	N/A	23.68	2	31	N/A
Total	7	77	60	37,441	231.49	60	19	3	423	94	1,461		25	245	110
Average	0.1	1.6	1.3	796.6	4.9	1.4	0.4	0.1	9.0	2.8	43.0	73	0.5	5.3	2.4

Notes:

All data derived from documents issued in the G7/8 leaders' names at each summit. N/A = not available.

Domestic political management: References by name to G7/8 member(s) that specifically give credit in the context of climate change; % indicates how many G7/8 members received compliments in the official documents, depending on the number of full members participating.

Deliberation: Number of references to climate change and the number of documents that contain references to climate change; unit of analysis is the paragraph; % refers to the percentage of words in all documents that refer to climate change.

Direction setting: Priority placement — number of references to climate change in the chapeau or chair's summary; unit of analysis is the sentence. Democracy and human rights: number of references in relation to climate change.

Decision making: # made — number of climate change commitments; assessed — number and percentage of climate change commitments assessed for compliance of the total made.

Delivery: Compliance score for all climate change commitments assessed for that summit.

Development of global governance: Inside — number of references to G7/8 institutions and ministerials related to climate change; outside — number of external multilateral organizations related to climate change; unit of analysis is the sentence.

Appendix D

G20 Climate Change Performance, 2008–2020

Summit	Domestic political management Compliments		Deliberation Words		Direction setting					Decision making	Delivery		Development of global governance Inside				Outside	
	#	%	#	%	Financial stability	Globalization	Priority placement	Democracy	Human rights	# made	Score (%)	% assessed	Ministerial created	Official-level body created	# references	# bodies	# references	# bodies
2008 Washington	0	0	64	1.7	0	0	0	0	1	0	–	–	0	0	0	0	0	0
2009 London	0	0	64	1.0	0	0	1	0	0	3	45	33 (1)	0	0	0	0	1	1
2009 Pittsburgh	1	5	911	9.7	0	0	4	0	0	3	93	33 (1)	4	0	2	2	10	5
2010 Toronto	1	5	838	7.4	0	0	0	1	0	3	71	100 (3)	0	0	0	0	3	3
2010 Seoul	2	10	2,018	12.7	0	0	2	1	0	8	53	50 (4)	5	3	10	7	20	11
2011 Cannes	2	10	1,167	8.2	0	0	0	1	0	8	69	37 (3)	2	0	4	2	11	7
2012 Los Cabos	0	0	1,160	9.1	0	0	0	1	0	6	80	50 (3)	1	5	8	3	6	5
2013 St Petersburg	1	5	1,697	5.9	0	0	1	0	0	11	42	27 (3)	0	3	6	5	10	7
2014 Brisbane	0	0	323	3.5	0	0	0	0	0	7	76	71 (5)	0	0	0	0	4	2
2015 Antalya	0	0	1,129	8	0	0	0	0	0	3	85	33 (1)	1	1	2	2	4	3
2016 Hangzhou	0	0	1,754	11	1	0	0	1	0	2	83	100 (2)	1	3	4	3	5	4
2017 Hamburg	0	0	5,255	15	0	0	1	1	1	22	71	23 (5)	0	11	11	5	26	9

2018 Buenos Aires	0	0	532	6	0	0	0	0	0	3	79	67 (2)	0	0	0	0	3	3
2019 Osaka	0	0	2,034	31	1	1	0	0	0	13	72	38 (5)	1	1	3	3	10	9
2020 Riyadh	0	0	679	4	2	1	0	0	0	3	N/A	N/A	0	0	2	2	4	2
Total	7	–	19,625	–	3	3	9	6	2	95	–	40 (38)	15	27	52	34	118	71
Average	0.5	0.0	1,308	8.9	0.2	0.2	0.6	0.4	0.1	6.3	69	56	1.0	1.8	3.5	2.3	7.9	4.7

Notes:

All data derived from documents issued in the G20 leaders' names at each summit. N/A = not available.

Domestic political management: References by name to G20 member(s) that specifically give credit in the context of climate change; % indicates how many G20 members received compliments in the official documents, depending on how many full members participating.

Deliberation: Number of references to climate change in the G20 leaders' documents; unit of analysis is the paragraph; % refers to the percentage of words in all documents that refer to climate change.

Direction setting: Priority placement—number of references to climate change in the chapeau or chair's summary; unit of analysis is the sentence. Financial stability, globalization, democracy and human rights: number of references in relation to climate change.

Decision making: # made—number of climate change commitments made.

Delivery: Compliance score for all climate change commitments assessed for that summit; % assessed—percentage of climate change commitments assessed for compliance of the total made; number in parentheses refers to number of commitments assessed.

Development of global governance: Inside—number of references to internal institutions made in relation to climate change; outside—number of external multilateral organizations related to climate change; unit of analysis is the sentence.

Appendix E

G7 Recognized Shock-Activated Vulnerabilities, 2015–2021

Subject	Total	2015 Elmau	2016 Ise-Shima	2017 Taormina	2018 Charlevoix	2019 Biarritz	2020 United States	2021 Cornwall
Shocks								
Health	76	2	5		1			68
Regional security	24	1	4	7	8	2	2	
Gender	16		2	1	10	1	1	1
Environment/ biodiversity	12				6			6
Economy	9							9
Terrorism	9			4	1	2	2	
Migration and refugees	5		2	1	2			
Cybersecurity	4				2	1	1	
Climate change	3							3
Food and agriculture	3	2		1				
Development	3							3
Human rights	3							3
Democracy	3							3
"Crisis/emergency"	3							3
Trade	2							2
Financial crisis	2	1	1		1			
Disasters	2	2						
Labour and employment	1							1
Crime and corruption	1							1
Development	1			1				
Democracy	1						1	
Total	183	8	14	15	31	6	7	103
Vulnerabilities								
Terrorism	6		6					
Health	2		2					
Energy	2	1	1					
Environment	2				2			
Crime and corruption	2		2					
Gender	1		1					
Infrastructure	1		1					
Financial crisis	1		1					
Cybersecurity	1		1					
Regional security	1	1						
Climate change	0							
Total	19	2	15	0	2	0	0	

Appendix F

G20 Recognized Shock-Activated Vulnerabilities, 2015–2020

Subject	Total	2015 Antalya	2016 Hangzhou	2017 Hamburg	2018 Buenos Aires	2019 Osaka	2020 Riyadh
Shocks							
Financial crisis	7		3	4			
Macroeconomic crisis	0						
Trade	1	1					
Labour and employment	1			1			
Migration and refugees	5	3	2				
Terrorism	3	2				1	
Health	38						38
Food and agriculture	0						
Climate change	0						
Total	55	6	5	5	0	1	38
Vulnerabilities							
Financial crisis	8	1	3	3	1		
Macroeconomic crisis	1		1				
Trade	0						
Labour and employment	0						
Migration and refugees	0						
Terrorism	1	1					
Health	8		1	2			5
Food and agriculture	1			1			
Crime and corruption	1			1			
Climate change	0						
Total	20	2	5	7	1	0	5

Appendix G

Physical Shock-Activated Vulnerabilities, 2014–2020

Year	Number of extreme weather events	Number of deaths	Number of injuries
G7 countries			
2014	51	556	22,301
2015	50	3,736	15,537
2016	47	693	4,467
2017	44	493	1,011
2018	48	857	51,231
2019	55	1,921	24,980
2020	48	4,868	794
Total	343	13,124	120,321
G20 countries			
2014	147	3,757	29,199
2015	149	7,898	19,853
2016	136	4,350	10,343
2017	131	4,125	9,334
2018	129	8,441	70,437
2019	146	5,696	29,252
2020	135	8,454	3,758
Total	974	42,721	172,176

Source: EM-DAT, CRED/UCLouvain, Brussels, Belgium. www.emdat.be (D. Guha-Sapir). Version of May 7, 2021; file created on May 7, 2021, at 23:50:28 CEST.

Note: Disasters recorded include extreme temperatures, floods, storms, typhoons, volcanic activity, earthquakes, landslides, wildfires and drought. Does not include the European Union as a whole.

Appendix H

G7 and G20 Relative Capability, 2014–2020

Gross domestic product (annual growth, %)

	2014	*2015*	*2016*	*2017*	*2018*	*2019*	*2020*
Argentina	−2.5	+2.7	−2.1	+2.8	−2.6	−2.1	−9.9
Australia	+2.5	+2.2	+2.8	+2.3	+2.9	+2.2	−0.3
Brazil	+0.5	−3.5	−3.3	+1.3	+1.3	+1.1	−4.1
Canada	+2.9	+0.7	+1.0	+3.2	+2.0	+1.7	−5.4
China	+7.4	+7.0	+6.8	+6.9	+6.7	+6.1	+2.3
France	+1.0	+1.1	+1.1	+2.3	+1.8	+1.5	−8.1
Germany	+2.2	+1.5	+2.2	+2.6	+1.3	+0.6	−4.9
India	+7.4	+8.0	+8.3	+7.0	+6.1	+4.2	−8.0
Indonesia	+5.0	+4.9	+5.0	+5.1	+5.2	+5.0	−2.1
Italy	0	+0.8	+1.3	+1.7	+0.9	+0.3	−8.9
Japan	+0.4	+1.2	+0.5	+2.2	+0.3	+0.7	N/A
Korea	+3.2	+2.8	+2.9	+3.2	+2.9	+2.0	−1.0
Mexico	+2.8	+3.3	+2.6	+2.1	+2.2	−0.1	−8.2
Russia	+0.7	−2.0	+0.2	+1.8	+2.5	+1.3	−3..0
Saudi Arabia	+3.7	+4.1	+1.7	−0.7	+2.4	+0.3	−4.1
South Africa	+1.8	+1.2	+0.4	+1.4	+0.8	+0.2	−7.0
Turkey	+4.9	+6.1	+3.3	+7.5	+3.0	+0.9	+1.8
United Kingdom	+2.6	+2.4	+1.9	+1.9	+1.3	+1.5	−9.8
United States	+2.5	+2.9	+1.6	+2.4	+2.9	+2.2	−3.5
European Union	+1.6	+2.3	+2.0	+2.8	+2.1	+1.6	−6.2

Source: World Bank Indicators (2021). "GDP Growth (Annual %)." Date accessed: August 28, 2021. https://databank.worldbank.org/indicator/NY.GDP.MKTP.KD.ZG/1ff4a498/Popular-Indicators#.

Note: N/A = not available.

Appendix I

G7 and G20 Emissions, 2014–2019

Member	2014	2015	2016	2017	2018	2019
Argentina	−0.69	+1.57	−7.49	+0.36	−0.55	1.7
Australia	+1.97	+0.05	+1.92	+7.93	−0.61	−0.90
Brazil	+2.83	−1.35	+6.62	+1.21	−2.80	+9.6
Canada	+0.82	−0.41	−13.44	+2.00	+2.37	+0.20[a]
China	+0.10	−0.05	+0.52	+1.79	+2.61	+2.6
European Union (27)	−4.65	+1.87	+11.21	+0.41	−2.02	−3.8
France	−8.36	+1.65	−0.27	+0.66	−2.46	−0.90
Germany	−4.71	+0.47	+0.30	−2.13	−3.31	−8.00[a]
India	+6.10	+0.46	+2.37	+4.22	+4.49	+1.8
Indonesia	+23.23	+2.68	−29.80	+0.88	+16.22	+4.66
Italy	−5.05	+2.94	+4.50	−1.41	−1.59	−1.00
Japan	−2.92	−2.88	+0.85	−1.82	−4.04	−4.30
Korea	−1.72	+3.51	+2.49	+2.06	+1.28	+5.05
Mexico	−1.33	+3.79	+1.99	−0.58	+0.10	+2.95
Russia	−0.18	−2.72	+7.09	+1.09	+2.72	−0.71
Saudi Arabia	+7.44	+4.92	+0.57	−1.22	−2.74	+5.85
South Africa	+2.33	−4.52	+0.15	+2.58	−0.16	+1.5
Turkey	+7.26	+1.83	+15.5	+9.69	−0.03	+2.88
United Kingdom	−7.52	−3.19	−4.62	−3.03	−2.02	−3.60[a]
United States	+0.36	−2.18	+1.61	−1.11	+3.22	−2.90
G7 average	−4.00	−2.22	+0.02	−0.80	−9.85	−3.04
G20 average	+0.76	+0.42	+0.10	+1.18	+0.66	+0.67

Sources:
2014–2018: Climate Watch (2020). "GHG Emissions." Washington, DC: World Resources Institute. Date accessed: July 14, 2021. https://www.climatewatchdata.org/ghg-emissions.
2019: Al Jazeera (2020). "Brazil's Carbon Emissions Rose in 2019 with Amazon Deforestation." November 7. https://www.aljazeera.com/news/2020/11/7/brazils-carbon-emissions-rose-in-2019-with -amazon-deforestation; Canada. Environment and Climate Change Canada (n.d.). "Greenhouse Gas Emissions." Last Modified: April 15, 2021. https://www.canada.ca/en/environment-climate-change /services/environmental-indicators/greenhouse-gas-emissions.html. European Environment Agency (2021). "Major Drop in EU's Greenhouse Gas Emissions in 2019, Official Data Confirms." May 31. https://www.eea.europa.eu/highlights/major-drop-in-eus-greenhouse; Goswami, Urmi (2019). "Carbon Dioxide Emissions Slow Down on Weak Growth, Less Coal Use," *Economic Times*, December 5. https://economictimes.indiatimes.com/news/economy/indicators/carbon-dioxide-emissions-slow -down-on-weak-growth-less-coal-use/articleshow/72360874.cms; Grant, Mikhail and Kate Larsen (2020). "Preliminary China Emissions Estimates for 2019." Rhodium Group, March 18. https:// rhg.com/research/preliminary-china-emissions-2019; International Energy Agency (2020). "Global CO2 Emissions in 2019." February 11. https://www.iea.org/articles/global-co2-emissions-in-2019;

Italian National Agency for New Technologies, Energy and Sustainable Economic Development (2020). "Climate: ENEA- 2019, Italy Towards 1% Drop in Greenhouse Gas Emissions." February 13. https://www.enea.it/en/news-enea/news/climate-enea-2019-italy-towards-1-drop-in-greenhouse-gas -emissions; Karombo, Tawanda (2020). "South Africa, Already the Continent's Biggest Polluter, Saw a Rise in Carbon Emissions Last Year." Quartz, December 15. https://qz.com/africa/1946022/south -africa-saw-a-rise-in-carbon-emissions-in-2019; Knoema (n.d.). "World Data Atlas: Argentina — CO2 Emissions." Date accessed: August 31, 2021. https://knoema.com/atlas/Argentina/CO2-emissions; Knoema (n.d.). "World Data Atlas: Indonesia — CO2 Emissions Per Capita." Date accessed: August 31, 2021. https://knoema.com/atlas/Indonesia/CO2-emissions-per-capita; Knoema (n.d.). "World Data Atlas: Mexico — CO2 Emissions." Date accessed: August 31, 2021. https://knoema.com/atlas/ Mexico/CO2-emissions; Knoema (n.d.). "World Data Atlas: Republic of Korea — CO2 Emissions." Date accessed: August 31, 2021. https://knoema.com/atlas/Republic-of-Korea/CO2-emissions; Knoema (n.d.). "World Data Atlas: Russian Federation — CO2 Emissions Per Capita." Date accessed: August 31, 2021. https://knoema.com/atlas/Russian-Federation/CO2-emissions-per-capita; Knoema (n.d.). "World Data Atlas: Saudi Arabia — CO2 Emissions." Date accessed: August 31, 2021. https://knoema.com/atlas/Saudi-Arabia/CO2-emissions; Knoema (n.d.). "World Data Atlas: Turkey — CO2 Emissions Per Capita." Date accessed: August 31, 2021. https://knoema.com/atlas/Turkey/ CO2-emissions-per-capita; Reuters (2020). "Australia's Greenhouse Gas Emissions Fall Slightly in 2019." May 28. https://www.reuters.com/article/us-australia-emissions-idUSKBN235088; Tapolsky, Stephanie (2020). "France Greenhouse Gas Emissions Decreased by 16.9% From 1990 Levels." Climate Scorecard, December 18. https://www.climatescorecard.org/2020/12/france-greenhouse-gas -emissions-decreased-by-16-9-from-1990-levels; United Kingdom. Department for Business, Energy and Industrial Strategy (2020). "2019 UK Greenhouse Gas Emissions, Provisional Figures." March 26. https://assets.publishing.service.gov.uk/government/uploads/system/uploads/attachment_data/file /875485/2019_UK_greenhouse_gas_emissions_provisional_figures_statistical_release.pdf.

Notes:
2014–2018: Sectors/subsectors: Total greenhouse gas emissions with land use, land change and forestry. Gases: Aggregate greenhouse gases (carbon dioxide, methane, nitrous oxide). Calculation: Percentage change from prior year.

[a] Carbon dioxide emissions only.

Appendix J
G7 and G20 Democracy Rank, 2014–2020

Member	2014	2015	2016	2017	2018	2019	2020	2015–2020
Argentina	52	50	49	48	47[a]	48	48	+2
Australia	9	9	10	8	9	9	9[a]	0
Brazil	44	51	51	49	50	52	49	+3
Canada	7	7	6[a]	6[a]	6[a]	7[a]	5	+2
China	144	136[a]	136[a]	139	130	153	151	−15
France	23	27	24[a]	29	29	20	24	+3
Germany	13	13	13	13	13	13	14	−1
India	27	35	32	42	41	51	53	−18
Indonesia	49	49	48	68	65	64	64[a]	−15
Italy	29	21	21[a]	21[a]	33	35	29	−8
Japan	20	23[a]	20	23[a]	22	24	21	+2
Korea	21	22	24[a]	20	21	23	23	−1
Mexico	57[a]	66	67	66[a]	71[a]	73	72	−6
Russia	132	132[a]	134	135	144[a]	134[a]	124	+8
Saudi Arabia	161	160[a]	159[a]	159[a]	159[a]	159[a]	156	+4
South Africa	30	37	39	41	40	40	45	−8
Turkey	98[a]	97	97	100	110	110	104	−7
United Kingdom	16	16	16	14	14	14	16	0
United States	19	20	21[a]	26	25	25	25	−5
G20 average	47	35	48	52	43	47	56	−21
G7 average	18	17	16	21	23	22	19	+2
BRICS	75	41	64	81	65	74	84	−43

Source: Economist Intelligence Unit (2017). "Democracy Index 2017: Free Speech under Attack." London. https://services.eiu.com/campaigns/democracy-index-2017-free-speech-under-attack.

Notes: Democracy scores and ranks are measured by the following indicators: electoral process and pluralism, functioning of government, political participation, political culture and civil liberties. Rankings for 2014–2019 cover 161 countries; rankings for 2020 cover 167 countries.
[a] Country is tied with another country, either inside or outside the G7/G20.

Appendix K

G7 and G20 Members' Climate Change Performance, 2014–2020

Country	2014	2015	2016	2017	2018	2019	2020
Argentina	41	48	48	36	46	34	46
Australia	57	60	59	57	57	55	54
Brazil	36	49	43	40	19	22	25
Canada	58	58	56	55	51	54	58
China	46	45	47	48	41	33	33
France	10	12	8	4	15	21	23
Germany	19	22	22	29	22	27	19
India	30	31	25	20	14	11	10
Indonesia	34	23	24	22	37	38	24
Italy	18	17	11	16	16	23	27
Japan	50	53	58	60	50	49	45
Korea	53	55	57	58	58	57	53
Mexico	20	18	28	28	27	25	32
Russia	56	56	53	53	53	52	52
Saudi Arabia	61	61	61	61	60	60	60
South Africa	39	37	38	32	48	39	37
Turkey	54	51	50	51	47	50	42
United Kingdom	5	6	5	6	9	8	5
United States	43	44	34	43	56	59	61
European Union	N/A	N/A	N/A	N/A	21	16	16
Average	38	39	38	38	37	37	36

Source: Germanwatch (2015–2020). "Climate Change Performance Index." Various editions. Bonn. https://germanwatch.org/en/CCPI.

Note: The years 2015, 2016, 2019, 2020 ranked 58 countries; 2017 ranked 57 countries; 2018 ranked 56 countries; 2019 and 2020 ranked 60 countries. N/A = not available.

Appendix L
Club at the Hub

Plurilateral summit institution	United States	Japan	Germany	France	United Kingdom	Italy	Canada	Russia	China	India	Brazil	South Africa	Mexico	Indonesia	Korea	Turkey	Australia	Saudi Arabia	Argentina
G7	1975	1975	1975	1975	1975	1975	1976	1998											
G20	2008	2008	2008	2008	2008	2008	2008	2008	2008	2008	2008	2008	2008	2008	2008	2008	2008	2008	2008
APEC	1993	1993					1993	1998	'91				1993	1993	1993		1993		
BRF						2017		2017	2017					2017		2017			2017
BRICS								2009	2009	2009	2009	2011							
CHOGM	1948				1948		1948			1948		1948					1948		
EAS	2011	2005						2011	2005	2005				2005	2005		2005		
ICO														1969		1969		1969	
NATO	1957		1957	1957	1957	1957	1957												
NSS	2010	2010	2010	2010	2010	2010	2010	2010	2010	2010	2010	2010	2010	2010	2010	2010	2010	2010	2010
OIF	1985			1985			1985												
Quad	2021	2021								2021							2021		
Total	9	6	4	5	5	5	7	7	6	6	3	4	3	6	4	4	6	3	3

Notes:

Indicates the year the G20 or G7 member joined the plurilateral summit organization. Blank = no membership.

Russia was a member of the G8 from 1998 until it was suspended in 2014.

Includes plurilateral summit institutions with G20 members from at least two transoceanic regions and operating in 2021. Excludes the following as they are land based, regional, trilateral or bilateral: Lusophones; India, Brazil and South Africa (IBSA); Shanghai Cooperation Organization; European Union; North American Leaders' Summit; and the Asia-Europe Meeting.

APEC = Asia-Pacific Economic Cooperation; founded 1993; annual summits.

BRF = Belt and Road Forum; founded 2015; summits every two years.

CHOGM = Commonwealth Heads of Government Meeting; founded 1948; summits every two years.

EAS = East Asian Summit; founded 2002; annual summits.

ICO = Islamic Conference Organization; founded 1969; annual summits.

NATO = North Atlantic Treaty Organization; founded 1957; 29 summits.

NSS = Nuclear Security Summit; founded 2010; four summits.

OIF = Organisation internationale de la Francophonie; founded 1970; summits every two years since 1986.

Quad = Quadrilateral Summit, 2021.

Bibliography

Note: Documents issued by the G7 and G20 leaders and their ministers are available on the G7 Information Centre at http://www.g7.utoronto.ca and the G20 Information Centre at http://www.g20.utoronto.ca.

AAP (2017). "Brazil Warns Climate Spat May Derail Talks." *SBS News*, November 11. https://www.sbs.com.au/news/brazil-warns-climate-spat-may-derail-talks.

Abbott, Kenneth W. and Duncan Snidal (2000). "Hard and Soft Law in International Governance." *International Organization* 54(3): 421–56.

Abe, Daisuke (2019). "Koizumi Admits to Japan's Coal Addiction, but Offers No Way Out." *Nikkei Asia*, December 12. https://asia.nikkei.com/Spotlight/Environment/Koizumi-admits-to-Japan-s-coal-addiction-but-offers-no-way-out.

Abe, Shinzo (2019a). "Defeatism About Japan Is Now Defeated." *Davos*, January 23. https://www.weforum.org/agenda/2019/01/abe-speech-transcript.

Abe, Shinzo (2019b). "With Great Support from You All, I Am Determined to Lead the Osaka Summit Towards Great Success." January. https://www.mofa.go.jp/policy/economy/g20_summit/osaka19/en/summit/message.

Abeysinghe, Achala, Brianna Craft and Janna Tenzing (2016). "The Paris Agreement and the LDCS: Analysing COP21 Outcomes from LDC Positions." International Institute for Environment and Development, London, March. http://pubs.iied.org/pdfs/10159IIED.pdf.

Abnett, Kate and Matthew Green (2021). "Global Temperatures Reached Record Highs in 2020, Say EU Scientists." *Reuters*, January 8. https://www.reuters.com/article/us-climate-change-temperature-idUSKBN29D0QQ.

Adesina, Akinwumi (2017). "Remarks by Akinwumi A. Adesina, President of the African Development Bank, at the G7 Summit, May 26–27, 2017, Taormina, Italy." May 27. https://www.afdb.org/fileadmin/uploads/afdb/Documents/Generic-Documents/SPEECH_ENGL_PRST_G7_Summit_Italy.pdf.

AFP (2018). "UN Climate Summit: The Three Key Outcomes You Need to Know." *The Journal*, December 16. https://www.thejournal.ie/un-climate-summit-three-key-outcomes-4398372-Dec2018.

African Development Bank (2019). "The African Development Bank Pledges US$ 25 Billion to Climate Finance for 2020–2025, Doubling Its Commitments." March 14. https://www.afdb.org/en/news-and-events/the-african-development-bank-pledges-us-25-billion-to-climate-finance-for-2020-2025-doubling-its-commitments-19090.

Agence France-Presse (2017a). "Donald Trump Leaves G7 Summit in Historic Climate Change Split." *The Journal*, May 27. https://www.thejournal.ie/g7-trump-climate-3412098-May2017/.

Agence France-Presse (2017b). "Glimmer of Hope for Climate Deal as 'Wide Open' Trump Pauses to Think." *The Standard*, May 17. https://www.thestandard. com.hk/section-news/section/11/183391/Glimmer-of-hope-for-climate-deal-as-'wide-open'-Trump-pauses-to-think.

Al-Khattaf, Eman (2020). "G20 Riyadh Summit Leads Efforts to Fight Climate Change." *Asharq Al-Awsat*, November 20. https://english.aawsat.com/home/article/2653611/g20-riyadh-summit-leads-efforts-fight-climate-change.

al-Sudairi, Mazen (2020). "Did the G20 Summit Matter." *Al Arabiya*, November 30. https://english.alarabiya.net/in-translation/2020/11/30/Did-the-G20-summit-matter-.

Alabi, Leke Oso (2021). "UK Still in Pain as US and China Rebound." *Financial Times*, January 2, p. 11.

Aluwaisheg, Abdel Aziz (2020). "Key Takeaways from an Unprecedented G20 Summit." *Arab News*, November 24. https://www.arabnews.com/node/1767781.

Ambumozhi, Venkatachalam (2018). "The G20 and Its Role in Driving Low-Carbon Transition through Private Finance." *Global Solutions Journal* 1(3): 80–91. December. https://www.global-solutions-initiative.org/wp-content/uploads/2019/09/GS18_Journal3_E-Reader.pdf.

Amelang, Sören (2017). "Germany to Miss Climate Targets 'Disastrously': Leaked Government Paper." *Climate Home News*, October 11. https://www.climatechangenews. com/2017/10/11/germany-miss-climate-targets-disastrously-leaked-government-paper.

America's Pledge (2017). "America's Pledge Phase 1 Report: States, Cities and Businesses in the United States Are Stepping up on Climate Action." November. https://assets. bbhub.io/dotorg/sites/28/2017/11/AmericasPledgePhaseOneReportWeb.pdf.

Anderson, Ashley and Heidi Huntington (2017). "Social Media, Science and Attack Discourse: How Twitter Discussions of Climate Change Use Sarcasm and Incivility." *Science Communication* 39(5): 598–620. https://doi.org/10.1177/1075547017735113.

Andreoni, Manuela and Ernesto Londoño (2021). "Brazil, Besieged by Covid, Now Faces a Severe Drought." *New York Times*, June. 19. https://www.nytimes.com/2021/06/19/world/americas/brazil-drought.html.

Andrione-Moylan, Alex, Wouters Jan and Oldani Chiara (2019). "Reflecting on the Future of Global Governance." In *The G7, Anti-Globalization and the Governance of Globalization*, Chiara Oldani and Jan Wouters, eds. Abingdon: Routledge, pp. 169–77. https://doi.org/10.4324/9780429506642.

Angus Reid Institute (2021). "Federal Politics: Trudeau Approval Sinks over Vaccination Rollout Delays, but National Political Picture Remains Static." February 22. https://angusreid.org/wp-content/uploads/2021/02/2021.02.19_Federal_Feb.pdf.

ANSA (2016). "Referendum Key to Future Political Credibility Says Renzi." July 4. https://www.ansa.it/english/news/2016/07/04/referendum-key-to-future-political-credibility-says-renzi_1b004dda-3a65-4351-b10c-05a4c41e3c8e.html.

Aose, Takeshi and Mayo Tomioka (2019). "気候変動対策にウイグル問題…Ｇ２０での議論求めデモ." ["Demonstrations Calling for G20 Discussion on Climate Change and Uighurs."] *Asahi Shimbun*, June 27. https://www.asahi.com/articles/ASM6W3R8PM6WPTIL00P.html.

Appunn, Kerstine, Ruby Russell and Ellen Thalman (2015). "Controversial Climate Summit Issues: Positions in Germany." *Clean Energy Wire*, November 20. https://www.cleanenergywire.org/factsheets/controversial-climate-summit-issues-positions-germany.

Asahi Shimbun(2019a). "Ｇ２０首脳宣言、「パリ協定言及なければ賛成せぬ」 マクロン大統領牽制." ["On G20 Leaders Declaration, President Macron Says 'Will Not Agree If the Paris Agreement Is Not Mentioned'."] June 27. https://www.asahi.com/articles/DA3S14071776.html.

Asahi Shimbun (2019b). "大阪ｇ２０閉幕 安倍外交の限界見えた." ["As the G20 Summit Ends, the Limits of Abe's Diplomacy Are Revealed."] June 30. Editorial. https://www.asahi.com/articles/DA3S14076118.html.

Balmer, Crispian and Gavin Jones (2021). "Former Ecb Chief Mario Draghi Forms New Italian Government, Unveils Cabinet." *Globe and Mail*, February 21. https://www.theglobeandmail.com/world/article-draghi-tells-italys-president-he-has-secured-enough-support-to-form/.

Banda, Maria (2018). "G7 Charlevoix: The G6's Missed Opportunity on Climate Change." *Open Canada*, June 15. https://www.opencanada.org/features/g7-charlevoix-g6s-missed-opportunity-climate-change.

Banos Ruiz, Irene (2018). "Can Poland End Its Toxic Relationship with Coal?" *Deutsche Welle*, November 23. https://www.dw.com/en/can-poland-end-its-toxic-relationship-with-coal/a-46356824.

Barber, Tony (2021a). "Cdu Stumbles as the 16-Year Merkel Era Draws to Its Close." *Financial Times*, March 16, p. 3.

Barber, Tony (2021b). "Democracies Must Put Their Houses in Order." *Financial Times*, March 4, p. 15.

Barnett, Sophie (2016). "The G7 and Gender Equality: Ise-Shima Edition." *NATO Association of Canada*, June 6. https://natoassociation.ca/the-g7-and-gender-equality-ise-shima-edition/.

Bartlett, Duncan (2021). "Prime Minister Suga Boldly Took His Place among the G7 Stars and Came Back with Some Big Wins for Japan." Japan Forward, June 14. https://japan-forward.com/prime-minister-suga-boldly-took-his-place-among-the-g7-stars-and-came-back-with-some-big-wins-for-japan.

Bauer, Steffen (2017). "COP23: Not Great, but Good Enough." *The Current Column (Blog)*. German Development Institute, November 23. https://www.die-gdi.de/en/the-current-column/article/cop23-not-great-but-good-enough/.

Bauer, Steffen, Axel Berger and Gabriela Iacobuta (2019). "With or without You: How the G20 Could Advance Global Action toward Climate-Friendly Sustainable Development." Briefing Paper 10/2019, German Development Institute. https://www.die-gdi.de/uploads/media/BP_10.2019.pdf.

Bauer, Steffen, Axel Berger and Gabriela Iacobuta (2020). "With or without You: How the G20 Could Advance Global Action toward Climate-Friendly Sustainable Development." *Global Solutions Journal* 5: 115–21. April. https://www.global-solutions-initiative.org/wp-content/uploads/2020/04/GSJ_issue5.pdf.

Bay Area News Group (2020). "Map: 33 People Killed in California Wildfires, 2020 Season." *Mercury News*, October 2. Updated November 18, 2020. https://www.mercurynews.com/2020/10/02/map-31-people-killed-in-california-wildfires-2020-season.

BBC Monitoring European (2019). "French President Sums up G20 Summit Discussion." Text of report by Presidency of the French Republic website on 29 June, July.

BBC News (2016). "Japan's Tsunami Debris: Five Remarkable Stores." March 9. https://www.bbc.com/news/world-asia-35638091.

BBC News (2019a). "Chile Cancels Climate and Apec Summits Amid Mass Protests." October 30. https://www.bbc.com/news/world-latin-america-50233678.

BBC News (2019b). "Climate Change: What Did Greta Thunberg Say at Cop25?" December 11. https://www.bbc.co.uk/newsround/50743328.

BBC News (2020a). "Australia Bushfires: Hundreds of Deaths Linked to Smoke, Inquiry Hears." May 26. https://www.bbc.com/news/world-australia-52804348.

BBC News (2020b). "In Pictures: Hurricanes Leave Hondurans Homeless and Destitute." November 29. https://www.bbc.com/news/world-latin-america-55064560.

Beaulieu, Annie (2017a). "A Focus on Sustainable Growth but No Mention of Tourism's Potential." *G20 Research Group*, July 8. https://www.g7g20.utoronto.ca/comment/170713-beaulieu.html.

Beaulieu, Annie (2017b). "A Sustainable Site for Canada's G7 Summit in 2018." *G7 Research Group*, May 27. https://www.g7.utoronto.ca/evaluations/2017taormina/beaulieu-charlevoix.html.

Berger, Axel, Andrew F. Cooper and Sven Grimm (2019). "A Decade of G20 Summitry: Assessing the Benefits, Limitations and Future of Global Club Governance in Turbulent Times." *South African Journal of International Affairs* 26(4): 493–504. https://doi.org/10.1080/10220461.2019.1705889.

Best, Emma, Taylor Grott, Philip Gazaleh et al. (2015). "The G20 Antalya Summit: Expectations, Results and the Road Ahead." *G20 Research Group*, November 16. https://www.g20.utoronto.ca/analysis/151116-analysts.html.

Betsill, Michele, Navroz K. Dubash, Matthew Paterson et al. (2015). "Building Productive Links between the UNFCCC and the Broader Global Climate Governance Landscape." *Global Environmental Politics* 15(2): 1–10. https://doi.org/10.1162/GLEP_a_00294.

Beyond (2016). "COP22: Outcomes and Reactions." November 18. https://www.beyond-magazine.com/cop22-outcomes-and-reactions/.

Bhatia, Rajiv (2020). "G20: What All Did Riyadh Achieve." *Financial Express*, December 3. https://www.financialexpress.com/opinion/g20-what-all-did-riyadh-achieve/2141959.

Bishop, Matthew Louis and Anthony Payne (2020). "Steering Towards Reglobalization: Can a Reformed G20 Rise to the Occasion." *Globalizations* 18(1): 120–40. https://doi.org/10.1080/14747731.2020.1779964.

Bittner, Jochen (2019). "The Greens Are Germany's Leading Political Party. Wait, What?" *New York Times*, June 19. https://www.nytimes.com/2019/06/19/opinion/greens-party-germany.html.

Blanchfield, Mike (2015). "Canada, Japan Blocking Consensus at G7 on Greenhouse Gas Reductions." *Globe and Mail*, June 8. https://www.theglobeandmail.com/news/politics/canada-japan-blocking-consensus-at-g7-on-greenhouse-gas-reductions/article24840166/.

Bland, Katrina (2018). "Canada's Sixth G7 Summit: Charlevoix." *G7 Research Group*, January 29. http://www.g7g20.utoronto.ca/comment/180129-bland.html.

Blinken, Anthony (2021). "Tackling the Crisis and Seizing the Opportunity: America's Global Climate Leadership." Speech to the Philip Merrill Environmental Centre, Annapolis, MD, U.S. Department of State, April 19. https://www.state.gov/secretary-antony-j-blinken-remarks-to-the-chesapeake-bay-foundation-tackling-the-crisis-and-seizing-the-opportunity-americas-global-climate-leadership/.

Boffey, Daniel (2019). "G7 Leaders Wait Nervously for Boris Johnson's Debut on the World Stage." *Guardian*, August 18. https://www.theguardian.com/world/2019/aug/18/g7-leaders-wait-nervously-for-boris-johnson-debut-biarritz.

Boffey, Daniel and Jennifer Rankin (2020). "EU Leaders Seal Deal on Spending and €750bn Covid-19 Recovery Plans." *Guardian*, July 21. https://www.theguardian.com/world/2020/jul/20/macron-seeks-end-acrimony-eu-summit-enters-fourth-day.

Böhme, Henrik (2015). "Opinion: Schloss Elmau Is the Last Chance for the G7." *DW*, June 4. https://www.dw.com/en/opinion-schloss-elmau-is-the-last-chance-for-the-g7/a-18496963.

Boykoff, Max, Rogelio Fernández-Reyes, Jennifer Katzung et al. (2021). "'This Is Not Your Grandparents' Climate'." *Media and Climate Change Observatory* 54. June. https://scholar.colorado.edu/concern/articles/ff365644n.

BP (2019). *Statistical Review of World Energy 2019*. 68th edition, London. https://www.bp.com/content/dam/bp/business-sites/en/global/corporate/pdfs/energy-economics/statistical-review/bp-stats-review-2019-full-report.pdf.

BP (2020). *Statistical Review of World Energy 2020*. 69th edition, London. https://www.bp.com/content/dam/bp/business-sites/en/global/corporate/pdfs/energy-economics/statistical-review/bp-stats-review-2020-full-report.pdf.

Britton, Bianca (2019). "Greta Thunberg Criticizes World Leaders' Climate Actions as They Meet at Cop25 to Discuss the Crisis." *CNN*, December 6. https://www.cnn.com/2019/12/06/europe/greta-thunberg-cop25-climate-crisis-intl/index.html.

Brocchieri, Federico (2015). "Matteo Renzi Announces Pre-COP21 Meeting in Italy." June 22. http://climatetracker.org/matteo-renzi-announces-pre-cop21-meeting-in-italy/.

Brown, Alexander (2021). "Boris Johnson Urges World Leaders to Work Together So the Whole Planet Is Vaccinated." *Scotsman*, February 19. https://www.scotsman.com/news/politics/boris-johnson-urges-world-leaders-to-work-together-so-the-whole-planet-is-vaccinated-3140857.

Brown, Garrett Wallace (2016). "Global Health Gets a Vital Injection at the G7 Japan Summit, but Not the Cure." *Policy Brief*, Global Policy, May. https://www.globalpolicyjournal.com//sites/default/files/inline/files/Brown%20-%20Global%20Health%20gets%20a%20Vital%20Injection%20at%20the%20G7%20Japan%20Summit,%20but%20not%20the%20Cure.pdf.

Brown, Gordon (2020). "G20 Nations Must Devise a Strategy for Post-Pandemic Growth." *Financial Times*, December 17, p. 17. https://www.ft.com/content/e6b2fced-7392-4624-8e4e-ff039e0020dd.

Buncombe, Andrew (2017). "Donald Trump Seems Happy to Destroy the Planet. Only China and India Can Save Us." *Independent*, May 28. http://www.independent.co.uk/voices/donald-trump-climate-change-paris-accord-destroy-planet-india-china-a7760731.html.

Byrne, Caitlin (2016). "China's G20 Summit Was Big on Show, but Short on Substance." *The Conversation*, September 5. https://theconversation.com/chinas-g20-summit-was-big-on-show-but-short-on-substance-64866.

Caballero, Paula, David Waskow and Christina Chan (2017). "4 Signs to Watch at COP23." World Resources Institute, October 27. https://www.wri.org/insights/4-signs-watch-cop23.

Cairns Post (2016). "Rivals Turn Allies in Climate Change Fight." September 5.

Canada. Environment and Climate Change Canada (2017a). "Canada and the United Kingdom Announce a Global Alliance on Coal Phase-Out." London, October 11. https://www.canada.ca/en/environment-climate-change/news/2017/10/canada_and_the_unitedkingdomannounceaglobalallianceoncoalphase-o.html.

Canada. Environment and Climate Change Canada (2017b). "Canada and the World Bank Group to Support the Clean Energy Transition in Developing Countries and Small Island Developing States." Paris, December 12. https://www.canada.ca/en/environment-climate-change/news/2017/12/canada_and_the_worldbankgrouptosupportthecleanenergytransitionin.html.

Canada. Environment and Climate Change Canada (2017c). "Ministerial Meeting on Climate Action Co-Chairs Summary." *Montreal*, September 16. https://www.canada.ca/en/environment-climate-change/news/2017/09/ministerial_meetingonclimateaction.html.

Canada. Prime Minister's Office (2017). "Prime Minister Unveils Themes for Canada's 2018 G7 Presidency." December 14. https://pm.gc.ca/en/news/news-releases/2017/12/14/prime-minister-unveils-themes-canadas-2018-g7-presidency.

Canadian Foreign Policy Journal (2018). "2018 Trudeau Report Card." Prepared by Norman Paterson School of International Affairs in partnership with iAffairs Canada, March. https://iaffairscanada.com/wp-content/uploads/2018/07/TrudeauReportCard2018.pdf.

Carbon Market Watch (2018). "COP24 Overshadowed by Market Failure as Countries Fail to Agree on Basic Accounting Principles and the Future of the CDM." December 15. https://carbonmarketwatch.org/2018/12/15/cop24-overshadowed-by-market-failure-as-countries-fail-to-agree-on-basic-accounting-principles-and-the-future-of-the-cdm.

Carmichael, Kevin (2018). "The G7 Meeting in Montreal Shows Relationships Matter." *Financial Post*, March 28. https://financialpost.com/news/economy/the-g7-meeting-in-montreal-shows-relationships-matter.

Carney, Mark (2021). *Value(S): Building a Better World for All*. Toronto: Penguin Random House.

Carrington, Damian (2021). "UN Global Climate Poll: 'The People's Voice Is Clear – They Want Action'." *Guardian*, January 21. https://www.theguardian.com/environment/2021/jan/27/un-global-climate-poll-peoples-voice-is-clear-they-want-action.

Casci, Mark (2021). "G7 Let Us Down on the Climate but Yorkshire Stands Ready to Deliver." *Yorkshire Post*, June 15. https://www.yorkshirepost.co.uk/business/g7-let-us-down-on-the-climate-but-yorkshire-stands-ready-to-deliver-mark-casci-3273843.

Cavendish, Camilla (2021). "Now Brexit Is 'Done', Britain Must Rebuild Trust with Europe." *Financial Times*, January 1. https://www.ft.com/content/e4b43ddc-7a37-4511-9b03-61c0b1625ed7.

CBC News (2017). "Canada Tops G7 in Latest IMF Estimate for 2017 Economic Growth in 2018." October 10. https://www.cbc.ca/news/business/imf-canada-economy-g7-1.4347737.

Central Asia News (2017). "Angela Merkel Says Europeans Must Rely on Themselves after Trump Tensions at G7 Meeting." May 29.

Champion, Marc, Ilya Arkhipov and Helene Fouquet (2018). "World Dysfunction Threatens to Derail the G-20." *Bloomberg*, November 26. Last modified: November 27, 2018. https://www.bloomberg.com/news/articles/2018-11-26/g-20-is-working-on-an-endangered-species-the-summit-communique.

Chan, Nicholas (2016). "The 'New' Impacts of the Implementation of Climate Change Response Measures." *Review of European, Comparative and International Environmental Law* 25(2): 228–37. https://doi.org/10.1111/reel.12161.

Chan, Sander, Idil Boran, Harro van Asselt et al. (2019). "Promises and Risks of Nonstate Action in Climate and Sustainability Governance." *WIREs Climate Change* 10(3): e572. https://doi.org/10.1002/wcc.572.

Chan, Sander, Clara Brandi and Steffen Bauer (2016). "Aligning Transnational Climate Action with International Climate Governance." *Review of European, Comparative and International Environmental Law* 25(2): 238–47. https://doi.org/10.1111/reel.12168.

Chazan, Guy (2018). "Angela Merkel's Plane to G20 Forced to Land Shortly after Take-Off." *Financial Times*, November 30. https://www.ft.com/content/79e1825e-f42b-11e8-ae55-df4bf40f9d0d.

Chazan, Guy and Stefan Wagstyl (2017). "Climate and Trade 'Dissonance' Hampers Meeting: Search for Consensus." *Financial Times*, July 8, p. 2.

China (2019). "Statement by China on Behalf of BASIC at the Opening Plenary of Cop25." *United Nations Climate Change*, December 2. https://www4.unfccc.int/sites/SubmissionsStaging/Documents/201912111926---STATEMENT%20BY%20CHINA%20ON%20BEHALF%20OF%20BASIC%20AT%20THE%20OPENING%20PLENARY%20OF%20COP25.pdf.

China's G20 Presidency (2015). "G20 Summit 2016, China." December 1. Beijing. http://www.g20.utoronto.ca/2016/xi-151201-en.pdf.

Christian Aid (2018). "Christian Aid Verdict on the Katowice Climate Summit: A C-Minus When They Needed Straight As." December 15. https://mediacentre.christianaid.org.uk/stronguchristian-aid-verdict-on-the-katowice-climate-summit-a-c-minus-when-they-needed-straight-asustrong/.

Civey (2018). "Civey/Der Spiegel Poll, March 20–27, 2018." https://www.electograph.com/2018/03/germany-civey-poll-for-der-spiegel.html.

Clark, Pilita (2016a). "Climate Battle Bears Early Fruit as Global Energy Emissions Stall." *Financial Times*, March 15. https://www.ft.com/content/ad0f58fa-eabe-11e5-bb79-2303682345c8.

Clark, Pilita (2016b). "COP21 Paris Climate Talks: Modi Tells Rich Nations of Their Duty to Lead Climate Change Fight." *Financial Times*, November 29. https://www.ft.com/content/929d0924-968e-11e5-9228-87e603d47bdc.

Clark, Pilita and Jim Pickard (2015). "UK Political Leaders in Joint Pledge on Climate Change." *Financial Times*, February 13. https://www.ft.com/content/a8ad6260-b37a-11e4-a45f-00144feab7de.

Clark, Pilita and Stefan Wagstyl (2017). "Ankara Threat to Pull out of Paris Accord Downplayed by Berlin: Climate Pact." *Financial Times*, July 10, p. 2.

Clémençon, Raymond (2016). "The Two Sides of the Paris Climate Agreement: Dismal Failure or Historic Breakthrough?" *Journal of Environment and Development* 25(1): 3–24. https://doi.org/10.1177/1070496516631362.

Climate Action 100+ (2017). "Global Investors Launch New Initiative to Drive Action on Climate Change by World's Largest Corporate Greenhouse Gas Emitters." Paris, December 12. https://www.climateaction100.org/news/global-investors-launch-new-initiative-to-drive-action-on-climate-change-by-worlds-largest-corporate-greenhouse-gas-emitters/.

Climate Action Network (2016). "Historic Breakthrough: 48 Countries Commit to Transition to 100% Renewable Energy." November 18. https://unfccc.int/files/meetings/marrakech_nov_2016/application/pdf/can_historicbreakthrough_release_en.pdf.

Climate Action Network (2018). "The Declaration of Human Rights Issue." *ECO*, December 10. https://eco.climatenetwork.org/cop24-eco7.

Climate Action Tracker (2021). "Global Update: Climate Summit Momentum." May 4. https://climateactiontracker.org/publications/global-update-climate-summit-momentum.

Climate Home News (2018). "Katowice Brief: Final Push for Paris." December 17. https://www.climatechangenews.com/2018/12/07/katowice-brief-final-push-paris.

Climate Vulnerable Forum (2016). "The Climate Vulnerable Forum Vision." Outcome document of the CVF High Level Meeting at Marrakech, Morocco, November 18. https://thecvf.org/activities/program/official-documents/marrakech-vision/.

CNN (2021). "2011 Japan Earthquake: Tsunami Fast Facts." April 14. https://www.cnn.com/2013/07/17/world/asia/japan-earthquake---tsunami-fast-facts/index.html.

Coalition of Finance Ministers for Climate Action (2019). "Cop25: Finance Ministers from over 50 Countries Join Forces to Tackle Climate Change." Madrid, December 9. https://www.financeministersforclimate.org/events/cop25-finance-ministers-over-50-countries-join-forces-tackle-climate-change.

Connolly, Kate (2015). "G7 Leaders Agree to Phase out Fossil Fuel Use by End of Century." *Guardian*, September 8. https://www.theguardian.com/world/2015/jun/08/g7-leaders-agree-phase-out-fossil-fuel-use-end-of-century.

Costa Rica. Ministry of the Environment and Energy (2020). "32 Leading Countries Set Benchmark for Carbon Markets with San Jose Principles." Originally published on December 14, 2019, San José, Costa Rica, January 6. https://cambioclimatico.go.cr/press-release-leading-countries-set-benchmark-for-carbon-markets-with-san-jose-principles.

Cowan, Micki (2018). "What Saskatchewan Stands to Lose — and Gain — by Not Signing the Federal Climate Change Plan." *CBC News*, March 2. https://www.cbc.ca/news/canada/saskatchewan/saskatchewan-climate-change-plan-1.4558216.

Darby, Meagan (2017). "Lead Diplomat: Bonn Climate Talks Must 'Restate Vision of Paris'." *Climate Home News*, October 3. https://www.climatechangenews.com/2017/10/03/lead-diplomat-bonn-climate-talks-must-restate-vision-paris.

Darby, Megan (2018a). "Armed with Faded Copies, Four Diplomats Write the Rules of the Paris Agreement." *Climate Home News*, September 7. https://www.climatechangenews.com/2018/09/07/armed-faded-copies-four-diplomats-write-rules-paris-climate-deal/.

Darby, Megan (2018b). "'Everyone Was Frustrated': US-China Stand-Off Holds up Climate Talks." *Climate Home News*, September 2. https://www.climatechangenews.com/2018/09/09/everyone-frustrated-us-china-stand-off-holds-climate-talks/.

Darby, Megan (2018c). "With Time Running out, Negotiators Inch Forwards on Paris Climate Rulebook." *Climate Home News*, August 3. https://www.climatechangenews.com/2018/08/03/time-running-negotiators-inch-forwards-paris-climate-rulebook.

Darby, Megan (2019). "Next UN Climate Summit Scheduled for December in Chile." *Climate Home News*, March 7. https://www.climatechangenews.com/2019/03/07/next-un-climate-summit-scheduled-december-chile/.

Darby, Meagan, Karl Mathiesen and Solia Apparicio (2018). "Bonn Morning Brief: We Simply Don't Have Any Time to Waste." *Climate Home News*, May 1. https://www.climatechangenews.com/2018/05/01/bonn-morning-brief-simply-dont-time-waste.

Davenport, Coral (2014). "A Climate Accord Based on Global Peer Pressure." *New York Times*, December 14. http://www.nytimes.com/2014/12/15/world/americas/lima-climate-deal.html.

Davutoğlu, Ahmet (2015). "Turkey's Vision for the G20." *Video, World Economic Forum*, Davos, January 21. https://www.weforum.org/events/world-economic-forum-annual-meeting-2015/sessions/turkeys-vision-g20.

Deheza, Mariana, Morgane Nicol, Vivian Dépoues et al. (2016). "COP22 in Marrakech: A Push for Accelerated Action by 2018." Climate Brief No. 43, Institute for Climate Economics, Paris. https://www.i4ce.org/download/cop22-in-marrakech-a-push-for-accelerated-action-by-2018/.

Dennis, Brady and Chico Harlan (2019). "'Tick Tock': UN Climate Talks End with Fresh Doubts over Global Unity." *Sydney Morning Herald*, December 15. https://www.smh.com.au/world/europe/un-climate-talks-drag-into-second-extra-day-20191215-p53k69.html.

Diplomat (2016). "G7 Falters on Growth Test." May 28. https://thediplomat.com/2016/05/
g7-falters-on-growth-test.

Dobson, Hugo (2020). "The UK, Prime Ministerial Leadership and Recent G7 Summits."
Presentation to "Recipes for G7 Summit Success for the 2021 UK Presidency," G7
Research Group, December 3. http://www.g7.utoronto.ca/conferences/2020-Recipes-
for-G7-Success.html.

Dodwell, David (2021). "As Joe Biden Works Quickly to Restore US Diplomacy, the
World Breathes a Sigh of Relief." *South China Morning Post*, February 20. https://
www.scmp.com/comment/opinion/article/3122260/biden-works-quickly-restore-us-
diplomacy-world-breathes-sigh-relief.

Doherty, Been, (2018). "Australia Likely to Use Controversial Kyoto Loophole to
Meet Paris Agreement." *Guardian*, December 12. https://www.theguardian.com/
environment/2018/dec/12/australia-likely-to-use-controversial-kyoto-loophole-to-
meet-paris-agreement.

Doig, Alison (2020). "Climate Ambition Summit 2020: A Good Start but Much More
Needed in the Year to COP26." *Energy and Climate Intelligence Unit*, December 16.
https://eciu.net/blog/2020/climate-ambition-summit-2020-a-good-start-but-much-
more-needed-in-the-year-to-cop26.

Downie, Christian (2015a). "Energy Governance: Can the G20 Drive Reform?" In *The G20
and the Future of International Economic Governance*, Mike Callaghan and Tristram
Sainsbury, eds. Sydney: NewSouth Publishing, pp. 119–31.

Downie, Christian (2015b). "Global Energy Governance in the G20: States, Coalitions and
Crises." *Global Governance* 21(3): 475–92. https://doi.org/10.1163/19426720-02103008.

Downie, Christian (2020). "Steering Global Energy Governance: Who Governs and What
Do They Do." *Regulation and Governance*, August. https://doi.org/10.1111/rego.12352.

Doyle, Alister (2018). "Logjam for Paris Climate 'Rule Book' as End-2018 Deadline
Looms." *Reuters*, May 10. https://www.reuters.com/article/uk-climatechange-accord-
idUKKBN1IB1XE.

DW (2016). "UN COP22 Closes with Work Plan to Implement Paris Climate Pact."
November 19. https://p.dw.com/p/2Svry.

Dwortzan, Mark (2017). "Monitoring Implementation of the Paris Agreement." *MIT
Joint Program on the Science and Policy of Global Change*, August 17. https://
globalchange.mit.edu/news-media/jp-news-outreach/monitoring-implementation-
paris-agreement.

Earth Negotiations Bulletin (2015a). "Summary of the Bonn Climate Change Conference:
1–11 June 2015." 12(648). June 14. https://enb.iisd.org/download/pdf/enb12638e.pdf.

Earth Negotiations Bulletin (2015b). "Summary of the Paris Climate Change Conference:
29 November–13 December 2015." 12(663). December 15. http://enb.iisd.org/
download/pdf/enb12663e.pdf.

Earth Negotiations Bulletin (2016). "Summary Report of the Marrakech Climate Change
Conference, 7–19 November 2016." 12(689). November 21. https://enb.iisd.org/
events/marrakech-climate-change-conference-november-2016/summary-report-7-18-
november-2016.

Earth Negotiations Bulletin (2017a). "Summary of the Bonn Climate Change Conference:
8–18 May 2017." 12(702). May 22. https://enb.iisd.org/download/pdf/enb12701e.pdf.

Earth Negotiations Bulletin (2017b). "Summary of the Fiji/Bonn Climate Change
Conference: 6–17 November 2017." 12(714). November 21. https://enb.iisd.org/
download/pdf/enb12714e.pdf.

Earth Negotiations Bulletin (2018). "Summary of the Katowice Climate Change Conference: 2–15 December 2018." 12(747). https://enb.iisd.org/download/pdf/enb12747e.pdf.

Economist (2021a). "Britain Has Lost the EU. Can It Find a Role?", January 2. https://www.economist.com/leaders/2021/01/02/britain-has-lost-the-eu-can-it-find-a-role.

Economist (2021b). "Britain Needs a Post-Brexit Foreign Policy." January 2. https://www.economist.com/briefing/2021/01/02/britain-needs-a-post-brexit-foreign-policy.

edie (2021). "COP26 Will Go Ahead in Person, Alok Sharma Confirms." *edie.net*, May 14. https://www.edie.net/news/9/COP26-will-go-ahead-in-person--Alok-Sharma-confirms/.

Edwardes-Evans, Henry (2021). "G7 Pledges to End Support for Unabated Coal by End of 2021." *S&P Global Platts*, June 13. https://www.spglobal.com/platts/en/market-insights/latest-news/coal/061321-g7-pledges-to-end-support-for-unabated-coal-by-end-of-2021.

EFE News Service (2016). "China Hails Paris Accord as Fresh Start for Climate Governance." November 4.

Ehley, Brianna and Alice Miranda Ollstein (2020). "Trump Announces U.S. Withdrawal from the World Health Organization." *Politico*, May 29. https://www.politico.com/news/2020/05/29/us-withdrawing-from-who-289799.

Eilperin, Juliet and Brady Dennis (2019). "Epa Chief Will Focus on Ocean Trash, Not Climate Change, at Upcoming Global Summit." *Washington Post*, June 10. https://www.washingtonpost.com/climate-environment/2019/06/10/epa-chief-will-focus-ocean-trash-not-climate-change-upcoming-global-summit/.

Eilperin, Juliet and Brady Dennis (2021). "U.S. Aims to Halve Emissions by 2030." *Washington Post*, April 21, p. A1.

Ekos Politics (2019). "Conservative Fortunes Waning as Liberal Fortunes Rise in Nearly Deadlocked Race." June 17. https://www.ekospolitics.com/index.php/2019/06/conservative-fortunes-waning-as-liberal-fortunes-rise-in-nearly-deadlocked-race.

El-Erian, Mohamed (2018). "Trump-Xi Will Make Us Forget This Weekend's G-20 Confab." *Bloomberg*, November 27. https://www.bloomberg.com/opinion/articles/2018-11-27/trump-xi-meeting-will-make-us-forget-this-weekend-s-g-20-summit.

Emorine, Hélène (2018). "G20 Summit, Overshadowed by Geopolitics, Still out to Prove Itself." *Open Canada*, December 5. https://opencanada.org/g20-summit-overshadowed-geopolitics-still-out-prove-itself.

Energy and Climate Intelligence Unit (n.d.). "What Is COP26, Who Will Attend It and Why Does It Matter." https://eciu.net/analysis/briefings/international-perspectives/what-is-cop26-who-will-attend-it-and-why-does-it-matter. Accessed: April 10, 2021.

Energy and Power (2016). "G20 Leaders Reaffirm Green Finance in Hangzhou." October 3. https://ep-bd.com/view/details/article/MTI0/title?q=g20+leaders+reaffirm+green+finance+in+hangzhou.

Engelbrekt, Kjell (2016). *High Table Diplomacy: The Reshaping of International Security Institutions*. Washington, DC: Georgetown University Press.

Environics Institute for Survey Research (2018). "Canada's World Survey 2018." In partnership with Bill Graham Centre for Contemporary International History, *Canadian International Council and SFU Public Square*, April. https://www.environicsinstitute.org/docs/default-source/project-documents/canada's-world-2018-survey/canada's-world-survey-2018---final-report.pdf.

Environment News Service (2019). "UN Climate Action Summit Yields Rich Funding Promises." September 25. https://ens-newswire.com/un-climate-action-summit-yields-rich-funding-promises.

Erdoğan, Recep Tayyip (2015). "Remarks at the Opening of the G20 Energy Ministers Meeting." *Unofficial Transcript*, Istanbul, October 2. http://www.g20.utoronto.ca/2015/151002-erdogan.html.

Erickson, Andrew S. and Gabriel Collins (2021). "Competition with China Can Save the Planet." *Foreign Affairs* 100(136–149). May/June. https://www.foreignaffairs.com/articles/united-states/2021-04-13/competition-china-can-save-planet.

European Commission (2021). "G7 Summit: President Von Der Leyen Outlines Key EU Priorities." June 11. https://ec.europa.eu/commission/presscorner/detail/en/ac_21_2969.

European Parliament (2019). "The European Parliament Declares Climate Emergency." Press release, November 29. https://www.europarl.europa.eu/news/en/press-room/20191121IPR67110/the-european-parliament-declares-climate-emergency.

European Union (n.d.). "Recovery Plan for Europe." https://ec.europa.eu/info/strategy/recovery-plan-europe_en. Accessed: July 30, 2021.

Evans, Simon (2014). "Who Is Donald Tusk and What Does He Think About Climate." *Carbon Brief*, September 2. https://www.carbonbrief.org/who-is-donald-tusk-and-what-does-he-think-about-climate.

Evans, Simon and Josh Gabbatiss (2019). "Cop25: Key Outcomes Agreed at the UN Climate Talks in Madrid." *Carbon Brief*, December 15. https://www.carbonbrief.org/cop25-key-outcomes-agreed-at-the-un-climate-talks-in-madrid.

Evans, Simon and Jocelyn Timperley (2017). "Bonn Climate Talks: Key Outcomes from the May 2017 UN Climate Conference." Carbon Brief, May 19. https://www.carbonbrief.org/bonn-climate-talks-key-outcomes-from-may-2017-climate-conference.

Evans, Simon and Jocelyn Timperley (2018). "COP 24: Key Outcomes Agreed at the UN Climate Talks in Katowice." *Carbon Brief*, December 18. http://www.carbonbrief.org/cop24-key-outcomes-agreed-at-the-un-climate-talks-in-katowice.

Fahim, Kareen (2020). "G-20 Leaders Close Riyadh Summit with Calls for Coordinated Response to Coronavirus." *Washington Post*, November 23. https://www.washingtonpost.com/world/g20-saudi-arabia-riyadh-coroanvirus-pandemic/2020/11/22/5d909c0a-2a81-11eb-9c21-3cc501d0981f_story.html.

Falkner, Robert (2016). "A Minilateral Solution for Global Climate Change? On Bargaining Efficiency, Club Benefits, and International Legitimacy." *Perspectives on Politics* 14(1): 87–101. https://doi.org/10.1017/S1537592715003242.

Farand, Chloé (2019). "What Is Article 6? The Issue Climate Negotiators Cannot Agree." *Climate Home News*, December 2. https://www.climatechangenews.com/2019/12/02/article-6-issue-climate-negotiators-cannot-agree/.

Fell, Tony (2018). "In Justin Trudeau's Canada, We Still Can't Get Anything Big Done." *Globe and Mail*, February 23. https://www.theglobeandmail.com/opinion/in-justin-trudeaus-canada-we-still-cant-get-anything-big-done/article38086398/.

Filliâtre, Pascale (2019). "24 Hours in a Palace: The Hôtel Du Palais in Biarritz." *Explore France*, October 29. https://ca.france.fr/en/biarritz-basque-country/list/24-hours-in-a-palace-the-hotel-du-palais-in-biarritz.

Financial Times (2019). "G20 Meets a Low Bar for Global Co-Operation." July 1, p. 16. Editorial.

Financial Times (2020). "Bearing the Burden of Coronavirus Debt." November 24, p. 16. Editorial.

Financial Times (2021a). "8 of the 10 Warmest Years on Record Have Occurred in the Past Decade." January, p. 16. 9–10.

Financial Times (2021b). "Biden Must First Repair Democracy at Home." January 9.

Financial Times (2021c). "Datawatch: Tackling Climate Change." April 22, p. 1.

Financial Times (2021d). "Earth Day Climate Goals Stepped Up." April 23, p. 1.

Financial Times (2021e). "Greens Freshen Stale Air of German Politics." April 22, p. 22.

Financial Times (2021f). "Time Is Running Short to Avert 'Hell on Earth'." August 14, p. 10.

Fiorino, Daniel J. (2018). *Can Democracy Handle Climate Change*. Cambridge, UK: Polity Press.

FiveThirtyEight (2021). "How Unpopular Is Donald Trump." Last modified: January 20, 2021. https://projects.fivethirtyeight.com/trump-approval-ratings.

Floyd, Rita (2015). "Global Climate Security Governance: A Case of Institutional and Ideational Fragmentation." *Conflict, Security and Development* 15(2): 119–46. https://doi.org/10.1080/14678802.2015.1034452.

Follain, John, Arne Delfs, Ilya Arkhipov et al. (2020). "Trump Tweets About Voter 'Fraud' During G-20 Opening Meeting." *Bloomberg*, November 21. https://www.bloomberg.com/news/articles/2020-11-21/trump-tweets-about-voter-fraud-during-g-20-s-opening-meeting.

Food and Agriculture Organization and United Nations Environment Programme (2020). "The State of the World's Forests 2020: Forests, Biodiversity and People.Food and Agriculture Organization," Rome. https://doi.org/10.4060/ca8642en.

Ford, James, Michelle Maillet, Vincent Pouliot et al. (2016). "Adaptation and Indigenous Peoples in the United Nations Framework Convention on Climate Change." *Climatic Change* 139(3): 429–43. https://doi.org/10.1007/s10584-016-1820-0.

Forsyth, James (2020). "Johnson's Levelling-up Mission Starts Now." *Times*, December 17. https://www.thetimes.co.uk/article/boris-johnsons-levelling-up-mission-starts-now-q38mz32pc.

Forsyth, James (2021). "Boris Johnson's Great Climate Change Challenge." *Times*, February 18. https://www.thetimes.co.uk/article/boris-johnsons-great-climate-change-challenge-s2vj7673b.

Fotheringham, Alasdair (2019). "Cop25: Climate Summit Kicks Off as Guterres Issues Stark Warning." December 2. https://www.aljazeera.com/news/2019/12/2/cop25-climate-summit-kicks-off-as-guterres-issues-stark-warning.

Fownes, Jennifer, Chao Yu and Drew Margolin (2018). "Twitter and Climate Change." *Sociology Compass* 12(6): e12587. https://doi.org/10.1111/soc4.12587.

France. Ministry for Europe and Foreign Affairs (n.d.a). "3rd Edition of the One Planet Summit: Africa's Pledge (Nairobi, 14 March 2019)." Paris. https://www.diplomatie.gouv.fr/en/french-foreign-policy/climate-and-environment/the-one-planet-movement/article/1st-edition-of-the-one-planet-summit-a-meeting-for-the-planet-paris-12-12-17. Accessed: July 4, 2021.

France. Ministry for Europe and Foreign Affairs (n.d.b). "One Planet Summit: A Meeting for the Planet (12 December 2017)." Paris. https://www.diplomatie.gouv.fr/en/french-foreign-policy/climate-and-environment/the-one-planet-movement/article/1st-edition-of-the-one-planet-summit-a-meeting-for-the-planet-paris-12-12-17. Accessed: July 4, 2021.

Freytag, Andreas and John Kirton (2017). "Pushed toward Partnership: Increasing Cooperation between the Bretton Woods Bodies." In *Palgrave Handbook of Inter-Organizational Relations in World Politics*, Rafael Biermann and Joachim Koops, eds. London: Palgrave Macmillan, pp. 569–90.

Friedman, Lisa (2017). "Trump Team to Promote Fossil Fuels and Nuclear Power at Bonn Climate Talks." *New York Times*, November 2. https://www.nytimes.com/2017/11/02/climate/trump-coal-cop23-bonn.html.

Friedman, Lisa (2021). "Biden Wants Leaders to Make Climate Commitments for Earth Day." *New York Times*, April 13. https://www.nytimes.com/2021/04/13/climate/biden-climate-change-diplomacy.html.

G7 France (2019a). "1st Sherpa Meeting in Lyon (4–5 February)." February 6. https://www.elysee.fr/en/g7/2019/02/06/1st-sherpa-meeting-in-lyon-4-5-february.

G7 France (2019b). "2nd Sherpa Meeting in Lille (16–18 April)." April 19. https://www.elysee.fr/en/g7/2019/04/19/2nd-sherpa-meeting-in-lille-16-18-april.

G7 France (2019c). "3rd Sherpa Meeting in Paris (13–14 June)." June 16. https://www.elysee.fr/en/g7/2019/06/16/3rd-sherpa-meeting-in-paris-13-14-june.

G7 France (2019d). "Biarritz Summit." January 1. https://www.elysee.fr/en/g7/2019/01/01/biarritz-summit.

G7 Italy (2017). "Second Meeting of the G7 Sherpas in Florence." March 21. https://www.g7italy.it/en/news/second-meeting-g7-sherpas-florence/index.html.

G7 Italy (n.d.). "Italian G7 Presidency 2017: Priorities." https://www.g7italy.it/en/priorities. Accessed: April 12, 2021.

G20 Germany (2016). "Priorities of the 2017 G20 Summit." December 1. Berlin. https://www.g20.utoronto.ca/2017/2016-g20-praesidentschaftspapier-en.pdf.

G20 Saudi Arabia (2019). "Overview of Saudi Arabia's 2020 G20 Presidency." Riyadh, December. https://www.g20.utoronto.ca/2020/2020-Presidency_Agenda-V5.pdf.

G20 Saudi Arabia (2020). "G20 Environment Ministers Promote a More Sustainable Future for All." September 17. https://www.g20.utoronto.ca/2020/2020-g20-environment-0917.html.

G20 Turkey (2014). "Turkish G20 Presidency Priorities for 2015." December 1. https://www.g20.utoronto.ca/2015/141201-turkish-priorities.html.

G77 (2015). "Statement on Behalf of the Group of 77 and China by Ambassador Nozipho Mxakato-Diseko from South Africa at the Open-Ended Consultation on Finance Process." Paris, December 2. https://www.g77.org/statement/getstatement.php?id=151202b.

Gabbatiss, Josh (2019). "Bonn Climate Talks: Key Outcomes from the June 2019 UN Climate Conference." *Carbon Brief*, July 1. https://www.carbonbrief.org/bonn-climate-talks-key-outcomes-from-june-2019-un-climate-conference.

Galey, Patrick and Marlowe Hood (2019). "UN Climate Talks Stagger Towards 'Mediocre' Outcome." *International Business Times*, December 15. https://www.ibtimes.com/un-climate-talks-stagger-towards-mediocre-outcome-2885442.

Geiling, Natasha (2015). "France and China Announce Major Climate Action Agreement." *Think Progress*, November 2. https://archive.thinkprogress.org/france-and-china-announce-major-climate-action-agreement-48b2da53e53c/.

Germany. Federal Government (2015). "Partners in Shouldering Global Responsibility." March 10. https://www.bundesregierung.de/breg-en/issues/europe/partners-in-shouldering-global-responsiblity-601490.

Germany. Federal Ministry for the Environment, Nature Conservation and Nuclear Safety (n.d.). "Cooperative Action under Article 6." *Carbon Mechanisms: Market-Based Climate Policy Mechanisms*. https://www.carbon-mechanisms.de/en/introduction/the-paris-agreement-and-article-6. Accessed: April 10, 2021.

Gigounas, George, Jesse Medlong and Elizabeth Callahan (2020). "Cop25's Key Outcome: Adoption of the San Jose Principles on Carbon Market Mechanisms: Takeaways for Business." *DLA Piper*, February 12. https://www.dlapiper.com/en/europe/insights/publications/2020/02/cop25s-key-outcome-adoption-of-the-san-jose-principles-on-carbon-market-mechanisms.

Giles, Chris and Robin Harding (2019). "Upbeat End to G20 Papers over Cracks in Global Co-operation." *Financial Times*, July 2, p. 3.

Gillis, Justin (2016). "In Zika Epidemic, a Warning on Climate Change." *New York Times*, February 21, p. 6. https://www.nytimes.com/2016/02/21/world/americas/in-zika-epidemic-a-warning-on-climate-change.html.

Giordana, Chiara (2021). "Boris Johnson Tells Angela Merkel to 'Mute' in First Virtual G7 Meeting." *Independent*, February 19. https://www.independent.co.uk/news/uk/politics/boris-johnson-g7-angela-merkel-mute-b1804765.html.

Glavin, Terry (2018). "Putin, Not Gender Equality, Needs to Be the Focus of Canada's G7 Summit." *National Post*, March 29. https://nationalpost.com/opinion/terry-glavin-putin-not-gender-equality-needs-to-be-the-focus-of-canadas-g7-summit.

Goldar, Amrita and Varsha Jain (2021). "Global Climate Finance Agenda: Can the G20 Pave the Way." *Financial Express*, January 19. https://www.financialexpress.com/opinion/global-climate-finance-agenda-can-the-g20-pave-the-way/2173546.

Goldenberg, Suzanne (2015). "How US Negotiators Ensured Landmark Paris Climate Deal Was Republican-Proof." *Guardian*, December 13.

Goodman, Matthew (2019). "Parsing the Osaka G20 Communiqué." *Center for Strategic and International Studies*, July 3. https://www.csis.org/analysis/parsing-osaka-g20-communique.

Gorrey, Megan (2018). "'Missed Opportunity': Australia's 'Difficult' Position in UN Climate Change Talks." *Sydney Morning Herald*, December 16. https://www.smh.com.au/environment/climate-change/missed-opportunity-australia-s-difficult-position-in-un-climate-change-talks-20181216-p50mkf.html.

Government of Canada (2016). "Pan-Canadian Framework on Clean Growth and Climate Change: Canada's Plan to Address Climate Change and Grow the Economy." https://www.canada.ca/content/dam/themes/environment/documents/weather1/20170125-en.pdf.

Graney, Emma, Marieke Walsh and Adam Radwanski (2021). "Canada to Set New Emissions Target as Trudeau, Other Leaders Gather for Climate Summit." *Globe and Mail*, April 21. https://www.theglobeandmail.com/business/article-canadian-climate-goals-under-pressure-ahead-of-biden-summit.

Green Party (2014). "General Election 2015: Greens Poll at Highest Numbers since 1989." London, August 7. https://www.greenparty.org.uk/news/2014/08/07/general-election-2015-greens-polling-at-highest-numbers-since-1989.

Greenpeace (2015). "People from the Philippines and Pacific Island Nations Sign Declaration to Hold Big Polluters Accountable." June 9. https://wayback.archive-it.org/9650/20200321220752/http://p3-raw.greenpeace.org/seasia/ph/press/releases/Philippines-and-Pacific-Island-nations-seek-Climate-Justice/.

Greenpeace (2016). "Marrakech: Renewed Determination by All, Leaps of Leadership by Some." November 18. https://wayback.archive-it.org/9650/20200401062544/http://p3-raw.greenpeace.org/international/en/press/releases/2016/COP22-closing-press-release/.

Grieger, Gisela (2015). "China and Climate Change Ahead of COP21." *November European Parliamentary Research Service*. http://www.europarl.europa.eu/RegData/etudes/ATAG/2015/572791/EPRS_ATA%282015%29572791_EN.pdf.

Gross, Anna and Jim Pickard (2020). "Johnson to Boost WHO Backing with £571m Vaccine Pledge." *Financial Times*, September 25. https://www.ft.com/content/32d5befc-90e8-43a5-b805-7dc357e30666.

Guardian (2018). "The Guardian View on COP24: While Climate Talks Continue, There Is Hope." December 16. https://www.theguardian.com/commentisfree/2018/dec/16/the-guardian-view-on-cop24-while-climate-talks-continue-there-is-hope.

Gupta, Joydeep and Tirthankar Mandal (2015). "Paris Climate Summit: How the Negotiating Blocs Work." *Third Pole*, November 28. https://www.thethirdpole.net/2015/11/28/climate-abcd-alignments-blocs-countries-divisions-2.

Gupta, Joyeeta (2016). "Climate Change Governance: History, Future and Triple-Loop Learning." *WIREs Climate Change* 7(2): 192–210. https://doi.org/10.1002/wcc.388.

Guterres, António (2018). "To Achieve Historic Paris Accord, Secretary-General Calls for Engaging, Investing, Partnering Ahead of 2019 Climate Summit." SG/SM/19388-ENV/DEV/1913, Katowice, Poland, December 4. https://www.un.org/press/en/2018/sgsm19388.doc.htm.

Guterres, António (2019a). "A Climate Change: An Unstoppable Movement Takes Hold." October 3. https://www.un.org/sg/en/content/sg/articles/2019-10-03/climate-change-unstoppable-movement-takes-hold.

Guterres, António (2019b). "Nature Striking Back with Fury, Secretary-General Tells Climate Action Summit, Calling 'Apocalyptic' Bahamas Destruction 'the Future: If We Do Not Act Now'." SG/SM/19757-ENV/DEV/1996, New York, September 23. https://www.un.org/press/en/2019/sgsm19757.doc.htm.

Guterres, António (2019c). "Secretary-General's Remarks at Opening Ceremony of UN Climate Change Conference Cop25." *As delivered*, Madrid, December 2. https://www.un.org/sg/en/content/sg/statement/2019-12-02/secretary-generals-remarks-opening-ceremony-of-un-climate-change-conference-cop25-delivered.

Guterres, António (2019d). "Statement by the UN Secretary-General António Guterres on the Outcome of Cop25." Madrid, December 15. https://unfccc.int/news/statement-by-the-un-secretary-general-antonio-guterres-on-the-outcome-of-cop25.

Guterres, António (2020a). "Remarks at the Climate Ambition Summit." *United Nations*, December 12. https://www.un.org/sg/en/content/sg/speeches/2020-12-12/remarks-the-climate-ambition-summit.

Guterres, António (2020b). "Secretary-General, in Message on Leaders' Pledge for Nature, Urges New Global Framework to Address Biodiversity-Loss Drivers, Citing 'Planetary Emergency'." SG/SM/20287, New York, September 28. https://www.un.org/press/en/2020/sgsm20287.doc.htm.

Guterres, António (2020c). "Secretary-General's Remarks to High-Level Roundtable on Climate Ambition." September 24. https://www.un.org/sg/en/content/sg/statement/2020-09-24/secretary-generals-remarks-high-level-roundtable-climate-ambition-delivered.

Hale, Thomas (2016). "'All Hands on Deck': The Paris Agreement and Nonstate Climate Action." *Global Environmental Politics* 16(3): 12–22. https://doi.org/10.1162/GLEP_a_00362.

Hall, Nina (2016). "The Institutionalisation of Climate Change in Global Politics." In *Environment, Climate Change and International Relations*, Gustavo Sosa-Nunez and Ed Atkins, eds. Bristol: E-International Relations Publishing, pp. 60–74. https://www.e-ir.info/2016/05/27/the-institutionalisation-of-climate-change-in-global-politics/.

Hammond, Andrew (2017). "Host Italy Promises a G-7 with Geopolitical Focus." *Business Times Singapore*, April 7. https://www.businesstimes.com.sg/opinion/host-italy-promises-a-g-7-with-geopolitical-focus.

Hannam, Peter (2019). "Tiny Nations Challenge Australia's Carbon 'Carryover Credits'." *Sydney Morning Herald*, June 30. https://www.smh.com.au/environment/climate-change/tiny-nation-s-challenge-australia-s-carbon-carryover-credits-20190630-p522n0.html.

Happaerts, Sander (2015). "Rising Powers in Global Climate Governance: Negotiating inside and Outside the UNFCCC." In *Rising Powers and Multilateral Institutions*,

Dries Lesage and Thijs Van de Graaf, eds. New York: Palgrave Macmillan, pp. 317–42.

Harris, Bryan (2021). "Brazil's Amazon Pledge Aims to Reset Relations." *Financial Times*, April 21, p. 4.

Harris, Bryan and Carolina Ribeirao Preto (2021). "Brazil's Worst Drought in a Century Deepens Covid-19 Misery." *Financial Times*, June 19, p. 2.

Harvey, Fiona (2020a). "Cop26: Boris Johnson Urged to Resist Calls to Postpone Climate Talks." *Guardian*, March 18. https://www.theguardian.com/environment/2020/mar/18/cop26-boris-johnson-urged-resist-calls-postpone-climate-talks-coronavirus.

Harvey, Fiona (2020b). "Covid-19 Economic Rescue Plans Must Be Green, Say Environmentalists." *Guardian*, March 24. https://www.theguardian.com/environment/2020/mar/24/covid-19-economic-rescue-plans-must-be-green-say-environmentalists.

Harvey, Fiona (2020c). "'We Must Use This Time Well': Climate Experts Hopeful after Cop26 Delay." *Guardian*, April 2. https://www.theguardian.com/environment/2020/apr/02/we-must-use-this-time-well-climate-experts-hopeful-after-cop26-delay-coronavirus.

Harvey, Fiona (2021a). "Joe Biden to Reveal US Emissions Pledge in Key Climate Crisis Moment." *Guardian*, April 19. https://www.theguardian.com/environment/2021/apr/19/joe-biden-to-reveal-us-emissions-pledge-in-key-climate-crisis-moment.

Harvey, Fiona (2021b). "UK Insists Cop26 Must Be Held in Person If Possible." *Guardian*, May 14. https://www.theguardian.com/environment/2021/may/14/uk-will-vaccinate-and-test-to-ensure-cop26-is-in-person-event.

Hatiar, Dominik (2015). "G7 Steps up as a Catalyst of Global Decarbonisation." *Policy Brief*, Global Policy, June. https://www.globalpolicyjournal.com/sites/default/files/Hatiar%20-%20G7%20Steps%20Up%20as%20a%20Catalyst%20of%20Global%20De-carbonisation.pdf.

He, Alex Xingqiang (2016). "China in Global Energy Governance: A Chinese Perspective." *International Organisations Research Journal* 11(1): 71–91. https://doi.org/10.17323/1996-7845/2016-01-71.

Heffron, Raphael (2018). "COP24 Shows Global Warming Treaties Can Survive the Era of the Anti-Climate 'Strongman'." *Conservation*, December 16. https://theconversation.com/cop24-shows-global-warming-treaties-can-survive-the-era-of-the-anti-climate-strongman-107786.

Heinberg, Richard (2015). "Renewable Energy after COP21: Nine Issues for Climate Leaders to Think About on the Journey Home." *Post Carbon Institute*, December 14. https://www.postcarbon.org/renewable-energy-after-cop21.

Henden, Amalie (2019). "G7 Summit 2019 Schedule: Full Itinerary as Boris Johnson Takes on Merkel and Macron." *Express*, August 28. https://www.express.co.uk/news/world/1169231/g7-summit-2019-schedule-itinerary-boris-johnson-angela-merkel-emmanuel-macron-brexit-news.

Heubaum, Harald and Frank Biermann (2015). "Integrating Global Energy and Climate Governance: The Changing Role of the International Energy Agency." *Energy Policy* 87: 228–39. https://doi.org/10.1016/j.enpol.2015.09.009.

High-Level Climate Champions (2016). "Marrakech Partnership for Global Climate Action." *United Nations Climate Change*, November 17. https://unfccc.int/files/paris_agreement/application/pdf/marrakech_partnership_for_global_climate_action.pdf.

Hodgson, Camilla (2021a). "Carbon Emissions Set to Surge in 2021." *Financial Times*, April 21, p. 4.

Hodgson, Camilla (2021b). "UN Rebukes 75 Nations over Climate Failures." *Financial Times*, February 27, p. 4.

Holder, Josh and Marc Santora (2021). "Hospitals in England Struggle as the Virus Maintains Its Deadly Grip." *New York Times*, January 15. https://www.nytimes.com/2021/01/15/world/hospitals-in-england-struggle-as-the-virus-maintains-its-deadly-grip.html.

Hook, Leslie (2019a). "Climate Talks Undone by 'Ghost from Past': Cop25." *Financial Times*, December 6, p. 3.

Hook, Leslie (2019b). "Japan Dilutes G20 Climate Pledge in Push to Win US Trade Favours." *Financial Times*, June 26, p. 1.

Hook, Leslie (2019c). "Record July Temperatures Put Focus on Climate Change." *Financial Times*, August 6, p. 4.

Hook, Leslie (2021a). "Last Year Was Joint Hottest on Record." *Financial Times*, January 9–10, p. 2.

Hook, Leslie (2021b). "Rich Nations' Failure to Cough up Cash Threatens to Derail UN Climate Summit." *Financial Times*, June 19, p. 1.

Hook, Leslie and Camilla Hodgson (2019). "Macron Tells World Leaders to Act Faster on Climate Change." *Financial Times*, September 24, p. 3.

Hook, Leslie and Camilla Hodgson (2021). "US Pledges to Halve Carbon Emissions." *Financial Times*, April 23, p. 4.

Hook, Leslie and Anjli Raval (2021). "IEA Calls for End to New Oil and Gas Exploration." *Financial Times*, May 19, p. 3.

Hopkins, Andrea (2018). "Majority of Canadians Say They Will Avoid U.S. Goods, Poll Finds." *Globe and Mail*, June 15. https://www.theglobeandmail.com/canada/article-vast-majority-of-canadians-say-they-will-avoid-us-goods-new-poll/.

Hua, Sha (2021). "U.S. And China Engage, Tentatively, on Climate Change-Update." *Wall Street Journal*, March 9. https://www.wsj.com/articles/u-s-and-china-engage-tentatively-on-climate-change-11615301108.

IANS (2017). "Cooperation, Terror to Figure on Modi's Europe Tour (Curtain Raiser." *Business Standard*, May 28. http://www.business-standard.com/article/news-ians/cooperation-terror-to-figure-on-modi-s-europe-tour-curtain-raiser-117052800313_1.html.

Ibbitson, John (2018). "Tory by-Election Landslide in Quebec More Than a Local Race." *Globe and Mail*, June 19. https://www.theglobeandmail.com/politics/article-tory-by-election-landslide-in-quebec-more-than-a-local-race/.

ICAEW (2021). "Sharma Takes the Reins on COP26 Preparations." London, January 12. https://www.icaew.com/insights/viewpoints-on-the-news/2021/jan-2021/sharma-takes-the-reins-on-cop26-preparations.

Ifop (2021a). "Les Indices de Popularité." *Un baromètre pour le Journal du Dimanche*, Paris, April 19. https://www.ifop.com/publication/les-indices-de-popularite-8.

Ifop (2021b). "Les Indices de Popularité: Février." *Un baromètre pour le Journal du Dimanche*, Paris, February. https://www.ifop.com/wp-content/uploads/2021/02/117823-Indices-de-popularit%C3%A9-F%C3%A9vrier-2021.pdf.

Initiative of 89 French Companies (2017). "French Business Climate Pledge: Our Common Ambition." Paris, December 12. https://www.pernod-ricard.com/en/download/file/fid/9303/.

Inside Climate News (2016). "Marrakech Climate Talks End on Positive Note Despite Trump Threat." November 21. https://insideclimatenews.org/news/21112016/paris-climate-change-agreement-cop-22-marrakech-donald-trump.

Intergovernmental Panel on Climate Change (2015). "Climate Change 2014: Synthesis Report. Contribution of Working Groups I, II and III to the Fifth Assessment Report of the Intergovernmental Panel on Climate Change." [Core writing team: R.K. Pachauri and L.A. Meyer, eds.], Geneva. https://www.ipcc.ch/report/ar5/syr/.

Intergovernmental Panel on Climate Change (2018). "Summary for Policymakers of IPCC Special Report on Global Warming of 1.5°C Approved by Governments." Incheon, October 8. https://www.ipcc.ch/2018/10/08/summary-for-policymakers-of-ipcc-special-report-on-global-warming-of-1-5c-approved-by-governments/.

Intergovernmental Panel on Climate Change (2021). "Climate Change Widespread, Rapid, and Intensifying." Geneva, August 9. https://www.ipcc.ch/2021/08/09/ar6-wg1-20210809-pr.

Intergovernmental Science-Policy Platform on Biodiversity and Ecosystem Services (2019). "Global Assessment Report on Biodiversity and Ecosystem Services of the Intergovernmental Science-Policy Platform on Biodiversity and Ecosystem Services." E.S. Brondizio, J. Settele, S. Díaz, and H.T. Ngo, eds. Intergovernmental Science-Policy Platform on Biodiversity and Ecosystem Services, Bonn. https://doi.org/10.5281/zenodo.3831673.

International Chamber of Commerce (2016). "ICC G20 Business Scorecard." 6th edition, Paris, December. https://iccwbo.org/content/uploads/sites/3/2017/01/ICC-G20-Business-Scorecard-December-2016.pdf.

International Energy Agency (2020). "World Energy Outlook 2020: Overview." Paris, October. https://www.iea.org/reports/world-energy-outlook-2020?mode=overview.

International Energy Agency (2021a). "After Steep Drop in Early 2020, Global Carbon Dioxide Emissions Have Rebounded Strongly." March, Paris. 2. https://www.iea.org/news/after-steep-drop-in-early-2020-global-carbon-dioxide-emissions-have-rebounded-strongly.

International Energy Agency (2021b). "Global Carbon Dioxide Emissions Are Set for Their Second-Biggest Increase in History." Paris, April 20. https://www.iea.org/news/global-carbon-dioxide-emissions-are-set-for-their-second-biggest-increase-in-history.

International Energy Agency (2021c). "Net Zero by 2050: A Roadmap for the Global Energy Sector." *Press Release*, Paris, May 18. https://www.iea.org/events/net-zero-by-2050-a-roadmap-for-the-global-energy-system.

International Institute for Sustainable Development (2019a). "Italy and Mexico Commit, Urge Others to Increase Climate Education Ahead of COP 26." *SDG Knowledge Hub*, December 12. https://sdg.iisd.org/news/italy-and-mexico-commit-urge-others-to-increase-climate-education-ahead-of-cop-26.

International Institute for Sustainable Development (2019b). "One Planet Summit Calls for Developing Renewable Energy, Protecting Biodiversity in Africa." *SDG Knowledge Hub*, March 19. https://sdg.iisd.org/news/one-planet-summit-calls-for-developing-renewable-energy-protecting-biodiversity-in-africa/.

International Institute for Sustainable Development (2020). "Climate Ambition Summit 2020." *SDG Knowledge Hub*, December 12. https://sdg.iisd.org/events/5th-anniversary-of-the-paris-agreement.

International Maritime Organization (2018). "UN Body Adopts Climate Change Strategy for Shipping." London, April 13. https://www.imo.org/en/MediaCentre/PressBriefings/Pages/06GHGinitialstrategy.aspx.

International Monetary Fund (2017a). "World Economic Outlook: Seeking Sustainable Growth." Washington, DC, October. https://www.imf.org/-/media/Files/Publications/WEO/2017/October/pdf/main-chapter/text.ashx.

International Monetary Fund (2017b). "World Economic Outlook Update: A Shifting Global Economic Landscape." International Monetary Fund, Washington, DC, January. https://www.imf.org/external/pubs/ft/weo/2017/update/01/. Accessed: September 2017.

International Monetary Fund (2021). "IMF Managing Director Kristalina Georgieva Calls for Strong G20 Policies to Counter 'Dangerous Divergence'." Press release no. 21/47, Washington, DC, February 26. https://www.imf.org/en/News/Articles/2021/02/26/pr2147-g20-imf-md-kristalina-georgieva-calls-strong-g20-policies-counter-dangerous-divergence.

Ipsos (2021). "Earth Day 2021: Globally People Wonder What's the Plan to Tackle Climate Change." April 22. https://www.ipsos.com/en-ca/earth-day-2021-globally-people-wonder-whats-plan-tackle-climate-change.

Irfan, Umair and Brian Resnick (2018). "Megadisasters Devastated America in 2017. And They're Only Going to Get Worse." *Vox*, March 26. https://www.vox.com/energy-and-environment/2017/12/28/16795490/natural-disasters-2017-hurricanes-wildfires-heat-climate-change-cost-deaths.

Irish, John (2017). "At G7, 'Pragmatist' Trump Leaves a Mark on New Boy Macron." *Reuters*, May 27. https://www.reuters.com/article/idUSKBN18N0V6.

Irish, John and Crispian Balmer (2017). "G7 Leaders Divided on Climate Change, Closer on Trade Issues." *Reuters*, May 25. https://www.reuters.com/article/idUSKBN18L2ZU.

Isaac, Joseph, Ronny Jumeau, Anisul Islam Mahmud et al. (2018). "When Will the World's Polluters Start Paying for the Mess They Made?" *Climate Home News*, May 2. https://www.climatechangenews.com/2018/05/02/will-worlds-polluters-start-paying-mess-made/.

Islam, Shada (2021). "Time to Dump the G7: It's a Relic of the Past." *EU Observer*, May 3. https://euobserver.com/opinion/151725.

Japan Ministry of Foreign Affairs (2018). "G7 Charlevoix Summit: Overview of the G7 Meeting." June 9. https://www.mofa.go.jp/ecm/ec/page4e_000855.html#section7.

Japan Times (2019). "Long Overdue Fight against Plastic Pollution." June 17. Editorial. https://www.japantimes.co.jp/opinion/2019/06/17/editorials/long-overdue-fight-plastic-pollution.

Japan Times (2020). "Prayers Offered for Victims of 2018 Hokkaido Quake." September 6. https://www.japantimes.co.jp/news/2020/09/06/national/2018-hokkaido-earthquake-anniversary/.

Jarvis, Oliver (2021). "World Insights: G7 Underlines Multilateralism to Tackle Global Challenges." *Xinhua*, February 20. https://www.xinhuanet.com/english/2021-02/20/c_139754364.htm.

Jewkes, Stephen (2021). "Italy's New PM Creates Green Superministry to Spur Economy with Eye on EU Recovery Fund." *Globe and Mail*, February 14. https://www.theglobeandmail.com/world/article-italy-gets-green-superministry-as-mario-draghi-eyes-eu-funds/.

Jiji Press (2019). "Japan Adopts Strategy to Achieve Paris Climate Goals." June 11. https://www.nippon.com/en/news/yjj2019061100564/japan-adopts-strategy-to-achieve-paris-climate-goals.html.

Jinnah, Sikina (2017). "Makers, Takers, Shakers, Shapers: Emerging Economies and Normative Engagement in Climate Governance." *Global Governance* 23(2): 285–306. https://doi.org/10.1163/19426720-02302009.

John, Tara, Arwa Damon, Ingrid Formanek et al. (2019). "Cop25 Was Meant to Tackle the Climate Crisis. It Fell Short." *CNN*, December 15. https://edition.cnn.com/2019/12/15/world/cop25-climate-change-intl/index.html.

Johnson, Boris (2021a). "PM Statement at the G7 Summit: 13 June 2021." Cornwall, UK, June 13. https://www.gov.uk/government/speeches/pm-statement-at-the-g7-summit-13-june-2021.

Johnson, Boris (2021b). "PM Statement to the House of Commons on G7 and NATO: 16 June 2021." London, June 16. https://www.gov.uk/government/speeches/pm-statement-to-the-house-of-commons-on-g7-and-nato-16-june-2021.

Johnston, Eric (2019a). "'Coal Japan' Threatens to Displace 'Cool Japan' as U.N. Climate Conference Set to Kick Off in Madrid." *Japan Times*, December 1. https://www.japantimes.co.jp/news/2019/12/01/national/united-nations-climate-conference-madrid-japan-shinjiro-koizumi-coal.

Johnston, Eric (2019b). "Plastic Waste Coming into Focus, but Can a Deal Be Reached?" *Japan Times*, June 27.

Jones, Gavin (2021). "G20 Fails to Agree on Climate Goals in Communique." *Reuters*, July 23. https://www.reuters.com/world/g20-loath-commit-climate-meeting-tussle-over-carbon-wording-2021-07-23.

Kahler, Miles (1992). "Multilateralism with Small and Large Numbers." *International Organization* 46(3): 681–708. http://www.jstor.org/stable/2706992.

Kahler, Miles (2017). "Domestic Sources of Transnational Climate Governance." *International Interactions* 43(1): 156–74. https://doi.org/10.1080/03050629.2017.1251687.

Kahn, Brian (2014). "2014 Officially Hottest Year on Record." *Scientific American*, January 5. https://www.scientificamerican.com/article/2014-officially-hottest-year-on-record.

Kami, Annie and Alan Rappeport (2020). "G20 Summit Closes with Little Progress and Big Gaps between Trump and Allies." *New York Times*, November 22. https://www.nytimes.com/2020/11/22/us/politics/g20-summit-trump.html.

Karlsson-Vinkhuyzen, Sylvia, Lars Friberg and Edoardo Saccenti (2016). "Read All About It!? Public Accountability, Fragmented Global Climate Governance and the Media." *Climate Policy* 17(8): 982–97. https://doi.org/10.1080/14693062.2016.1213695.

Karunungan, Renee Juliene (2016). "Marrakech Climate Talks Thrown into Confusion by US Election." *Asia Sentinel*, November 12. https://www.asiasentinel.com/p/marrakech-climate-talks-confusion-us-election.

Kaul, Inge (2017). "For the G20: Let's Return to the Original Idea." *German Development Institute*, July 20. https://blogs.die-gdi.de/2017/07/20/for-the-g20-lets-return-to-the-original-idea.

Kayagil, Alev (2015). "The Turkish Presidency and the W20." *G20 Research Group*, November 27. http://www.g20.utoronto.ca/analysis/151127-kayagil.html.

Keohane, Robert (2015). "The Global Politics of Climate Change: Challenge for Political Science." *Political Science and Politics* 48(1): 19–26. https://doi.org/10.1017/S1049096514001541.

Keohane, Robert O. and David G. Victor (2016). "Cooperation and Discord in Global Climate Policy." *Nature Climate Change* 6(June): 570–75. https://doi.org/10.1038/nclimate2937.

Kingdom of Saudi Arabia (2015). "The Intended Nationally Determined Contribution of the Kingdom of Saudi Arabia under the UNFCCC." Riyadh, November. https://www4.unfccc.int/sites/ndcstaging/PublishedDocuments/Saudi%20Arabia%20First/KSA-INDCs%20English.pdf.

Kinkartz, Sabine (2015). "G7 Pledges to Intensify Development Cooperation." *DW*, June 8. https://p.dw.com/p/1FdkP.

Kirton, John (2013). *G20 Governance for a Globalized World*. Farnham: Ashgate.

Kirton, John (2015a). "The G20 Antalya Summit's Substantial Success." *G20 Research Group*, November 16. http://www.g20.utoronto.ca/analysis/151116-kirton-participation.html.

Kirton, John (2015b). "A Summit of Significant Success: The G7 at Elmau." *G7 Research Group*, June 8. http://www.g7.utoronto.ca/evaluations/2015elmau/kirton-performance. html.

Kirton, John (2016a). *China's G20 Leadership*. Abingdon: Routledge.

Kirton, John (2016b). "The G7 Ise-Shima Summit's Solid, Security-Centred Success." *G7 Research Group*, May 27. http://www.g7.utoronto.ca/evaluations/2016shima/kirton-shima.html.

Kirton, John (2016c). "The Hangzhou Summit's First Small Step to Control Climate Change." *G20 Research Group*, August 31. http://www.g7g20.utoronto.ca/comment/ 160831-kirton.html.

Kirton, John (2016d). "President Xi's Priorities for the G20's Hangzhou Summit." *G20 Research Group*, August 30. http://www.g7g20.utoronto.ca/comment/160830-kirton. html.

Kirton, John (2016e). "Prospects for Hangzhou." In *G20 China: The 2016 Hangzhou Summit*, John Kirton and Madeline Koch, eds. London: Newsdesk Media, pp. 78–79. http://www.g7g20.utoronto.ca/books/g20hangzhou2016.pdf.

Kirton, John (2016f). "Prospects for Italy's G7 Taormina Summit 2017." *G7 Research Group*, October 7. http://www.g7.utoronto.ca/evaluations/2017taormina/kirton-prospects.html.

Kirton, John (2017a). "From Collision to Co-operation: Prospects for the Hamburg Summit." *Lecture at the Bavarian School of Public Policy, Technical University of Munich, G20 Research Group*, June 26. http://www.g20.utoronto.ca/analysis/170626-kirton-prospects.pdf.

Kirton, John (2017b). "A G7 Summit of Solid Security Success: Prospects for Togetherness with Trump at Taormina." *G7 Research Group*, May 23. http://www.g7.utoronto.ca/ evaluations/2017taormina/kirton-performance.html.

Kirton, John (2017c). "A Summit of Significant Success: The G20 at Hamburg." *G20 Research Group*, July 11. http://www.g20.utoronto.ca/analysis/170711-kirton-performance.html.

Kirton, John (2018a). "Canadians' Strong Support for Canada's National Interests and Distinctive National Values." Presentation to the Canadian International Council, Toronto, April 16. http://www.g8.utoronto.ca/scholar/kirton-cfp-2018.html.

Kirton, John (2018b). "A G7 Summit of Significant Success at Charlevoix 2018." *G7 Research Group*, July 13. http://www.g7.utoronto.ca/evaluations/2018charlevoix/ kirton-performance-full.html.

Kirton, John (2018c). "Implementing G20 Summit Commitments." Italian Institute for International Political Studies, Rome, November 13. https://www.ispionline.it/en/ pubblicazione/implementing-g20-summit-commitments-21670.

Kirton, John (2018d). "Japan's 2019 G20 Osaka Summit Agenda." *G20 Research Group*, December 1. http://www.g20.utoronto.ca/analysis/181201-kirton-osaka.html.

Kirton, John (2018e). "Japan's Priority Agenda for Its G20 Osaka Summit 2019." *G20 Research Group*, December 1. http://www.g20.utoronto.ca/analysis/181201-kirton-osaka-priorities.html.

Kirton, John (2018f). "Public Support for National Interests and Distinctive National Values in Canadian Foreign Policy." Paper prepared for a presentation at the *Canadian International Council's launch of Canada's 2018 World Survey Report, G7 Research Group*, April 16. http://www.g7.utoronto.ca/scholar/kirton-cfp-2018.html.

Kirton, John (2018g). "A Significant Security-Centred Success: Prospects for the G7 Charlevoix Summit in 2018." Paper prepared for a presentation at the Paul H. Nitze School for Advanced International Studies, Washington, DC, G7 Research Group,

April 18. http://www.g7.utoronto.ca/evaluations/2018charlevoix/kirton-prospects-2018-04-18.html.

Kirton, John (2018h). "Strengthening Economic, Energy and Food Security through Canada's G7 and G20 Summits from Kananaskis 2002 to Charlevoix 2018." Lecture prepared for the Ranchmen's Club, Calgary, February 7. http://www.g7.utoronto.ca/scholar/kirton-ranchmen-2018.html.

Kirton, John (2018i). "A Summit of Solid Success: The 2018 G20 Buenos Aires Summit." *G20 Research Group*, December 1. http://www.g20.utoronto.ca/analysis/181201-kirton-solid-success.html.

Kirton, John (2019a). "The G7 Biarritz Summit: A Substantial Success." *G7 Research Group*, August 26. http://www.g7utoronto.ca/evaluations/2019biarritz/kirton-performance.html.

Kirton, John (2019b). "Promising Prospects for Planetary Preservation at Saudi Arabia's G20 in 2020." *G20 Research Group*, December 6. http://www.g20.utoronto.ca/analysis/191206-kirton.html.

Kirton, John (2019c). "Prospects for the Biarritz Summit." In *G7 France: The 2019 Biarritz Summit*, John Kirton and Madeline Koch, eds. London: GT Media, pp. 26–27. http://bit.ly/G7France.

Kirton, John (2019d). "Prospects for the Osaka Summit." In *G20 Japan: The 2019 Osaka Summit*, John Kirton and Madeline Koch, eds. London: GT Media, p. 38. http://bit.ly/G20Japan.

Kirton, John (2020a). "Confronting the Crises." In *G7 USA: The 2020 Virtual Year*, John Kirton and Madeline Koch, eds. London: GT Media, p. 26. http://bit.ly/g7usa.

Kirton, John (2020b). "A Historic Year for G7 Summitry." In *G7 USA: The 2020 Virtual Year*, John Kirton and Madeline Koch, eds. London: GT Media, p. 11. http://bit.ly/g7usa.

Kirton, John (2020c). "A Small Short-Term Success at the G20's Riyadh Summit." *G20 Research Group*, November 22. http://www.g20.utoronto.ca/analysis/201122-kirton-performance.html.

Kirton, John (2020d). "Steady as She Goes: G20 Finance Ministers and Central Bank Governors in July 2020." *G20 Research Group*, July 18. http://www.g20.utoronto.ca/analysis/200718-kirton-finance.html.

Kirton, John (2020e). "The United States' Cooperative Leadership in G7 and G20 Governance." *SAIS Review of International Affairs* 40(1): 103–16. doi: 10.1353/sais.2020.0009.

Kirton, John (2021a). "G7 Summitry Builds Back Better under Boris Johnson in 2021." *G7 Research Group*, February 14. http://www.g7.utoronto.ca/evaluations/2021cornwall/kirton-G7-builds-back-better.html.

Kirton, John (2021b). "Joe Biden's Quad Summit Back's Global Britain's G7 Ones." *G7 Research Group*, March 13. http://www.g7.utoronto.ca/evaluations/2021cornwall/kirton-quad.html.

Kirton, John (2021c). "The Promising Performance of G7 Foreign and Development Ministers in May 2021." *G7 Research Group*, May 13. http://www.g7.utoronto.ca/evaluations/2021cornwall/kirton-foreign-and-development-ministers.html.

Kirton, John and Madeline Koch, eds. (2015). *G20 Turkey: The 2015 Antalya Summit*. London: Newsdesk Media. http://www.g7g20.utoronto.ca/books/g20antalya2015.pdf.

Kirton, John and Ella Kokotsis (2015). *The Global Governance of Climate Change: G7, G20 and UN Leadership*. Farnham: Ashgate.

Kirton, John and Marina Larionova (2018a). "Accountability for Effectiveness in Global Governance." In *Accountability and Effectiveness in Global Governance*, John Kirton and Marina Larionova, eds. Abdingdon: Routledge, pp. 3–22.

Kirton, John and Marina Larionova, eds. (2018b). *Accountability for Effectiveness in Global Governance*. Abdingdon: Routledge.

Kirton, John and Alissa Wang (2021). "China's Global Leadership through G20 Compliance." *Chinese Political Science Review* 1–41. https://doi.org/10.1007/s41111-021-00177-2.

Kirton, John and Brittaney Warren (2017). "G20 Insights: T20 Recommendations Realized." *G20 Research Group*, November 3. http://www.g20.utoronto.ca/analysis/t20-2017-recommendations-realized.html.

Kirton, John and Brittaney Warren (2018a). "G20 Climate Change Governance: Performance, Prospects, Proposals." Paper prepared for the conference on "Prospects and Possibilities for Japan's 2019 G20 Osaka Summit," Soka University, December 10, Tokyo. http://www.g20.utoronto.ca/biblio/G20_Climate_Change_Governance_181210.pdf.

Kirton, John and Brittaney Warren (2018b). "Making History in Halifax: The Performance of the G7 Environment Ministers' Meeting in 2018." *G7 Research Group*, September 27. http://www.g7.utoronto.ca/evaluations/kirton-warren-making-history-in-halifax.pdf.

Kirton, John and Brittaney Warren (2020a). "A Fragile First Step? G7 and G20 Governance of Climate Change, the Environment, Health and Indigenous Peoples." Paper prepared for the Convention of the 2020 International Studies Association (scheduled for March 25–28, 2020; cancelled because of COVID-19), April 5. http://www.g7.utoronto.ca/scholar/kirton-warren-isa-2020.pdf.

Kirton, John and Brittaney Warren (2020b). "From Silos to Synergies: G20 Governance of Climate Change, Digitalization and the SDGs." Paper prepared at the conference on "The Future of Global Economic Governance in a Digitalized World," Center for International Institutions Research, Russian Presidential Academy of National Economy and Public Administration, October 7–8, Moscow.

Kirton, John and Brittaney Warren (2020c). "Selected Steps against a Comprehensive Threat: The G20 Leaders' Statement on COVID-19." *G20 Research Group*, March 26. http://www.g20.utoronto.ca/analysis/200326-kirton-warren.html.

Kirton, John, Ella Kokotsis and Aurora Hudson (2018). "Controlling Climate Change through G7/8, G20 and UN Leadership." In *Accountability and Effectiveness in Global Governance*, Marina Larionova and John Kirton, eds. Abingdon UK: Routledge.

Kirton, John, Ella Kokotsis and Brittaney Warren (2019a). "G7 Governance of Climate Change: The Search for Effectiveness." In *The G7, Anti-Globalization and the Governance of Globalization*, Chiara Oldani and Jan Wouters, eds. Abingdon UK: Routledge, pp. 90–126.

Kirton, John, Julia Tops and Angela Min Yi Hou (2019b). "Final Compliance with Commitments of the 2018 G7 Halifax Environment Ministers Meeting." *G7 Research Group*, May 3. http://www.g7.utoronto.ca/evaluations/2018-G7-emm-final-compliance.pdf.

Kirton, John, Brittaney Warren and Madeline Koch (2017). "Improving G20 Compliance to Control Climate Change." Paper prepared for "The G20 and the Global Environment, Climate and Energy Agenda," Bavarian School of Public Policy, Technical University of Munich, June 28, Munich. http://www.g20.utoronto.ca/biblio/kirton-warren-koch-climate-2017.pdf.

Kizzier, Kelley, Kelly Levin and Mandy Rambharos (2019). "What You Need to Know About Article 6 of the Paris Agreement." World Resources Institute, December 2. https://www.wri.org/blog/2019/12/article-6-paris-agreement-what-you-need-to-know.

Klett, Steven (2017). "Is Trump Closer to Being Impeached Now That He's Being Investigated for Obstruction of Justice?" *International Business Times*, June 17. https://www.ibtimes.com/trump-closer-being-impeached-now-hes-being-investigated-obstruction-justice-2553752.

Koch, Madeline (2019). "China's B20 and T20 Leadership in the G20." Presentation to the International Symposium on Connecting the World and the Future, Shanghai Academy of Global Governance and Area Studies, November 8–9, Shanghai. http://www.g20.utoronto.ca/biblio/koch-china-b20-t20-leadership.html.

Kokotsis, Eleanore (1999). *Keeping International Commitments: Compliance, Credibility, and the G7, 1988–1995*. New York: Garland.

Krasner, Stephen (1983). *International Regimes*. Ithaca, NY: Cornell University Press.

Krupp, Fred (2019). "G20 Leaders Must Take the Lead." *China Daily*, June 17. https://www.chinadailyhk.com/articles/234/17/148/1560753850174.html.

Kudo, Yashushi (2016). "What's on the Agenda for the 2016 G7 Summit." *Council of Councils*, May 20. https://www.cfr.org/councilofcouncils/global-memos/whats-agenda-2016-g7-summit.

Kulik, Julia (2015). "G20 Gender Equality Performance in Antalya: Meagre Action, More Accountability." *G20 Research Group*, December 3. http://www.g20.utoronto.ca/analysis/151203-research-gender.html.

Kulik, Julia (2016). "G7 Performance on Gender Equality at Ise-Shima." *G7 Research Group*, June 7. http://www.g8.utoronto.ca/evaluations/2016shima/kulik-gender.html.

Kyodo News (2019a). "Abe Cabinet Support Rate Falls to 47.6% Amid Pension System Controversy." *Japan Times*, June 17. https://www.japantimes.co.jp/news/2019/06/17/national/politics-diplomacy/cabinet-support-rate-falls-47-6-amid-pension-system-controversy-telephone-poll-finds.

Kyodo News (2019b). "Japan Gets 'Fossil of the Day' Award for Hanging on to Coal-Fired Power." *Japan Times*, December 4. https://english.kyodonews.net/news/2019/12/1f42d6542d87-japan-gets-fossil-prize-for-hanging-on-to-coal-fired-power.html.

Kyodo News (2021). "Kumamoto Commemorates 5th Anniv. of Deadly Quakes." 14 April. https://english.kyodonews.net/news/2021/04/df302b75a32f-kumamoto-commemorates-5th-anniv-of-deadly-quakes.html.

Le Quéré, Corinne, Robert B. Jackson, Matthew W. Jones et al. (2020). "Temporary Reduction in Daily Global CO2 Emissions During the COVID-19 Forced Confinement." *Nature Climate Change* 10(7): 647–53. https://doi.org/10.1038/s41558-020-0797-x.

Levitz, Stephanie (2018). "Fewer Than Half of Canadians Hold an Open View of the World, Poll on Populism Finds." *Toronto Star*, January 22. https://www.thestar.com/news/canada/2018/01/22/fewer-than-half-of-canadians-hold-open-view-of-the-world-poll-on-populism-finds.html.

Lewis, Leo (2018). "Thousands Stranded at Japan Airport after Typhoon." *Financial Times*, September 4. https://www.ft.com/content/42f0045c-b059-11e8-8d14-6f049d06439c.

Li, Jing (2017a). "Dispute over Pre-2020 Climate Action 'Risks Repeat of Copenhagen'." *Climate Home News*, November 10. https://www.climatechangenews.com/2017/11/10/dispute-pre-2020-climate-action-risks-repeat-copenhagen.

Li, Jing (2017b). "Is China the Leader UN Climate Talks Need?" *Climate Home News*, November 1. https://www.climatechangenews.com/2017/11/01/china-leader-un-climate-talks-need.

Li, Jing and Meagan Darby (2017). "Tensions Emerge over Timetable for Raising Climate Ambition." *Climate Home News*, November 3. https://www.climatechangenews.com/2017/11/03/tensions-emerge-timetable-raising-climate-ambition.

Liberia. Executive Mansion (2015). "President Sirleaf Calls on G7 Nations to Change the Way They Do Business; Change Mindset and a New Paradigm for Development." June 8. https://emansion.gov.lr/2press.php?news_id=3307.

Lockett, Hudson and Benjamin Parkin (2021). "Rupee Retreats as New Covid Wave Threatens India's Economic Recovery." *Financial Times*, April 20, p. 10.

Londoño, Ernesto and Lisa Friedman (2018). "Brazil Backs out of Hosting 2019 Climate Change Meeting." *New York Times*, November 28. https://nyti.ms/2zsrnv1.

Lövbrand, Eva, Mattias Hjerpe and Björn-Ola Linnér (2017). "Making Climate Governance Global: How UN Climate Summitry Comes to Matter in a Complex Climate Regime." *Environmental Politics* 26(4): 580–99. https://doi.org/10.1080/09644016.2017.1319019.

Lowe, Jaime (2021). "'If You Move out Here, You Make a Deal with Nature': Life in a Fire-Prone Canyon." *New York Times*, June 19. https://www.nytimes.com/2021/06/19/us/topanga-canyon-wildfires.html.

Luce, Edward (2021). "Biden's Pivot to China." *Financial Times*, June 19, p. 5.

Luckhurst, Jonathan (2019). "The G20 Osaka Legacy, from Global Summitry to the Korean DMZ." *Future of Globalisation (Blog)*. German Development Institute, July 3. https://blogs.die-gdi.de/2019/07/03/the-osaka-legacy-from-g20-summitry-to-dmz.

Macdonald, Ted (2020). "How Broadcast TV Networks Covered Climate Change in 2019." Media Matters for America, Washington, DC, February 27. https://www.mediamatters.org/broadcast-networks/how-broadcast-tv-networks-covered-climate-change-2019.

Macdonald, Ted (2021). "How Broadcast TV Networks Covered Climate Change in 2020." Media Matters for America, Washington, DC, March 10. https://www.mediamatters.org/broadcast-networks/how-broadcast-tv-networks-covered-climate-change-2020.

Macron, Emmanuel (2018a). "France's Priorities for the G7." In *G7 Canada: The 2018 Charlevoix Summit*, John Kirton and Madeline Koch, eds. London: GT Media, pp. 10–11. http://bit.ly/G7Charlevoix2018.

Macron, Emmanuel (2018b). "United Nations General Assembly: Speech by President Emmanuel Macron." Ministère de l'Europe et des affaires étrangères, France, September 25. https://www.diplomatie.gouv.fr/en/french-foreign-policy/united-nations/news-and-events/united-nations-general-assembly/unga-s-73rd-session/article/united-nations-general-assembly-speech-by-president-emmanuel-macron-25-09-18.

Macron, Emmanuel (2021). "Leaders Summit on Climate: Speech by French President Emmanuel Macron." Speech to the Leaders Summit on Climate, Paris, April 22. https://www.elysee.fr/en/emmanuel-macron/2021/04/22/leaders-summit-on-climate-speech-by-french-president-emmanuel-macron.

Maisonnave, Fabiano (2017a). "Nicaragua Joined Paris Pact in Bid for Top Climate Fund Appointment: Sources." *Climate Home News*, November 9. https://www.climatechangenews.com/2017/11/09/nicaragua-joined-paris-pact-bid-top-climate-fund-appointment-sources.

Maisonnave, Fabiano (2017b). "Rich Countries 'Trying to Turn Climate Funds into World Bank'." *Climate Home News*, November 20. https://www.climatechangenews.com/2017/11/20/rich-countries-trying-turn-climate-fund-world-bank.

Malo, Sebastien (2018). "Brazil Green Groups Prepare Climate-Change Contingency Plan." *Reuters*, December 11. https://www.reuters.com/article/us-climatechange-accord-brazil/brazil-green-groups-prepare-climate-change-contingency-plan-idUSKBN1OA2BM.

Malpass, David (2021). "Remarks by World Bank Group President David Malpass at Session 1 of the G20 Finance Ministers and Central Bank Governors Meeting." February 26. https://www.worldbank.org/en/news/speech/2021/02/26/remarks-by-world-bank-group-president-david-malpass-at-the-g20-finance-ministers-and-central-bank-governors-meeting.

Mann, Michael E. (2021). "The G7 Can't Compromise with the Climate: Opinion." *Newsweek*, June 25. https://www.newsweek.com/g7-cant-compromise-climate-opinion-1603969.

Manson, Katrina and Leslie Hook (2021). "Blinken Says US Must Take on China in Green Energy." *Financial Times*, April 21, p. 4.

Manson, Katrina, Guy Chazan and George Parker (2021). "Biden Declares 'America Is Back' in Pledge to Repair NATO Alliance." *Financial Times*, February 20, p. 1.

Mariani, Sarah (2019). "The Role of Youth in the 2018 Charlevoix Summit." *G7 Research Group*, April 12. http://www.g7g20.utoronto.ca/comment/190412-mariani.html.

Martens, Catherine (2018). "French President Emmanuel Macron: Is the Honeymoon Over?" *Deutsche Welle*, September 3. https://www.dw.com/en/french-president-emmanuel-macron-is-the-honeymoon-over/a-45329639.

Martin, Dan (2019). "Who Needs the G20? Question Gets Louder in Osaka." *AFP*, June 29. https://sg.news.yahoo.com/needs-g20-gets-louder-osaka-093153344.html.

Mathiesen, Karl (2017a). "'Don't Wake the Bear': Fragile Climate Talks Begin in Bonn." *Climate Home News*, November 6. https://www.climatechangenews.com/2017/11/06/dont-wake-bear-fragile-climate-talks-begin-bonn.

Mathiesen, Karl (2017b). "Pope Francis Gives Trump a Climate Change Message." *Climate Home News*, May 24. http://www.climatechangenews.com/2017/05/24/pope-francis-gives-trump-climate-change-message.

Mathiesen, Karl (2017c). "US Climate Talks Delegation to Be Led by under Secretary Thomas Shannon." *Climate Home News*, October 17. https://www.climatechangenews.com/2017/10/17/us-climate-talks-delegation-lead-secretary-thomas-shannon.

Mathiesen, Karl (2018a). "Draft G20 Statement Waters Down Paris Climate Commitment." *Climate Home News*, November 26. https://www.climatechangenews.com/2018/11/26/draft-g20-statement-waters-paris-climate-commitment.

Mathiesen, Karl (2018b). "Paris Agreement Fight Could Push US out Permanently, Warn Top Obama Officials." *Climate Home News*, December 3. https://www.climatechangenews.com/2018/12/03/paris-agreement-fight-push-us-permanently-warn-top-obama-officials.

Mathiesen, Karl and Chloé Farand (2019). "Chile Pulls out of Hosting Cop25 Climate Talks Amid Civil Unrest." *Climate Home News*, October 30. https://www.climatechangenews.com/2019/10/30/chile-pulls-hosting-cop25-climate-talks.

Mathiesen, Karl and Jing Li (2017a). "China Flexes Its Muscle as Climate Talks End with Slow Progress." *Climate Home News*, November 17. https://www.climatechangenews.com/2017/11/17/china-flexes-muscle-climate-talks-make-slow-progress.

Mathiesen, Karl and Jing Li (2017b). "Trump's 'Top Priority' at Climate Talks: Protecting an Obama Legacy." *Climate Home News*, November 14. https://www.climatechangenews.com/2017/11/14/trump-priority-climate-talks-no-soft-option-china.

Mathiesen, Karl and Mantoe Phakathi (2017). "Rich Countries Not Talking Climate Finance Seriously, Say African Officials." Climate Home News, November. https://www.climatechangenews.com/2017/11/09/rich-countries-not-talking-climate-change-seriously-say-african-officials.

Mathiesen, Karl, Meagan Darby and Sara Stefanini (2018). "Countries Breathe Life into the Paris Climate Agreement." *Climate Home News*, December 15. https://www.climatechangenews.com/2018/12/15/countries-breathe-life-paris-climate-agreement.

Matthews, Jessica (2021). "Present at the Re-Creation?: U.S. Foreign Policy Must Be Remade, Not Restored." *Foreign Affairs*, March/April, pp. 10, 12–16. https://www.foreignaffairs.com/articles/united-states/2021-02-16/present-re-creation.

May, Theresa (2018). "PM Speaks at Commonwealth Press Conference." Prime Minister's Office, UK Government, April 20. https://www.gov.uk/government/speeches/pm-speaks-at-commonwealth-press-conference-20-april-2018.

McCarthy, Joe and Erica Sánchez (2019). "The 'Attenborough Effect' Is Causing Plastic Pollution to Plummet." *Global Citizen*, April 17. https://www.irishtimes.com/news/environment/attenborough-effect-praised-for-reduced-use-of-plastic-1.3945951.

McGrath, Matt (2014). "UN Members Agree Deal at Lima Climate Talks." *BBC News*, December 14. https://www.bbc.com/news/science-environment-30468048.

McGrath, Matt (2020). "Climate Change: US Formally Withdraws from Paris Agreement." November 4. https://www.bbc.com/news/science-environment-54797743.

McGrath, Matt (2021). "Climate Change: Net Zero Targets Are 'Pie in the Sky'." *BBC News*, April 1. https://www.bbc.com/news/science-environment-56596200.

McManis, Alex (2020). "Coping It in Madrid: Why Australia's Stance at Cop25 Was So Widely Condemned." *Australian Outlook*, January 16. https://www.internationalaffairs.org.au/australianoutlook/coping-it-in-madrid-why-australias-stance-at-cop25-was-so-widely-condemned.

McTague, Tom and Peter Nicholas (2020). "The World Order That Donald Trump Revealed." *Atlantic*, October 20. https://www.theatlantic.com/international/archive/2020/10/donald-trump-foreign-policy/616773.

Médecins Sans Frontières (2016). "G7 Fail to Address the Biggest Threats to Global Health." May 27. https://www.msf.org/g7-fail-address-biggest-threats-global-health.

Messenger, Michael (2020). "COVID Solutions? Invite Kids to the Adult Table." *Hamilton Spectator*, November 30. https://www.hamiltonnews.com/opinion-story/10279113-covid-solutions-invite-kids-to-the-adult-table.

Mezouar, Salaheddine (2016). "Informal Consultations on the First Session of the Conference of the Parties Serving as the Meeting of the Parties to the Paris Agreement (CMA 1) and on Item 4 of the Agenda of COP 22." Further revised elements of outcomes for CMA 1 and COP 22. Proposal by the President, November 14. https://unfccc.int/files/meetings/marrakech_nov_2016/application/pdf/announcement_on_cma_cop_president_further_revised_elements_141116_1000hrs.pdf.

Michael, Prince of Liechtenstein (2016). "Buying Time with Other People's Money." June 3. Geopolitical Intelligence Services. https://www.gisreportsonline.com/buying-time-with-other-peoples-money,politics,1872.html.

Michaelowa, Katharina and Axel Michaelowa (2017). "Transnational Climate Governance Initiatives: Designed for Effective Climate Change Mitigation." *International Interactions* 43(1): 129–55. https://doi.org/10.1080/03050629.2017.1256110.

Miharu, Mitsuki (2020). "内閣支持率・不支持率の平均. 1月24〜26日に実施された日経新聞の世論調査と、1月25〜26日に実施された朝日新聞の世論調査を反映しました。." ["Average Approval/Disapproval Rate of the Cabinet. Reflecting the Opinion Poll of the Nikkei Newspaper Conducted from January 24-26 and the Opinion Poll of the Asahi Shimbun Conducted from January 25-26."] *Twitter*, January, 27. https://twitter.com/miraisyakai/status/1221760463412682753.

Millan Lombrana, Laura and Hayley Warren (2020). "A Pandemic That Cleared Skies and Halted Cities Isn't Slowing Global Warming." *Bloomberg*, May 8. https://www.bloomberg.com/graphics/2020-how-coronavirus-impacts-climate-change.

Miller, Jake (2017). "Mattis Says Trump 'Wide Open' on Paris Climate Accord." *CBS News*, May 27. https://www.cbsnews.com/news/defense-secretary-james-mattis-trump-wide-open-paris-climate-change-accord/.

Milmand, Oliver and Vivian Ho (2020). "California Wildfires Spawn First 'Gigafire' in Modern History." *Guardian*, October 6. https://www.theguardian.com/us-news/2020/oct/06/california-wildfires-gigafire-first.

Mirza, Asad (2021). "G-7 Summit: Not Much to Cheer." *Navhind Times*, June 16. https://www.navhindtimes.in/2021/06/16/opinions/opinion/g-7-summit-not-much-to-cheer.

Miyake, Kuni (2019). "What a Difference a Day Makes after the G20." *Japan Times*, July 1. https://www.japantimes.co.jp/opinion/2019/07/01/commentary/japan-commentary/difference-day-makes-g20/.

Mohieldin, Mahmoud (2021). "Good Words but Not Major Action at the G7." *Ahram Online*, June 16. https://english.ahram.org.eg/News/414301.aspx.

Morning Consult (2021). "Global Leader Approval Rating Tracker." Last modified: August 5, 2021. https://morningconsult.com/form/global-leader-approval/#section-53.

Morris, Harvey (2020). "Concerted Global Action Needed to Help the Poor." *China Daily*, November 27. https://www.chinadaily.com.cn/a/202011/27/WS5fc04a57a31024ad0ba96c53.html.

Morrow, Adrian (2017). "Trump Changes Course on NAFTA, Drops Threat to Back Out." *Globe and Mail*, April 26. https://www.theglobeandmail.com/report-on-business/economy/trump-nafta-us-canada-mexico/article34818889.

Morton, Adam (2019). "About 100 Countries at UN Climate Talks Challenge Australia's Use of Carryover Credits." *Guardian*, December 9. https://www.theguardian.com/environment/2019/dec/09/about-100-countries-at-un-climate-talks-challenge-australias-use-of-carryover-credits.

Mulroney, David (2018). "Trudeau Is Delivering the Foreign Policy Canadians Deserve." *Globe and Mail*, February 25. https://www.theglobeandmail.com/opinion/trudeau-is-delivering-the-foreign-policy-canadians-deserve/article38103174.

Nanos (2018). "Liberals 41, Conservatives 29, NDP 16, Green 8 in Latest Nanos Federal Tracking." April 17. https://nanos.co/wp-content/uploads/2018/04/Political-Package-2018-04-13-R.pdf.

Nanos (2019). "Conservatives 33, Liberals 33, NDP 17, Green 10, People's 1 in Latest Nanos Federal Tracking." *Nanos Weekly Tracking*, ending June 21, 2019, June 25. https://www.nanos.co/wp-content/uploads/2019/06/Political-Package-2019-06-21-FR3.pdf.

Nanos (2021). "Nanos Ballot: Liberals 36, Conservatives 29, NDP 19, Green 7, BQ 7 Heading into Federal Budget Week." *Nanos Weekly Tracking*, ending April 16, 2021, April 20. https://nanos.co/wp-content/uploads/2021/04/Political-Package-2021-04-16-FR-with-tabs.pdf.

Nardelli, Alberto (2021). "Boris Johnson Plans a Virtual Meeting of G-7 Leaders Next Month." *Bloomberg*, January 12. https://www.bloomberg.com/news/articles/2021-01-12/u-k-s-johnson-plans-a-virtual-meeting-of-g-7-leaders-next-month.

NASA Earth Observatory (2016). "2015 Was the Hottest Year on Record." January 22. https://earthobservatory.nasa.gov/images/87359/2015-was-the-hottest-year-on-record.

NASA Global Climate Change: Vital Signs (2019). "2018 Fourth Warmest Year in Continued Warming Trend, According to NASA, NOAA." February 6. https://

climate.nasa.gov/news/2841/2018-fourth-warmest-year-in-continued-warming-trend-according-to-nasa-noaa.

Nature (2020). "Delaying COP26 Is Not a Reason to Delay Climate Action." 582: 7. June 2. Editorial. https://www.nature.com/articles/d41586-020-01621-0.

Naylor, Tristen (2020). "All That's Lost: The Hollowing of Summit Diplomacy in a Socially Distanced World." *Hague Journal of Diplomacy* 15(4): 583–98. https://doi.org/10.1163/1871191X-BJA10041.

Neslen, Arthur (2017a). "Syria to Join Paris Agreement, Isolating US." *Climate Home News*, November 11. https://www.climatechangenews.com/2017/11/07/syria-join-paris-agreement-isolating-us/.

Neslen, Arthur (2017b). "US 'No Position' on How Much Humans Are Changing Climate, Says Trump Envoy." *Climate Home News*, December 13. https://www.climatechangenews.com/2017/12/13/world-leaders-can-change-us-position-paris-says-trump-climate-advisor.

NHK (2019). "G7 Summit in France to Discuss Inequality." *NHK World News*. https://web.archive.org/web/20190820225916/https://www3.nhk.or.jp/nhkworld/en/news/20190818_03/.

Nikkei Asia (2018). "APEC Discord Reflects a Global Cooperation System in Crisis." *Nikkei Asia*, November 20. https://asia.nikkei.com/Opinion/The-Nikkei-View/APEC-discord-reflects-a-global-cooperation-system-in-crisis.

Nikkei News (2019). "マクロン氏、気候変動「後退防げた」G20首脳宣言." ["On G20 Leaders Declaration, Macron Says 'retreat was prevented' on Climate Change."] June 29. https://www.nikkei.com/article/DGXMZO46774060Z20C19A6FF8000.

Noelle, Selin (2016). "Teaching and Learning from Environmental Summits: COP 21 and Beyond." *Global Environmental Politics* 16(3): 31–40. https://doi.org/10.1162/GLEP_a_00364.

Norton, Andrew (2021). "The G7 Must Collaborate with China on Climate: Letters." *Financial Times*, June 18, p. 16.

O'Malley, Nick (2021). "World Must Break Its 'Deadly Addiction' to Coal, Say UN Chief." *Sydney Morning Herald*, March 3. https://www.smh.com.au/environment/climate-change/world-must-break-its-deadly-addiction-to-coal-says-un-chief-20210302-p57754.html.

Obama, Barack (2020). *A Promised Land.* New York: Crown Publishers.

Obayashi, Yuka (2019). "G20 Agrees to Tackle Ocean Plastic Waste." *Reuters*, June 16. https://www.reuters.com/article/us-g20-japan-energy-environment/g20-agrees-to-tackle-ocean-plastic-waste-idUSKCN1TH0B3.

OECD Ministerial Council (2017a). "Making Globalisation Work: Better Lives for All." Organisation for Economic Co-operation and Development, Paris, June 7–8. https://www.oecd.org/mcm/documents/2017-ministerial-council-statement.htm.

OECD Ministerial Council (2017b). "Statement of the Chair of Mcm 2017: International Trade, Investment and Climate Change." Organisation for Economic Co-operation and Development, Paris, June 7–8. https://www.oecd.org/mcm/documents/C-MIM-2017-18-EN.pdf.

Oldani, Chiara and Jan Wouters (2019). *The G7, Anti-Globalism and the Governance of Globalization.* Abingdon: Routledge.

One Planet Summit (2017). "12 International Commitments: Stepping up for Finance Adaptation and Resilience to Climate Change." December 12. Paris. https://www.diplomatie.gouv.fr/IMG/pdf/oneplanetsummit-dp-engagements-en_cle88d5e4.pdf.

One Planet Summit (2018a). "One Planet Summit 2018 Results in Reconfirmed and Fresh Climate Action." September 27. https://www.oneplanetsummit.fr/sites/default/files/2018-09/Fact%20Sheet%20Annoucements%20-FINAL%20V5-Vdef_0.pdf.

One Planet Summit (2018b). "Review of the Commitments." September. https://www.oneplanetsummit.fr/sites/default/files/2018-09/OneplanetSummit_ReviewOfThe Commitments_VGB_1.pdf.

One Planet Summit (2019). "Africa Pledge: Chair's Summary." Nairobi, March 14. https://www.oneplanetsummit.fr/sites/default/files/2019-03/AFRICA%20PLEDGE%20-%20VE-%2019-03.pdf.

Onyango, Protus (2019). "Uhuru Leads Other Presidents to Call for a Healthy Environment, Billions Pledged for Green Economy." *Business Standard*, March 15. https://www.standardmedia.co.ke/nairobi/article/2001316668/donors-pledge-sh47tr-to-fight-climate-change.

Organisation for Economic Co-operation and Development (2017a). "OECD Gdp Growth Falls to 0.4% in First Quarter of 2017." Press release, Paris, May 22. https://www.oecd.org/newsroom/gdp-growth-first-quarter-2017-oecd.htm.

Organisation for Economic Co-operation and Development (2017b). "OECD, France and Mexico Launch 'Paris Collaborative on Green Budgeting' at One Planet Summit." Paris, December 12. https://www.oecd.org/env/cc/one-planet-summit-paris-collaborative-on-green-budgeting-december-2017.htm.

Oroschakoff, Kalina and Paola Tamma (2018). "World's Nations Agree on Rules to Implement Paris Climate Deal." *Politico*, December 15. https://www.politico.eu/article/worlds-nations-agree-on-rules-to-implement-paris-climate-deal.

Oxfam (2016). "Rich Countries Turn a Blind Eye to the Needs of Climate-Vulnerable Countries at Marrakech Conference." November 16. https://www.oxfam.org/en/press-releases/rich-countries-turn-blind-eye-needs-climate-vulnerable-countries-marrakech.

Palacio, Ana (2019). "The Twilight of the Global Order." *Gulf Times*, September 4. https://www.gulf-times.com/Story/640717.

Palencia, Gustavo and Ismael Lopez (2020). "Storm Death Toll Rises in Central America as Honduran Leader Pleads for Help." *Reuters*, November 19. https://www.reuters.com/article/storm-iota-idINKBN27Z2K0.

Paravicini, Giulia (2017). "Angela Merkel: Europe Must Take 'Our Fate' into Own Hands." *Politico*, May 28. https://www.politico.eu/article/angela-merkel-europe-cdu-must-take-its-fate-into-its-own-hands-elections-2017/.

Paris Europlace (2017). "Acceleration! The Financial Industry Fully Committed to Fighting Climate Change." Paris, December 11. https://www.paris-europlace.com/en/file/2939/download?token=YUeis-bf.

Parker, George (2021). "Johnson Prepares 'Global Britain' Relaunch in Wake of Brexit." *Financial Times*, January 3. https://www.ft.com/content/a377fbc1-df1c-4de9-8660-5862b51ecd7f.

Patrick, Stewart (2015). "The G7 Summit: An Exclusive Club-but a Global Role." (Blog). *Council on Foreign Relations*. June 3. https://www.cfr.org/blog/g7-summit-exclusive-club-global-role.

Patrick, Stewart (2020). "Up in the Air: Ten Global Summits That Will Test Joe Biden in 2021." December 23. Council of Councils, Washington, DC. https://www.cfr.org/councilofcouncils/global-memos/air-ten-global-summits-will-test-joe-biden-2021-0.

Pauw, Pieter, Kennedy Mbeva and Harro van Asselt (2019). "Subtle Differentiation of Countries' Responsibilities under the Paris Agreement." *Palgrave Communications* 5(86). https://doi.org/10.1057/s41599-019-0298-6.

Payne, Sebastian and Jim Pickard (2021). "Behind Johnson's Green Conversion." *Financial Times*, January 27, p. 15. https://www.ft.com/content/24b55395-5e95-403f-9ef3-76b74f3e9960.

Permanent Mission of France to the United Nations in New York (n.d.a). "Financing the Fight against Climate Change." New York. Last modified: April 1, 2020. https://onu.delegfrance.org/Financing-the-fight-against-climate-change.

Permanent Mission of France to the United Nations in New York (n.d.b). "One Planet Summit." New York. Last modified: April 18, 2021. https://onu.delegfrance.org/One-Planet-Summit.

Perry, Claire (2018). "Britain Backs Strong Rules to Bring the Paris Agreement to Life." *Climate Home News*, December 4. https://www.climatechangenews.com/2018/12/04/britain-backs-strong-rules-bring-paris-agreement-life.

Pew Research Center (2019). "Climate Change and Russia Are Partisan Flashpoints in Public's Views of Global Threats." July 30. Washington, DC. https://www.people-press.org/2019/07/30/climate-change-and-russia-are-partisan-flashpoints-in-publics-views-of-global-threats.

Phakathi, Mantoe (2017). "Four Overlooked Issues at the Bonn Climate Talks." *Climate Home News*, November 17. https://www.climatechangenews.com/2017/11/17/four-overlooked-issues-bonn-climate-talks.

Pickrell, John (2020). "Bush-Fire Smoke Linked to Hundreds of Deaths." March 24. https://doi.org/10.1038/d41586-020-00886-9.

Politico (2014). "Juncker's Ten Priorities." November 17. https://www.politico.eu/article/junckers-ten-priorities.

Postel-Vinay, Karoline (2020). "Return of the Crisis Group." In *G7 USA: The 2020 Virtual Year*, John Kirton and Madeline Koch, eds. London: GT Media, pp. 124–25. http://bit.ly/g7usa.

Poushter, Jacob and Dorothy Manevich (2017). "Globally, People Point to ISIS and Climate Change as Leading Security Threats." Pew Research Center, August 1. https://www.pewresearch.org/global/2017/08/01/globally-people-point-to-isis-and-climate-change-as-leading-security-threats/.

PTI (2015). "BASIC Nations Calls for Roadmap on Climate Financing by Rich Nations." *Economic Times*, December 2. https://economictimes.indiatimes.com/news/politics-and-nation/basic-nations-calls-for-roadmap-on-climate-financing-by-rich-nations/articleshow/50009070.cms.

Purvis, Nigel (2015). "The White House's COP 21 Goals: Less Climate Idealism, More Political Realism." *Guardian*, December 10. http://www.theguardian.com/commentisfree/2015/dec/10/white-house-cop-21-goals-less-climate-idealism-more-political-realism.

Putnam, Robert and Nicholas Bayne (1987). *Hanging Together: Co-operation and Conflict in the Seven-Power Summit*. 2nd ed. London: Sage Publications.

Radwanski, Adam (2021). "Federal Budget 2021: With Their Budget's Green-Recovery Plans, the Liberals Place a Big Bet on Large Industry." *Globe and Mail*, April 19. https://www.theglobeandmail.com/politics/article-with-their-budgets-green-recovery-plans-the-liberals-place-a-big-bet/.

Raftery, Adrian E., Alec Zimmer, Dargan Frierson et al. (2017). "Less Than 2 °C Warming by 2100 Unlikely." *Nature Climate Change* 7: 637–41. https://doi.org/10.1038/nclimate3352.

Rappeport, Alan (2021). "Out of Trump's Shadow, World Bank President Embraces Climate Fight." *New York Times*, April 9. https://www.nytimes.com/2021/04/09/us/politics/david-malpass-world-bank-climate.html.

Rapson, Jessica (2020a). "Increasing the Impact of the G7." In *G7 USA: The 2020 Virtual Year*, John Kirton and Madeline Koch, eds. London: GT Media, pp. 126–27. http://bit.ly/g7usa.

Rapson, Jessica (2020b). "Using Data to Improve Compliance." In *G20 Saudi Arabia: The 2020 Riyadh Summit*, John Kirton and Madeline Koch, eds. London: GT Media, pp. 170–72. http://bit.ly/g20saudi.

Rashmi, Rajani Ramjan (2020). "Global Climate Change: Challenges for India." *G20 Digest* 2(1): 21–28. http://ris.org.in/sites/default/files/G20%20June-August%202020.pdf.

Reguly, Eric (2017). "Who's Hot and Who's Not Ahead of the 'Mini G7' in Italy." *Globe and Mail*, May 12. https://www.theglobeandmail.com/report-on-business/economy/economic-insight/whos-hot-and-whos-not-ahead-of-the-mini-g7-in-italy/article34969117/.

Reidmiller, David, Christopher W. Avery, David R. Easterling et al. (2018). "Impacts, Risks, and Adaptation in the United States: Fourth National Climate Assessment, Volume II." United States Global Change Research Program, Washington, DC. https://nca2018.globalchange.gov/.

Reuters (2021). "Germany's Merkel Hopes for G7 Infrastructure Plans in 2022." June 13. https://www.reuters.com/world/china/germanys-merkel-hopes-g7-infrastructure-plans-2022-2021-06-13/.

Reuters Staff (2017a). "After Summits with Trump, Merkel Says Europe Must Take Fate into Own Hands." *Reuters*, May 28. https://www.reuters.com/article/us-germany-politics-merkel-idUSKBN18O0JK.

Reuters Staff (2017b). "France Awards Climate Grants to U.S.-Based Scientists on Summit Eve." December 11. https://www.reuters.com/article/us-climatechange-summit-idUSKBN1E523J.

Rich, Motoko and Ben Dooley (2021). "Powerful Quake Hits Japan, Evoking a Worrisome Memory." *New York Times*, February 13. https://www.nytimes.com/2021/02/13/world/asia/earthquake-japan-fukushima.html.

Richards, Lisa and Nigel Brew (2020). "2019–20 Australian Bushfires—Frequently Asked Questions: A Quick Guide." March 12. Parliament of Australia, Canberra. https://www.aph.gov.au/About_Parliament/Parliamentary_Departments/Parliamentary_Library/pubs/rp/rp1920/Quick_Guides/AustralianBushfires.

Rinaldi, Augusto Leal and Patricia Nabuco Martuscelli (2016). "The BRICS on Climate Change Global Governance." *Meridiano 47 Journal of Global Studies* 17(October). https://periodicos.unb.br/index.php/MED/article/view/5270.

Robert, Aline (2016). "Global Climate Action Pushing Ahead, Despite Slim Pickings at COP22." November 21. https://www.euractiv.com/section/climate-environment/news/global-climate-action-pushing-ahead-despite-slim-pickings-at-cop22/.

Roberts, Timmons, Mara Dolan, Angelica Arellano et al. (2017). "US Governor Elections Inspire Hope During UN Climate Talks." *Climate Home News*, November 8. https://www.climatechangenews.com/2017/11/08/us-governor-elections-inspire-hope-un-climate-talks.

Robertson, Colin (2018). "Canada Should Use Its G7 Meeting to Tackle the Putin Problem." *Globe and Mail*, March 29. https://www.theglobeandmail.com/opinion/article-canada-should-use-its-g7-meeting-to-tackle-the-putin-problem.

Rose, Michel and Steve Holland (2021). "America Is Back with Biden, France's Macron Says." *Reuters*, June 12. https://www.reuters.com/world/china/america-is-back-with-biden-frances-macron-says-2021-06-12.

Rowlatt, Justin (2021). "COP26: Greta Thunberg Says Glasgow Summit Should Be Postponed." *BBC News*, April 9. https://www.bbc.com/news/uk-scotland-56686163.

Sachs, Jeffrey (2021). "We Don't Need the G7." *Project Syndicate*, June 16. https://www.jeffsachs.org/newspaper-articles/yl549dx2g4pc6e3p52fl677s8bk5aa.

Sainsbury, Tristram and Hannah Wurf (2016). "The G20 and the Future of Energy Governance." *International Organisations Research Journal* 11(1): 7–21. doi: http://doi.org/10.17323/1996-7845/2016-01-7.

Samuelson, Kate (2016). "Donald Trump Must Respect 'Irreversible' Paris Climate Deal, French President Hollande Says." *Time*, November 15. http://time.com/4571421/francois-hollande-donald-trump-paris-agreement/.

Sankei News (2019). "綱渡りだったg20宣言 温暖化wto・改革で対立." ["G20 Declaration on Tightrope: Division over Global Warming and WTO Reform."] June 30. https://www.sankei.com/article/20190630-3JGV27PQQ5J6BNNIBO6ZCSHYZI.

Santos, Marcelo (2017). "Global Justice and Environmental Governance: An Analysis of the Paris Agreement." *Revista Brasileira de Política Internacional* 60(1): e008. https://doi.org/10.1590/0034-7329201600116

Saudi Press Agency (2019). "First G20 Sherpa Meeting Held in Riyadh." December 6. https://www.spa.gov.sa/viewfullstory.php?lang=en&newsid=2007583.

Sauer, Natalie (2018). "The Paris Agreement Rulebook Explained." *Climate Home News*, December. https://www.climatechangenews.com/2018/12/05/paris-agreement-rulebook-explained.

Sauer, Natalie (2019). "Chile's 'Blue Cop' Will Push Leaders to Protect Oceans to Heal Climate." *Climate Home News*, April 25. https://www.climatechangenews.com/2019/04/25/chiles-blue-cop-will-push-leaders-protect-oceans-heal-climate.

Sauer, Natalie and Meagan Darby (2019). "Bonn Climate Talks End with Saudis and Brazil Defiant." *Climate Home News*, June 27. https://www.climatechangenews.com/2019/06/27/bonn-climate-talks-end-saudis-brazil-defiant.

Savio, Roberto (2021). "Italy and the Dubious Honour of Chairing the G20." January 11. http://www.ipsnews.net/2021/01/italy-dubious-honor-chairing-g20.

Savorskaya, E.V. (2016). "The European Union in Global Climate Governance: To Paris and Beyond." *Sravnitel'naa Politika [Comparative Politics]* 7(3): 71–84. https://doi.org/10.18611/2221-3279-2016-7-3(24)-71-84.

Scheer, Roddy and Doug Moss (2012). "What Causes Ocean Dead Zones." *EarthTalk (Blog)*. Scientific American. September 25. https://www.scientificamerican.com/article/ocean-dead-zones.

Schulz, Florence (2019). "German Ministers Urge EU to Lead the Way at Cop25 in Madrid." *EurActiv.com*, November 15. https://www.euractiv.com/section/climate-environment/news/german-ministers-urge-eu-to-lead-the-way-at-cop25-in-madrid.

Scott, Shirley (2015). "Implications of Climate Change for the UN Security Council: Mapping the Range of Potential Policy Responses." *International Affairs* 91(6): 1317–33. https://doi.org/10.1111/1468-2346.12455.

Seetharaman, Raghavan (2021). "G7 Has Brought Sustainable Value Creations." *Gulf Times*, June 15. https://www.gulf-times.com/story/694152/G7-has-brought-sustainable-value-creations.

Seifter, Andrew, Denise Robbins and Kevin Kalhoefer (2016). "Study: How Broadcast Networks Covered Climate Change in 2015." *Media Matters for America*, March 7. https://www.mediamatters.org/rush-limbaugh/study-how-broadcast-networks-covered-climate-change-2015.

Sengupta, Somini (2019). "U.N. Climate Talks End with Few Commitments and a 'Lost' Opportunity." *New York Times*, December 15. https://www.nytimes.com/2019/12/15/climate/cop25-un-climate-talks-madrid.html.

Sengupta, Somini (2020). "Intense Arctic Wildfires Set a Pollution Record." *New York Times*, July 7. https://www.nytimes.com/2020/07/07/climate/climate-change-arctic-fires.html.

Sengupta, Somini and Lisa Friedman (2019). "At U.N. Climate Summit, Few Commitments and U.S. Silence." *New York Times*, September 23. https://www.nytimes.com/2019/09/23/climate/climate-summit-global-warming.html.

Sethi, Nitin (2018). "Climate Finance by Rich Nations Must for Any Deal in Katowice: Negotiator." *Business Standard*, December 5. https://www.business-standard.com/article/current-affairs/climate-finance-by-rich-nations-must-for-any-deal-in-katowice-negotiator-118120501205_1.html.

Sethi, Nitin and Kumar Sambhav Shrivastava (2017). "India Prepares for Clashes with Developed World at UN Climate Talks." *Climate Home News*, November 2. https://www.climatechangenews.com/2017/11/02/india-prepares-clashes-developed-world-un-climate-talks.

Shalal, Andrea (2020). "Biden Should Seek Early G20 Meeting, Former U.S. Officials Say." *Reuters*, November 10. https://www.reuters.com/article/us-usa-election-g20-idUSKBN27Q3BT.

Sharma, Alok (2020a). "COP26 President Alok Sharma at Launch of COP26 Private Finance Agenda." Speech at the launch of the COP26 Private Finance Agenda, United Kingdom, Cabinet Office, London, February 27. https://www.gov.uk/government/speeches/cop26-president-alok-sharma-at-launch-of-cop26-private-finance-agenda.

Sharma, Alok (2020b). "COP26 President's Closing Remarks at Climate Ambition Summit 2020." December 12. https://www.gov.uk/government/speeches/cop26-presidents-closing-remarks-at-climate-ambition-summit-2020.

Sharma, Alok (2021a). "Bringing the Benefits of Clean Growth and Resilient Economies to Countries around the World." Speech at the Valedictory session of the World Sustainable Development Summit, United Kingdom, Cabinet Office, London, February 16. https://www.gov.uk/government/speeches/bringing-the-benefits-of-clean-growth-and-resilient-economies-to-countries-around-the-world.

Sharma, Alok (2021b). "COP26 President Addresses UN Member States." United Kingdom, Cabinet Office, February 8. https://www.gov.uk/government/speeches/cop26-president-addresses-un-member-states.

Sharma, Alok (2021c). "COP26 President Speech at UN Climate Change Conference." Press conference to close the 2021 Subsidiary Bodies Session, United Kingdom, Cabinet Office, June 17. https://www.gov.uk/government/speeches/cop26-president-speech-at-un-climate-change-conference.

Sharma, Alok (2021d). "Urging All Countries to Commit to Phase out Coal Power Ahead of COP26." Speech at the Powering Past Coal Alliance's London Climate Action Week European roundtable, United Kingdom, Cabinet Office, June 30. https://www.gov.uk/government/speeches/urging-all-countries-to-commit-to-phase-out-coal-power-ahead-of-cop26--2.

Sharma, Alok (2021e). "We Must Help Poorer Countries Tackle Climate Change." *Financial Times*, April 19, p. 17.

Sharma, Shubhi (2016). "Marrakech, Morocco: The COP of Action." *Akshay Urja* 10(3): 12–15. December. https://mnre.gov.in/img/documents/uploads/45051531c8864acc81fbccee443b3ed3.pdf.

Shear, Michael D. (2017). "Trump Will Withdraw U.S. From Paris Climate Agreement." *New York Times*, June 1. https://www.nytimes.com/2017/06/01/climate/trump-paris-climate-agreement.html.

Siddique, Abu and Meagan Darby (2017). "No Finance Plan for Climate Change Victims in Draft UN Decision." *Climate Home News*, November 14. https://www.climatechangenews. com/2017/11/14/no-finance-plan-climate-change-victims-draft-un-decision.

Singhai, Rajrishi (2021). "The Green Concerns of Accomplished Central Bankers." *Mint*, Delhi, January 3. https://www.livemint.com/opinion/columns/the-green-concerns-of-accomplished-central-bankers-11609689387610.html.

Slaughter, Anne-Marie and Gordon LaForge (2021). "Opening up the Order: A More Inclusive International System." *Foreign Affairs*, pp. 154–62. https://www.foreignaffairs. com/articles/world/2021-02-16/opening-order.

Slaughter, Steven (2020). *The Power of the G20: The Politics of Legitimacy in Global Governance*. Abingdon UK: Routledge.

Slezak, Michael (2016). "Marrakech Climate Talks: Giving the Fossil Fuel Lobby a Seat at the Table." *Guardian*, November 6. https://www.theguardian.com/environment/2016/ nov/07/marrakech-climate-talks-giving-the-fossil-fuel-lobby-a-seat-at-the-table.

Smee, Jess (2019). "G7 Summit — French Meeting Takes Aim at Rising Inequality." *Global Policy*, July 10. https://www.globalpolicyjournal.com/blog/10/07/2019/g7-summit-french-meeting-takes-aim-rising-inequality.

Snower, Dennis (2017). "The G20 Summit Was More Successful Than You Think." *G20 Insights*, July 11. https://www.g20-insights.org/2017/07/11/g20-summit-successful-think.

Sobel, Mark and Matthew Goodman (2020). "Biden Should Call for an Early G20 Summit." Center for Strategic and International Studies, November 10. https://www. csis.org/analysis/biden-should-call-early-g20-summit.

Speigel (2018). "Union Verliert, AfD Legt Zu." ["Union Loses, AfD Increases.]" March 27. https://www.spiegel.de/politik/deutschland/cdu-csu-verliert-afd-legt-deutlich-zu-spon-wahltrend-a-1200091.html.

Stanway, David (2019). "China's Climate 'Ambition' Pledge Could Lead to Tougher CO2 Targets: Experts." *Reuters*, July 2. https://www.reuters.com/article/us-climatechange-china-idUSKCN1TX0SY.

Stauffer, Caroline and Scott Squires (2018). "Europeans Say G20 Members Agree to Reform WTO in Draft Communique." *Reuters*, December 1. https://www.reuters.com/ article/uk-g20-argentina-communique-idUKKCN1O03AD.

Stavins, Robert N. and Robert C. Stowe (2010). "What Hath Copenhagen Wrought? A Preliminary Assessment." *Environment Magazine*, May/June, pp. 8–14. https://www.hks. harvard.edu/publications/what-hath-copenhagen-wrought-preliminary-assessment.

Steer, Andrew (2017). "$70 Trillion Investment Pool Goes Greener." World Resources Institute, November 15. https://www.wri.org/blog/2017/11/70-trillion-investment-pool-goes-greener.

Stevenson, Hayley (2021). "Reforming Global Climate Governance in an Age of Bullshit." *Globalizations* 18(1): 86–102. https://doi.org/10.1080/14747731.2020.1774315.

Stone, Jon (2021). "Climate Experts Urge Boris Johnson to Make International Emissions Agreements Legally Binding in UK." *Independent*, April 19. https://www.independent. co.uk/climate-change/news/climate-international-agreements-legally-binding-cop26-b1833972.html.

Strauss, Delphine and Jonathan Wheatley (2021). "World Bank Warns Jab Delays Risk Growth Hit." *Financial Times*, January 6, p. 2.

Stroeve, J. C., T. Markus, L. Boisvert et al. (2014). "Changes in Arctic Melt Season and Implications for Sea Ice Loss." *Geophysical Research Letters* 41(4): 1216–25. https:// doi.org/10.1002/2013GL058951.

Suzuki, Toshiyuki (2019). "Japan to Propose Int'l Framework on Plastic Waste in Oceans at G-20 Meet." *Mainichi*, June 7. https://mainichi.jp/english/articles/20190607/p2a/00m/0na/017000c.

Szalay, Eva and Colby Smith (2021). "Blue Sweep of Congress Adds to Dollar Gloom." *Financial Times*, January 8, p. 10. https://www.ft.com/content/9e58d2fb-37c5-432e-b4c6-dcf862d43b7d.

Taipei Times (2021). "G7 Pushes Back against China, Supports Taiwan." June 14. https://www.taipeitimes.com/News/front/archives/2021/06/14/2003759130.

Takenaka, Kyoshi and Elaine Lies (2018). "Abe's Support Rate Falls to 39% as Demonstrations Continue." *Japan Today*, March 17. https://japantoday.com/category/politics/Abe%27s-support-rate-falls-to-39-as-demonstrations-continue.

Thunberg, Greta (2019). "Transcript: Greta Thunberg's Speech at the U.N. Climate Action Summit." *Transcribed by NPR Staff, NPR*, September 23. https://www.npr.org/2019/09/23/763452863/transcript-greta-thunbergs-speech-at-the-u-n-climate-action-summit.

Tiberghien, Yves (2019). "Surprising Momentum but Meagre Outcomes from the G7." *East Asia Forum*, August 31. https://www.eastasiaforum.org/2019/08/31/surprising-momentum-but-meagre-outcomes-from-the-g7.

Timperley, Jocelyn (2017). "COP23: Key Outcomes Agreed at the UN Climate Talks in Bonn." *Carbon Brief*, November 19. https://www.carbonbrief.org/cop23-key-outcomes-agreed-un-climate-talks-bonn.

Timperley, Jocelyn (2019a). "Australia and Brazil Carbon Credits Will Put 1.5c out of Reach, 31 Countries Say." *Climate Home News*, December 14. https://www.climatechangenews.com/2019/12/14/australia-brazil-carbon-credits-will-put-1-5c-reach-ten-countries-say.

Timperley, Jocelyn (2019b). "Brazil Fights Attempt to Cancel Its Old Carbon Credits." *Climate Home News*, October 11. https://www.climatechangenews.com/2019/10/11/brazil-fights-attempt-cancel-old-carbon-credits.

Timperley, Jocelyn (2019c). "Cop25: What Was Achieved and Where to Next." *Climate Home News*, December 16. https://www.climatechangenews.com/2019/12/16/cop25-achieved-next.

Timperley, Jocelyn and Rosamund Pearce (2017). "Mapped: Where Multilateral Climate Funds Spend Their Money." *Carbon Brief*, November 6. https://www.carbonbrief.org/mapped-where-multilateral-climate-funds-spend-their-money.

Trudeau, Justin (2016a). "PM Trudeau Holds a Press Conference Following the G20 Summit in Hangzhou." September 5. https://pm.gc.ca/en/videos/2016/09/05/pm-trudeau-holds-press-conference-following-g20-summit-hangzhou.

Trudeau, Justin (2016b). "Working Together for a Better World." In *G20 China: The 2016 Hanghou Summit*, John Kirton and Madeline Koch, eds. London: Newsdesk Media, pp. 30–31. http://www.g7g20.utoronto.ca/books/g20hangzhou2016.pdf.

Trump, Donald (2018). "Remarks by President Trump to the World Economic Forum." Speech to the World Economic Forum, Davos, Switzerland, White House, Washington, DC, January 26. https://trumpwhitehouse.archives.gov/briefings-statements/remarks-president-trump-world-economic-forum.

Tubiana, Laurence, Emmanuel Guérin and Joss Garman (2019). "COP26: A Roadmap for Success." Chatham House, September 23. https://accelerator.chathamhouse.org/article/cop26-a-roadmap-for-success.

Tusk, Donald (2019). "Speech by President Donald Tusk at the UN Climate Action Summit." New York, September 23. https://www.consilium.europa.eu/en/press/press-releases/2019/09/23/speech-by-president-donald-tusk-at-the-un-climate-action-summit.

Twidale, Susanna and Matthew Green (2019). "Britain to Become First G7 Country with Net Zero Emissions Target." *Reuters*, June 11. https://www.reuters.com/article/us-climate-change-britain/britain-to-become-first-g7-country-with-net-zero-emissions-target-idUSKCN1TC2QI.

UCAR Center for Science Education (n.d.). "Predictions of Future Global Climate." University Corporation for Atmospheric Research. https://scied.ucar.edu/learning-zone/climate-change-impacts/predictions-future-global-climate. Accessed: July 29, 2011.

UN News (2018). "Assembly President Launches New Initiative to Purge Plastics and Purify Oceans." United Nations, December 4. https://news.un.org/en/story/2018/12/1027571.

United Kingdom. Cabinet Office (2021a). "COP26 President Visits India to Welcome Climate Leadership." February 17. https://www.gov.uk/government/news/cop26-president-visits-india-to-welcome-climate-leadership.

United Kingdom. Cabinet Office (2021b). "UK Releases Presidency Programme for Major Climate Summit in Glasgow." *Press Release*, July 7. https://www.gov.uk/government/news/uk-releases-presidency-programme-for-major-climate-summit-in-glasgow.

United Kingdom. Department for Business, Energy and Industrial Strategy (2019). "UK Becomes First Major Economy to Pass Net Zero Emissions Law." London, June 27. https://www.gov.uk/government/news/uk-becomes-first-major-economy-to-pass-net-zero-emissions-law.

United Kingdom. Department for Business, Energy and Industrial Strategy (2021). "UK Enshrines New Target in Law to Slash Emissions by 78% by 2035." *Press Release*, April 20. https://www.gov.uk/government/news/uk-enshrines-new-target-in-law-to-slash-emissions-by-78-by-2035.

United Kingdom. Foreign, Commonwealth and Development Office (2021). "New Global Coalition Launched to Address Impacts of Climate Change." January 25. https://www.gov.uk/government/news/new-global-coalition-launched-to-address-impacts-of-climate-change.

United Kingdom. Her Majesty's Treasury (2021). "Chancellor Prioritises Climate Change and Urged Support for Vulnerable Countries in First UK G7 Finance Meeting." February 12. http://www.g7.utoronto.ca/finance/210212-finance.html.

United Kingdom Presidency of the UN Climate Conference (2020). "New Dates Agreed for COP26 United Nations Climate Change Conference." *UK COP26*, London, May 29. https://ukcop26.org/new-dates-agreed-for-cop26-united-nations-climate-change-conference.

United Kingdom. Prime Minister's Office (2020). "PM: 'Climate Action Cannot Be Another Victim of Coronavirus'." September 24. https://www.gov.uk/government/news/pm-climate-action-cannot-be-another-victim-of-coronavirus.

United Kingdom. Prime Minister's Office (2021a). "PM Call with Chancellor Merkel: 8 February 2021." London, February 9. https://www.gov.uk/government/news/pm-call-with-chancellor-merkel-8-february-2021.

United Kingdom. Prime Minister's Office (2021b). "PM Call with Prime Minister Suga of Japan." London, February 16. https://www.gov.uk/government/news/pm-call-with-prime-minister-suga-of-japan-16-february-2021.

United Kingdom. Prime Minister's Office (2021c). "PM Call with United Nations Secretary-General." February 16. https://www.gov.uk/government/news/pm-call-with-united-nations-secretary-general-16-february-2021.

United Kingdom. Prime Minister's Office (2021d). "UK to Donate 100 Million Coronavirus Vaccine Doses." June 11. https://www.gov.uk/government/news/uk-to-donate-100-million-coronavirus-vaccine-doses.

United Nations (2020a). "Secretary-General Supports Delay of Climate Change Conference Amid COVID-19 Pandemic, Spotlighting Vulnerability to Existential Threats | Meetings Coverage and Press Releases." SG/SM/20030, New York, April 1. https://www.un.org/press/en/2020/sgsm20030.doc.htm.

United Nations (2020b). "UN Secretary-General to Convene High-Level Roundtable on Climate Action During General Assembly Week to Showcase Solutions and Progress." September 21. https://www.un.org/sustainabledevelopment/blog/2020/09/un-secretary-general-to-convene-high-level-roundtable-on-climate-action-during-general-assembly-week-to-showcase-solutions-and-progress/.

United Nations (2020c). "UN Secretary-General's Roundtable Shows Tide Turning for Climate Action as Governments, Businesses and Investors Ramp up Ambition." September 24. https://www.un.org/sites/un2.un.org/files/climateroundtable-closing-pressrelease.pdf.

United Nations (2021). "Secretary-General Calls Latest IPCC Climate Report 'Code Red for Humanity', Stressing 'Irrefutable' Evidence of Human Influence." *Press Release*, August 9. https://www.un.org/press/en/2021/sgsm20847.doc.htm.

United Nations (n.d.). "2019 Climate Action Summit." https://www.un.org/en/climatechange/2019-climate-action-summit. Accessed: July 22, 2021.

United Nations Climate Change (2016). "Marrakech Action Proclamation for Our Climate and Sustainable Development." Marrakech, November 17. https://unfccc.int/files/meetings/marrakech_nov_2016/application/pdf/marrakech_action_proclamation.pdf.

United Nations Climate Change (2017). "One Planet Summit: Finance Commitments Fire-up Higher Momentum for Paris Climate Change Agreement." December 12. https://cop23.unfccc.int/news/one-planet-summit-finance-commitments-fire-up-higher-momentum-for-paris-climate-change-agreement.

United Nations Climate Change (2018). "Countries Inch Forward on Paris Agreement Implementation as Bangkok Climate Talks Close." September 9. https://unfccc.int/news/countries-inch-forward-on-paris-agreement-implementation-as-bangkok-climate-talks-close.

United Nations Climate Change (2019a). "Cop25 Will Take Place in Madrid from 2 to 13 December 2019." November 1. https://unfccc.int/news/cop25-will-take-place-in-madrid-from-2-to-13-december-2019.

United Nations Climate Change (2019b). "President Sebastián Piñera and Minister Carolina Schmidt Launch Cop25 Climate Change Summit." April 11. https://unfccc.int/news/president-sebastian-pinera-and-minister-carolina-schmidt-launch-cop25-climate-change-summit.

United Nations Climate Change (2020a). "Climate Ambition Summit Builds Momentum for COP26." *Press Release*, December 12. https://unfccc.int/news/climate-ambition-summit-builds-momentum-for-cop26.

United Nations Climate Change (2020b). "COP26 Postponed." *UN Climate Press Release*, April 1. https://unfccc.int/news/cop26-postponed.

United Nations Climate Change (2021). "'Climate Commitments Not on Track to Meet Paris Agreement Goals' as NDC Synthesis Report Is Published." *Bonn*, February 26. https://unfccc.int/news/climate-commitments-not-on-track-to-meet-paris-agreement-goals-as-ndc-synthesis-report-is-published.

United Nations Environment Programme (2015). "UNEP Executive Director Achim Steiner Welcomes G7 Leaders' Declaration." June 8. https://www.unenvironment.org/news-and-stories/press-release/unep-executive-director-achim-steiner-welcomes-g7-leaders.

United Nations Environment Programme (2017). "UN Environment and Bnp Paribas Partner to Bring Private Capital to Sustainable Projects in Emerging Countries." Paris, December 12. https://www.unep.org/news-and-stories/press-release/un-environment-and-bnp-paribas-partner-bring-private-capital.

United Nations Framework Convention on Climate Change (2014). "Further Advancing the Durban Platform." Report of the Ad Hoc Working Group on the Durban Platform for Enhanced Action, Conference of the Parties Twentieth Session, Lima, Peru. FCCC/CP/2014/L.14, December 13. https://unfccc.int/resource/docs/2014/cop20/eng/l14.pdf.

United Nations Framework Convention on Climate Change (2017). "Talanoa Dialogue." Annex II: Informal Note by the Presidencies of COP 22 and COP 23. FCCC/CP/2017/L.13, November 17. https://cop23.com.fj/wp-content/uploads/2017/12/Informal-Note-on-the-Approach-to-the-Talanoa-Dialogue.pdf.

United Nations Framework Convention on Climate Change (2018). "Addendum. Part Two: Action Taken by the Conference of the Parties at Its Twenty-Third Session." *Report of the Conference of the Parties on its twenty-third session, held in Bonn from 6 to 18 November 2017*. FCCC/CP/2017/11/Add.1, February 8. https://unfccc.int/sites/default/files/resource/docs/2017/cop23/eng/11a01.pdf.

United Nations General Assembly (2020). "United Nations Summit on Biodiversity: Summary of the President of the General Assembly." September 30. New York. https://www.un.org/pga/75/wp-content/uploads/sites/100/2020/11/Summary-Biodiversity-Summit-4-November-clearedFINAL-002.pdf.

United States Department of State (2021). "U.S.-China Joint Statement Addressing the Climate Crisis." Washington, DC, April 17. https://www.state.gov/u-s-china-joint-statement-addressing-the-climate-crisis.

United States Department of the Treasury (2021a). "Letter from Treasury Secretary Janet L. Yellen to G20 Colleagues." February 25. https://home.treasury.gov/news/press-releases/jy0034.

United States Department of the Treasury (2021b). "Statement from Treasury Department Spokesperson on Secretary Janet L. Yellen's Participation in the G7 Finance Ministers and Central Bank Governors Meeting." Washington, DC, February 12. https://home.treasury.gov/news/press-releases/jy0025.

Urban 20 (2021). "Urban 20 Calls on G20 to Empower Cities to Ensure a Green and Just Recovery." June 17. https://www.urban20.org/wp-content/uploads/2021/06/U20-2021-Communique-Final.pdf.

Vainshtein, Annie (2020). "'Hidden Cost' of Wildfire Smoke: Stanford Researchers Estimate up to 3,000 Indirect Deaths." *San Francisco Chronicle*, September 24. https://www.sfchronicle.com/california-wildfires/article/Hidden-cost-of-wildfire-smoke-Stanford-15595754.php.

Van de Graaf, Thijs and Harro van Asselt (2017). "Introduction to the Special Issue: Energy Subsidies at the Intersection of Climate, Energy, and Trade Governance." *International Environmental Agreements* 17(3): 313. https://doi.org/10.1007/s10784-017-9359-8.

Vashist, Sanjay (n.d.). "Chennai Floods : Why India Must Support 1.5°C Goal at Paris Climate Summit." *Climte Action Network South Asia*. https://cansouthasia.net/chennai-floods-why-india-must-support-1-5c-goal-at-paris-climate-summit. Accessed: April 10, 2021.

Verhaegen, Soetkin, Jan Aart Scholte and Jonas Tallberg (2021). "Explaining Elite Perceptions of Legitimacy in Global Governance." *European Journal of International Relations* 27(2): 622–50. https://doi.org/10.1177/1354066121994320.

Vetter, David (2020). "Climate Change Experts: Here's Why a Virtual COP26 Won't Work." *Forbes*, April 8. https://www.forbes.com/sites/davidrvetter/2020/04/08/climate-change-experts-heres-why-a-virtual-cop26-wont-work/.

Viscidi, Lisa and Nate Graham (2019). "Brazil Was a Global Leader on Climate Change. Now It's a Threat." *Foreign Policy*, January 4. https://foreignpolicy.com/2019/01/04/brazil-was-a-global-leader-on-climate-change-now-its-a-threat/.

Volcovici, Valierie (2019). "At Climate Talks, Japan's Koizumi Confronts Critics over Coal." *Reuters*, December 11. https://www.reuters.com/article/us-climate-change-accord-japan/at-climate-talks-japans-koizumi-confronts-critics-over-coal-idUSKBN1YF2KG.

Volker, Kurt (2014). "Where's NATO's Strong Response to Russia's Invasion of Crimea." *Foreign Policy*, March 18. https://foreignpolicy.com/2014/03/18/wheres-natos-strong-response-to-russias-invasion-of-crimea/.

Walker, Derek (2020). "How the Biden-Harris Administration Can Restore and Strengthen U.S. Climate Leadership." *Environmental Defense Fund*, December 26. http://blogs.edf.org/climate411/2020/12/21/how-biden-harris-administration-can-restore-and-strengthen-climate-leadership.

Walker, Peter (2021). "Alok Sharma to Work Full-Time on Cop26 Climate Conference Preparation." *Guardian*, January 8.

Wang, Yi (2016). "Strive to Achieve Ten Results from the G20 Hangzhou Summit." Beijing, May 26. http://www.fmprc.gov.cn/mfa_eng/zxxx_662805/t1367533.shtml.

Warren, Brittaney (2019a). "G7 Environment Ministers' Meetings, Commitments and Compliance, 1992–2018." *G7 Research Group*, January 9. http://www.g7.utoronto.ca/evaluations/warren-environment-ministers-meetings-190109.pdf.

Warren, Brittaney (2019b). "G7 Performance on Climate Change." In *G7 France: The 2019 Biarritz Summit*, John Kirton and Madeline Koch, eds. London: GT Media, pp. 48–49. http://bit.ly/G7France.

Warren, Brittaney (2020). "G20 Performance on Climate Change." In *G20 Saudi Arabia: The 2020 Riyadh Summit*, John Kirton and Madeline Koch, eds. London: GT Media, pp. 122–23. http://bit.ly/g20saudi.

Warren, Brittaney (2021a). "G7 Performance on Climate Change." In *G7 UK: The 2021 Cornwall Summit*, John Kirton and Madeline Koch, eds. London: GT Media, pp. 68–69. http://bit.ly/ukg721.

Warren, Brittaney (2021b). "The G7's Pre-Cornwall Summit: Is the G7 Acting Like Its House Is on Fire." *G7 Research Group*, February 19. http://www.g7.utoronto.ca/evaluations/2021cornwall/warren-february-summit.html.

Warren, Brittaney (2021c). "Global Change Governance Off to a Strong Start in 2021." *G7 Research Group*, February 18.

Warren, Brittaney (2021d). "Improving G7 Performance on Climate Change." G7 Research Group, April 12. http://www.g7.utoronto.ca/evaluations/warren-performance-climate-change-210412.html.

Warren, Brittaney and John Kirton (2021). "Holding the G7 Environment Ministers' Meeting to the Highest Standard." *G7 Research Group*, May 19. http://www.g7.utoronto.ca/evaluations/2021cornwall/warren-kirton-emm.html.

Waskow, David, Yamide Dagnet, Eliza Northrop et al. (2018). "At Climate Talks, a 'Bicycle' Built for Bonn." *World Resources Institute*, May 1. https://www.wri.org/blog/2018/05/climate-talks-bicycle-built-bonn.

Waterfield, Bruno and Tom Kington (2017). "G7 Riven by White House Doubts over Free Trade and Curbing Carbon Emissions." *Times*, May 26. https://www.thetimes.co.uk/article/g7-riven-by-white-house-doubts-over-free-trade-and-climate-change-rpnkkt6rc.

Watts, Jonathan and Niko Kommenda (2020). "Coronavirus Pandemic Leading to Huge Drop in Air Pollution." *Guardian*. https://www.theguardian.com/environment/2020/mar/23/coronavirus-pandemic-leading-to-huge-drop-in-air-pollution.

White, Edward and Leslie Hook (2021). "US-China Climate Pledge Boosts Hopes for Global Emissions Deal." *Financial Times*, April 19, p. 1.

White House (2019). "Press Briefing by Acting Chief of Staff Mick Mulvaney." October 17. https://trumpwhitehouse.archives.gov/briefings-statements/press-briefing-acting-chief-staff-mick-mulvaney.

White House (2020). "Remarks by President Trump in Press Briefing, August 10, 2020." August 11. https://trumpwhitehouse.archives.gov/briefings-statements/remarks-president-trump-press-briefing-august-10-2020.

White House (2021b). "President Biden Invites 40 World Leaders to Leaders Summit on Climate." March 26. https://www.whitehouse.gov/briefing-room/statements-releases/2021/03/26/president-biden-invites-40-world-leaders-to-leaders-summit-on-climate.

White House (2021b). "Readout of President Joseph R. Biden, Jr. Call with Prime Minister Narendra Modi of India." Washington, DC, February 8. https://www.whitehouse.gov/briefing-room/statements-releases/2021/02/08/readout-of-president-joseph-r-biden-jr-call-with-prime-minister-narendra-modi-of-india.

Widakuswara, Patsy (2021). "G-7 Summit Kicks Off with 'Build Back Better' Message." *VOA News*, June 11. https://www.voanews.com/europe/g-7-summit-kicks-build-back-better-message.

Wike, Richard, Bruce Stokes, Jacob Poushter et al. (2017). "U.S. Image Suffers as Publics around World Question Trump's Leadership." Pew Research Center, Washington, DC, June 26. https://www.pewresearch.org/global/2017/06/26/u-s-image-suffers-as-publics-around-world-question-trumps-leadership/.

Winsor, Morgan (2015). "G7 Summit 2015: Buhari Requests Help for Boko Haram in Nigeria, Economy in 'Wish List' for Group of Seven Leaders." *International Business Times*, June 8. https://www.ibtimes.com/g7-summit-2015-buhari-requests-help-boko-haram-nigeria-economy-wish-list-group-seven-1956460.

Witz, Alexandra (2020). "The Arctic Is Burning Like Never before: and That's Bad News for Climate Change." *Nature* 585: 336–37. September 10. https://doi.org/10.1038/d41586-020-02568-y.

Wolf, Martin (2021a). "The Economic Recovery Masks a Global Divide." *Financial Times*, April 21, p. 15.

Wolf, Martin (2021b). "The G20 Has Failed to Meet Its Challenges." *Financial Times*, July 14, p. 17.

Wolf, Martin (2021c). "The Healing of Democracies Starts at Home." *Financial Times*, June 23, p. 17.

Worland, Justin (2016). "Donald Trump Promises to Cut Regulation on 'Phony' Environmental Issues." *Time*. https://time.com/4349309/donald-trump-bismarck-energy-speech/.

Worland, Justin (2017). "50 World Leaders Will Discuss Climate Change in Paris. Donald Trump Wasn't Invited." *Time*, December 11. https://time.com/5058736/climate-change-macron-trump-paris-conference.

World Bank (2017). "World Bank Group Announcements at One Planet Summit." Paris, December 12. https://www.worldbank.org/en/news/press-release/2017/12/12/world-bank-group-announcements-at-one-planet-summit.

World Bank (2019). "World Bank Group Announcements at One Planet Summit." Washington, DC, March 13. https://www.worldbank.org/en/news/press-release/2019/03/13/world-bank-group-announcements-at-one-planet-summit.

World Economic Forum (2015). "Global Risks 2015." Davos. http://www3.weforum.org/docs/WEF_Global_Risks_2015_Report15.pdf.

World Economic Forum (2018). "Global Risks Report 2018." Davos. http://www3.weforum.org/docs/WEF_GRR18_Report.pdf.

World Economic Forum (2020). "Burning Planet: Climate Fires and Political Flame Wars Rage." January 15. https://reports.weforum.org/global-risks-report-2020/press-release.

World Economic Forum (2021). "Global Risks Report 2021." Davos. https://www.weforum.org/reports/the-global-risks-report-2021.

World Health Organization (2020a). "Disease Outbreak News – Japan." January 17. https://www.who.int/emergencies/disease-outbreak-news/item/2020-DON237.

World Health Organization (2020b). "Novel Coronavirus (2019-Ncov): Situation Report 3." January 23. https://www.who.int/docs/default-source/coronaviruse/situation-reports/20200123-sitrep-3-2019-ncov.pdf.

World Health Organization (2020c). "Novel Coronavirus (2019-Ncov): Situation Report 54." March 14. https://www.who.int/docs/default-source/coronaviruse/situation-reports/20200314-sitrep-54-covid-19.pdf.

World Health Organization (2020d). "Weekly Operational Update on COVID-19." November 20. https://www.who.int/publications/m/item/weekly-operational-update-on-covid-19---20-november-2020.

World Meteorological Organization (2016). "The Global Climate 2011–2015: Heat Records and High Impact Weather." Press Release Number 14, November 8. https://public.wmo.int/en/media/press-release/global-climate-2011-2015-hot-and-wild.

World Meteorological Organization (2017). "One Planet Summit Drives Forward Climate Action." Geneva, December 12. https://public.wmo.int/en/media/news/one-planet-summit-drives-forward-climate-action.

World Meteorological Organization (2020). "New Climate Predictions Assess Global Temperatures in Coming Five Years." Geneva, July 8. https://public.wmo.int/en/media/press-release/new-climate-predictions-assess-global-temperatures-coming-five-years.

World Meteorological Organization (2021a). "Climate Change Indicators and Impacts Worsened in 2020." Press release no. 19042021, Geneva, April 19. https://public.wmo.int/en/media/press-release/climate-change-indicators-and-impacts-worsened-2020.

World Meteorological Organization (2021b). "New Climate Predictions Increase Likelihood of Temporarily Reaching 1.5 °C in Next 5 Years." Press release 27052021, Geneva, May 27. https://public.wmo.int/en/media/press-release/new-climate-predictions-increase-likelihood-of-temporarily-reaching-15-%C2%B0c-next-5.

Wouters, Jan and Sven Van Kerchhoven (2017). "The Role of the EU in the G7 in Times of Anti-Globalization." Paper prepared for the conference on "G7 and Global Governance in an Age of Anti-globalization," LUISS Carli Guido University, May 22, Rome.

Wuebbles, Donald J., David W. Fahey and Kathy A. Hibbard (2017). "Climate Science Special Report: Fourth National Climate Assessment, Volume I." United States Global Change Research Program, Washington, DC. https://science2017.globalchange.gov.

Wyns, Arthur (2018). "Poland Presidency under Criticism in Bangkok." (Blog). ClimateTracker.org., September 4. http://climatetracker.org/poland-president-under-criticism-in-bangkok.

Xi, Jinping (2016a). "A New Starting Point for China's Development: A New Blueprint for Global Growth." Opening Remarks to the B20, Summit Hangzhou, September 3. http://www.g20.utoronto.ca/2016/160903-xi.html.

Xi, Jinping (2016b). "Towards an Innovative, Invigorated, Interconnected and Inclusive World Economy." In *G20 China: The 2016 Hangzhou Summit*, John Kirton and

Madeline Koch, eds. London: Newsdesk Media, p. 3. http://www.g7g20.utoronto.ca/books/g20hangzhou2016.pdf.

Xi, Jinping (2019). "Working Together to Deliver a Brighter Future for Belt and Road Cooperation." Keynote Speech at the Opening Ceremony of the Second Belt and Road Forum for International Cooperation, Beijing, April 26. https://www.fmprc.gov.cn/mfa_eng/zxxx_662805/t1658424.shtml.

Xi, Jinping (2021). "For Man and Nature: Building a Community of Life Together." Remarks at the Leaders Summit on Climate, Beijing, April 22. https://www.fmprc.gov.cn/mfa_eng/zxxx_662805/t1870852.shtml.

Xinhua (2017). "Climate Action Continues Despite U.S. Withdrawal: COP22 President." *China Daily*, June 3. https://www.chinadaily.com.cn/world/2017-06/03/content_29605999.htm.

Yale Environment 360 (2018). "It's Official: 2017 Was the Second Hottest Year on Record." *Yale School of the Environment*, January 4. https://e360.yale.edu/digest/its-official-2017-was-the-second-hottest-year-on-record.

Yeo, Sophie and Karl Mathiesen (2017). "Despite Trump, US Climate Team to Forge on with Paris Deal." *Climate Home News*, November 1. https://www.climatechangenews.com/2017/11/01/despite-trump-us-climate-team-forge-paris-deal.

Yeung, Peter (2016). "The Paris Climate Agreement Has 'Failed' Poor Countries." *Independent*, May 16. https://www.independent.co.uk/environment/climate-change/paris-climate-agreement-report-oxfam-a7030446.html.

Yomiuri Shimbun (2019). "Strive to Maintain G7's Significance, Framework for Global Cooperation." *Japan News*, August 28. Editorial.

YouGov (2018). "Yougov/the Times Survey Results." March 29. https://d25d2506sfb94s.cloudfront.net/cumulus_uploads/document/o3oayi8z58/TimesResults_180327_VI_Trackers_W.pdf.

Zhao, Minghao (2016). "Climate Change Is Both a Burden and an Opportunity for Economic Change." *Global Times*, September 14. https://www.globaltimes.cn/content/1004618.shtml.

Index

284   *Index*

For Product Safety Concerns and Information please contact our EU
representative GPSR@taylorandfrancis.com
Taylor & Francis Verlag GmbH, Kaufingerstraße 24, 80331 München, Germany

www.ingramcontent.com/pod-product-compliance
Lightning Source LLC
Chambersburg PA
CBHW052120230326
41598CB00080B/3917